Embedded Networking
with CAN and CANopen

Embedded Networking with CAN and CANopen

Embedded Systems Academy Inc.

http://www.esacademy.com

Embedded Networking with CAN and CANopen
by Olaf Pfeiffer, Andrew Ayre and Christian Keydel

Revised First Edition

Published by
Embedded Systems Academy Inc.
1250 Oakmead Parkway, Sunnyvale, CA 94085, USA
http://www.esacademy.com

Formerly published by
Copperhill Technologies Corporation, Greenfield, MA, USA
RTC Books, San Clemente, CA, USA

Printed in the United States of America
ISBN 978-0-692-74087-3
formerly ISBN 978-0-9765116-2-5 and ISBN 0-929392-78-7
First Printing November 2003

We welcome your comments. Email us at author@canopenbook.com. Errata and clarifications will be posted to www.CANopenBook.com.

About this Book

"Few things in life are less efficient than a group of people trying to write a sentence. The advantage of this method is that you end up with something for which you will not be personally blamed."
Scott Adams

Being three authors, we divided *Embedded Networking with CAN and CANopen* into three parts so that each of us could focus on one of them.

Part One "Using CANopen" (Chapters 1 through 4) by Olaf Pfeiffer focuses on CANopen up to the system integrator level. Any technician or engineer that needs to be able to configure and/or maintain a CANopen network will find the required knowledge to do so in this part. The last chapter in this part contains a step-by-step example of a network configuration and test cycle.

Part Two "CANopen Engineering" (Chapters 5 and 6) by Christian Keydel is for engineers that either need to have a detailed knowledge of how CAN and CANopen work or that will be developing their own CANopen devices. Different implementation methods are introduced and compared with each other.

Part Three "CANopen Reference" (Appendices) by Andrew Ayre is a pure reference section for all CANopen users. Key elements of CANopen are summarized in a way that allows for quick look-up. The core of this part is an Object Dictionary reference listing all Object Dictionary entries specified by [CiADS301] and [CiADS302].

In this book we will often use text boxes to provide the reader with additional personal opinions, recommendations, experiences, goals and objectives. Although not always critical to the topic under discussion these texts often provide additional insight that might help the reader better understand how or why something was specified or implemented.

Be sure to visit the companion website, www.CANopenBook.com for additional resources, examples, downloads and much more.

Contributions and Acknowledgments

"Love and work are the cornerstones of our humanness."
Sigmund Freud

This project would not have been possible without the support of Cyndi, Irena, Leah and Katja.

Furthermore, the authors would like to thank the many persons and companies that helped with the realization of this book project. The companies Vector CANTech, Philips Semiconductors and Schneider Electric provided us with many CAN and CANopen related software and hardware products which enabled us to add several real-world examples where appropriate.

Valuable contributions came from Holger Zeltwanger, William Seitz and Thilo Schumann from the CiA, the CAN in Automation user's and manufacturer's group. Special thanks to the CiA for providing the glossary of CANopen terms.

Michael B. Simmonds of Quantum Design kindly allowed us to use parts of his paper "Customizing CANopen for Use in an Automated Laboratory Instrument" for a real-world customized CANopen implementation example.

Additional feedback was provided by Jürgen Baumgartner, John Dammeyer, Jürgen Klüser, Paul Lukowicz and Axel Wolf.

And finally we would like to thank Craig Choisser and all the others at The RTC Group for their help and drive to turn a long-term virtual project into a real book. As Yoda said: "Do, or do not. There is no 'try'."

Andrew Ayre
Christian Keydel
Olaf Pfeiffer

November 2003

Contents

Preface

by William E. Seitz

General Manager, CAN in Automation North America

The Controller Area Network, commonly known as CAN, was originally designed for use in automobiles. By virtue of its massive adoption by automakers worldwide, low-cost microcontrollers with CAN controller interfaces are available from over twenty manufacturers, making CAN a mainstream network technology. Moreover, CAN has migrated into many non-automobile applications over the last ten years creating a requirement for an open, standardized higher-layer protocol that provides a reliable message exchange system along with a means to detect, configure and operate nodes.

Several higher-layer CAN protocols emerged such as SAE J1939, DeviceNet and CANopen. While each protocol has its own special purpose, CANopen is the most popular higher-layer protocol for embedded networking applications – those networks that are completely hidden within a machine or cell – and is found in over twenty vertical markets such as transportation, medical, industrial machinery, building automation and military, just to name a few.

Embedded Networking with CAN and CANopen is one of the most useful books embedded network designers can own – whether they are just starting out or have years of experience. Arranged in three easy-to-read parts, *Embedded Networking with CAN and CANopen* introduces the reader to CAN and characterizes its flexibility in over twenty vertical industries. Subsequent chapters take the reader through a stepwise description of CAN and CANopen standards from the perspective of the embedded systems engineer. There are also sections devoted to a small set of mandatory functionality and a large set of optional functions that illustrate the extent of customization available in the CAN and CANopen standards.

Key topics include requirements for understanding embedded networking, code and communications, underlying CAN technology, selecting CAN controllers, conformance testing and application specific examples of popular device profiles used to implement designs. The last part of the book is devoted to reference information and frequently asked questions (FAQs) that facilitate quick reference to standards and methods outlined in the book.

Written by leading CAN and CANopen technology consultants, *Embedded Networking with CAN and CANopen* has been especially written for CANopen developers and integrators, providing them with the ability to see ahead and even implement functionality that is currently not available yet as CAN and CANopen standards.

This book is a must for CAN laymen, developers and integrators who want to learn more about CAN and its wide range of applications in embedded control systems.

History of CAN and CANopen

by Holger Zeltwanger

Managing Director, CAN in Automation

In February of 1986, Robert Bosch introduced the CAN (Controller Area Network) serial bus system at the SAE congress in Detroit. In mid-1987, Intel delivered the first stand-alone CAN controller chip, the 82526. Shortly thereafter, Philips Semiconductors introduced the 82C200. Today, almost every new passenger car manufactured in Europe is equipped with at least one CAN network. Also used in other types of vehicles, from trains to ships, as well as in industrial controls, CAN is one of the most dominating bus protocols. To date, chip manufacturers have produced and sold more than 500 million CAN devices in total.

Although CAN was originally developed to be used in passenger cars, the first applications came from other market segments. Especially in northern Europe, CAN was already very popular even in its early days. At the beginning of 1992, users and manufacturers established the CAN in Automation (CiA) international users and manufacturers association. One of the first tasks of the CiA was the specification of the CAN Application Layer (CAL). Although the CAL approach was academically correct and it was possible to use it in industrial applications, every user needed to design a new profile because CAL was a true application layer. Since 1993 and within the scope of the Esprit project ASPIC, a European consortium led by Bosch had been developing a prototype of what would become CANopen, the CAL-based profile for embedded networking in production cells. In 1995, CiA released the completely revised CANopen communications profile. The CANopen profile family defines a framework for programmable systems as well as different device, interface and application profiles. This is an important reason why whole industry segments (e.g. printing machines, maritime applications, medical systems, etc.) decided to use CANopen during the late 1990s.

In the early 1990s, engineers at the US mechanical engineering company Cincinnati Milacron started a joint venture together with Allen-Bradley and Honeywell Microswitch regarding a control and communications project based on CAN. However, after a short while important project members changed jobs and the joint venture fell apart. But Allen-Bradley and Honeywell continued the work separately. This led to the two higher layer protocols 'DeviceNet' and 'Smart Distributed System' (SDS), which are quite similar, at least in the lower communication layers. In early 1994, Allen-Bradley turned the DeviceNet specification over to the Open DeviceNet Vendor Association (ODVA), which boosted the popularity of DeviceNet. Honeywell failed to go a similar way with SDS, which makes SDS look more like an internal solution by Honeywell Microswitch. DeviceNet was developed especially for factory automation and therefore presents itself as a direct opponent to protocols like Profibus-DP and Interbus. Providing off-the-shelf plug-and-play functionality, DeviceNet has become the leading bus system in this particular market segment in the US.

With DeviceNet and CANopen, two standardized (EN 50325) application layers are now available, addressing different markets. DeviceNet is optimized for factory automation and CANopen is especially well suited for embedded networks in all kinds of machine controls. This has made proprietary application layers obsolete; the necessity to define application-specific application layers is history (except, perhaps, for some specialized high-volume embedded systems).

Of course, the more than 50 semiconductor vendors who have implemented CAN modules into their micro-controllers and ASICs are mainly focused on the automotive industry. Since the mid-1990s, Infineon Technologies (formerly Siemens) and Motorola have shipped large quantities of CAN controllers to European passenger car manufacturers. As a next wave, Far Eastern semiconductor vendors have also offered CAN controllers since the late 1990s. Since 1992, Mercedes-Benz has been using CAN in their high-end passenger cars. Now nearly all new European passenger cars are equipped with several networks, with some high-end cars implementing up to five CAN networks.
Although the CAN protocol is now 15 years old, it is still being enhanced. In the last two years an ISO task force defined a protocol for a time-triggered transmission of CAN messages. The TTCAN extension will add about five to ten years to the lifetime of CAN. Considering CAN is still at the beginning of a global market penetration, even conservative estimates show further growth for this bus system for the next ten to fifteen years. This is underlined by the fact that the US and Far Eastern car manufacturers are just starting to use CAN in the production of their vehicles. Furthermore, new potentially high-volume applications are in the pipeline – not only in passenger cars but also entertainment, domestic appliances and automatic building doors, among many others.

Several enhancements regarding the approval for different safety-relevant and safety-critical applications can be expected for the higher-layer protocols (HLP). The German professional association BIA and the German safety standards authority TÜV have already certified some of the proprietary CAN-based safety systems. CANopen-Safety and DeviceNet Safety are the first standardized CAN solutions to earn a tentative TÜV approval. Approval of the CANopen framework for maritime applications by one of the leading classification societies worldwide, Germanischer Lloyd, is in preparation. Among other things, this specification defines the automatic switchover from a CANopen network to a redundant bus system.

In the future, CiA members will define several CANopen application profiles. An application profile specifies all device interfaces used in a specific application. This includes direct communication between dedicated devices overcoming the master/slave PDO communication as usual in standard device profiles. The first CANopen application profiles will be for automatic building doors, lift control systems, road construction machinery and light railways.

Part One: Using CANopen

1 Understanding Embedded Networking Requirements

"Everything should be as simple as it is, but not simpler."
Albert Einstein

The intention of this first section is to lay a foundation of knowledge required to truly understand the terminology and issues typical to embedded systems and networked embedded systems. It was written with newcomers to embedded systems and embedded networking in mind.

Readers with experience in this field should double check to see if all the terms explained in Section 1.1.3 are familiar to them.

For additional reading material, the reference section at the end of the book lists books about embedded systems; see in particular [Barr99], [Berger01], [Ganssle00] and [Ganssle03]. For additional information on process control see [Stenerson02].

1.1 Embedded Networking for Beginners

Objective

In this chapter we describe the basic terms and technologies involved with "embedded networking" from a generic point of view, without getting into the details of how CANopen relates to them. If you have a lot of experience in both embedded systems and computer networks (preferably with real-time requirements), feel free to skip this chapter for now. If during further reading you detect knowledge gaps, come back to this chapter for a "memory refresh."

Here we cover generic networking terms such as serial networks, master, slave, server, client, producer, consumer, point-to-point, multicast, broadcast, message triggering, time driven, event driven and change-of-state (COS). In addition, we will also look at terms and technologies used in embedded or industrial control systems, involving things like automation systems, fieldbuses, real-time and performance requirements.

1.1.1 What is "Embedded Networking"?

Since the introduction of the personal computer the semiconductor components receiving most of the attention by the media are the main processor (CPU) and the memory. One of the first things every computer user learns is that CPU performance and memory size continuously increase with time. Both improvements are based on technological changes and enhancements that allow chip manufacturers to pack more and more transistors into the same silicon area. So although chips are the same size and same price as in the past, greater performance becomes available.

Unfortunately, the effects these technological improvements have had on the other side of the scale receive far less attention, even though the consequences are reaching far into our everyday lives. On this other side of the scale chip manufacturers can build "low-performance" microcontrollers ever smaller and smaller. This not only makes them cheaper, but also brings down their power consumption.

As a consequence, intelligent electronics get embedded into more everyday products. Parents know that there are hardly any toys these days that do not have some electronics built in. Further examples of microcontrollers used in "embedded systems" are kitchen appliances, any sort of audio equipment, phones, and computer peripherals such as modems, printers, keyboards, etc.

The trend towards more affordable microcontrollers results in embedded systems which utilize several microcontrollers. Typically there is a need for a communication channel between those controllers embedded in a system, hence "embedded networking." Typical examples of such "multi-controller" embedded systems with communication requirements are cars and trucks, household appliances, lift/elevator systems and a whole array of industrial machinery.

1.1.2 Communication in the Automation Pyramid

Industrial automation applications as used on factory floors contain most of the elements applicable to embedded systems and embedded networking. Looking at an embedded networking system from the industrial angle not only helps us to understand basic communication requirements, but this model can also easily be adapted to a variety of embedded systems.

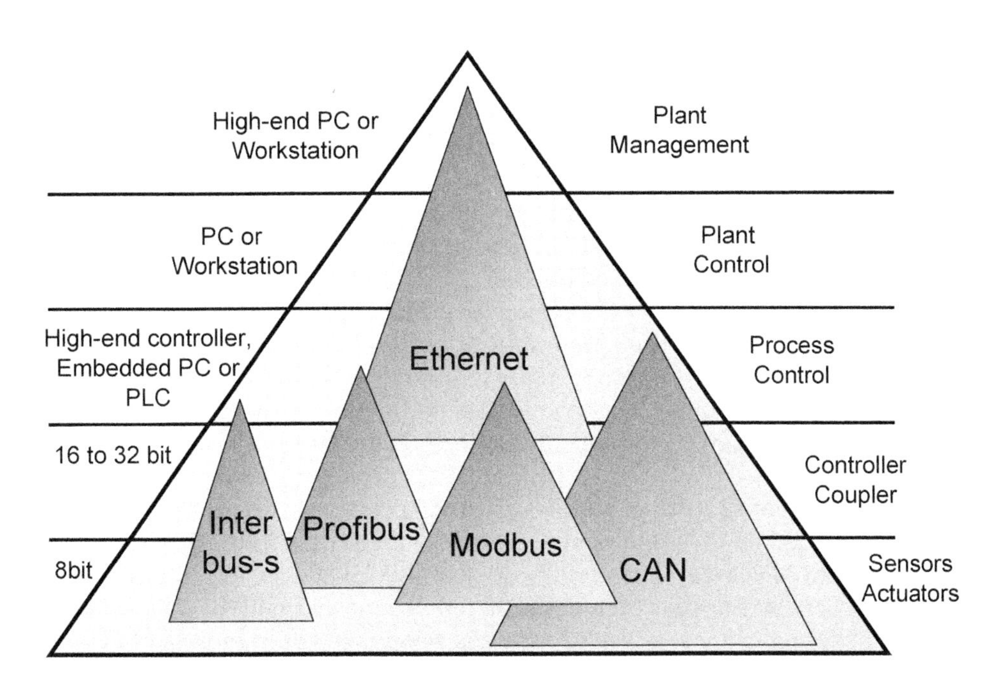

Figure 1.1 Communication in the Automation Pyramid

The automation pyramid symbolizes the different control levels and number of computerized systems in a factory automation system. On top of the pyramid there are a few workstations handling the management of one or multiple plants, followed by levels with more workstations or PCs controlling specific sections of the manufacturing process.

The three lower levels are those implemented in a complex machine or production / manufacturing cell. The Sensor and Actuator level contains simple sensors (contact sensors, distance sensors, temperature sensors, etc.) and actuators (hydraulics, drives, etc.) used in the process. The Controller level implements direct control loops between sensors and actuators. For example, the sensor input from a rotary encoder could be used by a controller to calculate new outputs for the actuator; i.e. an electric motor / drive.

The Process Control level combines several systems from the Controller level. So a production machine typically has one Process Control system that has individual sub-systems on the Controller level. For example, such subsystems might first feed material into a machine, then work on/with the material and then pass the product on to the next step in the production process.

Figure 1.1 shows a variation of the industrial automation pyramid. It symbolizes the hierarchy in an industrial automation system. At the bottom is the Sensor and Actuator level with input and output elements that directly read switches and sensors like current speed of a conveyor belt, RPM values from anything rotating or a current temperature. Typical actuators include hydraulic or pneumatic elements or electrical motors, which in industrial lingo are usually referred to as drives. Being at the bottom of the pyramid also symbolizes that in any installation these elements or modules are used in the highest quantity, compared to modules in the layers above.

Due to the higher quantity, these modules are often price sensitive, as the price-per-module is multiplied by the large number of devices required. In the not-so-distant past of some 20 to 30 years ago, communication at this level was not computerized, meaning that every sensor or actuator was directly connected with its own set of wires to the next higher control level. Today, the trend is to equip more and more sensors and actuators with a networking interface.

However, since single components on the lowest level are needed in large quantities, cost is still a major issue. Equipping simple sensors that just report a single or a few values with high-performance processors and high-end network adapters like Ethernet is simply not an option. Other technologies typically based on serial buses have been used for years because they can be handled by some of the lowest performance (but most affordable) microcontrollers and microprocessors. CAN – the Controller Area Network – is just one of many contenders in the field of networking technologies that are suitable to reach into the lowest level of the automation pyramid.

The next layer up is the Controller level. In this level controllers are used to collect all the inputs, perform some sort of control algorithm and transmit the appropriate commands back to the actuators, the outputs.

The next layers of the automation pyramid are of only limited concern for embedded networking. With each level up in the pyramid the performance of systems needs to be higher, as it needs to handle multiple systems from the layer below. The communication requirements become more significant in the upper levels as more bandwidth is required to handle all the accumulated information coming from the multiple systems in the layers below. In these levels, interfaces to embedded networks are only used if a direct link to the lowest levels is required.

1.1.2.1 Placing embedded systems into the automation pyramid

Embedded systems using multiple microcontrollers and any sort of communication between them can often be directly compared to the lowest levels of the automation pyramid. There will be some sensors and actuators for the inputs and outputs and some sort of controller. Sometimes there might be truly distributed control (in which case the controller functionality is divided between modules) however the basic model and its consequences still applies. The closer a module is to the Sensor and Actuator level (or to the inputs and outputs), the more cost-sensitive it is and the more basic the communication requirements.

> As an example of placing an embedded communication system into the automation pyramid let's have a look at a fully automated shuttle train (as found at many airports) with a focus on the doors.
>
> On the Sensor and Actuator level there is a whole array of signals. Sensors detect not only the current status of a door (is it open, closed or something in-between), they also detect what happens around it. Is something "in" an open door or are passengers (too) close to the door?
>
> The sensors report their findings to the controller level, probably one control module in each passenger car supervising all the doors of this car.

The control modules of each car report up to the Process level – some sort of control module controlling the entire train.

Any communication to higher levels would go beyond "embedded networking" as it would leave the "embedded system" of the automated train. At the Plant Control level there would probably be some wireless communication to a station where all the trains in the system are controlled.

1.1.3 Terminology used in Embedded Networking

The following is a collection of terms and their explanations that are frequently used in conjunction with computer networks, especially those related to industrial automation or to embedded systems.

1.1.3.1 Fieldbuses, Serial Buses

The term "fieldbus" originates from bus systems used in the production field of a manufacturing or processing plant. By itself the term is generic, meaning that by simply referring to a "fieldbus" one cannot determine which exact type of fieldbus is used.

Unfortunately the "Foundation Fieldbus" is sometimes referred to as "Fieldbus" which can lead to confusion since the Foundation Fieldbus is simply one particular "brand" of fieldbus, comparable to DeviceNet, Profibus, Interbus, Modbus and others.

Many fieldbuses are based on a serial bus, meaning that data is transmitted over the fieldbus on a bit-by-bit basis.

1.1.3.2 Arbitration, Token-ring, Multi-master, CSMA/CD

Most computerized communication systems require a method for avoiding collisions. A collision occurs if two or more nodes transmit at the same time and thus destroy each other's messages. In other words, when may a particular node transmit something, and for how long?

One method would be to pass a token (could be a specific message) from one node to another (forming a logical ring). Only nodes that currently have the token may transmit something to the network. Once a node is done, it passes on the token and remains silent until it gets the token again.

In a multi-master environment nodes may transmit at any time and collisions are resolved immediately upon detection. The method used by Ethernet is Carrier Sense Multiple Access with Collision Detection (CSMA/CD). In short it means that each node listens to the network (carrier sense) and may transmit at any time (multiple access). If a collision is detected in Ethernet, a jamming sequence is started which destroys the message for all nodes participating in the communication. After a random time delay the transmitting nodes will re-try.

The problem with such a communication scheme is that the jamming sequence destroys bandwidth (no data can be transmitted during the jamming) and message delays are not deterministic, making the overall response or transmit times hard to predict.

CAN uses a modified version of CSMA/CD with Collision Avoidance (CA). Instead of a jamming sequence, CAN resolves collisions by priority so that in a collision the message with the higher priority gets access to the network. This process is described in detail in Chapter 5, Section 5.2.8.

1.1.3.3 Input, Output

When looking at a single communication node, one could argue that there are inputs and outputs to both the application (sensors and actuators) and the network – an application input is transmitted "out" to the network, so it could be considered an output (to the network).

To avoid confusion, all control systems consider "input" and "output" as they relate to the application, not to the network.

An input signal comes from the application, typically from a sensor and goes into the control system. If a network is used, the signal gets transferred via the network to another node, either a master or directly to another output node.

An output signal goes to the application, typically to an actuator, and comes from the control system. If a network is used, an output module receives the signal from either the master or another input module.

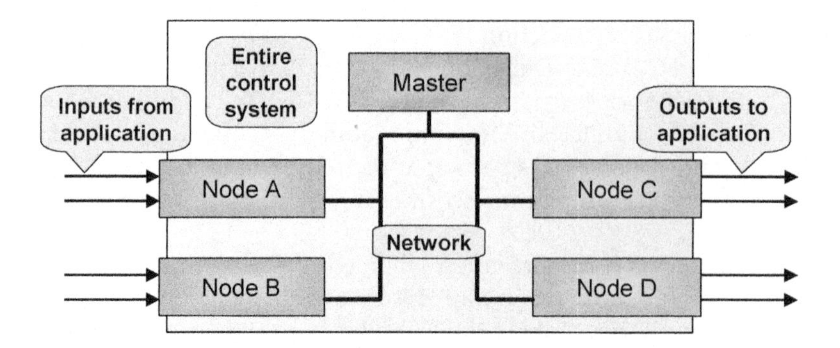

Figure 1.2 Inputs and Outputs - Traditional

The traditional control system in Figure 1.2 has 2 input nodes (A and B) and 2 output nodes (C and D). As long as a master is involved the inputs and outputs are fairly clear - all the input nodes transmit their data to the master and the master transmits the calculated output data to the output nodes.

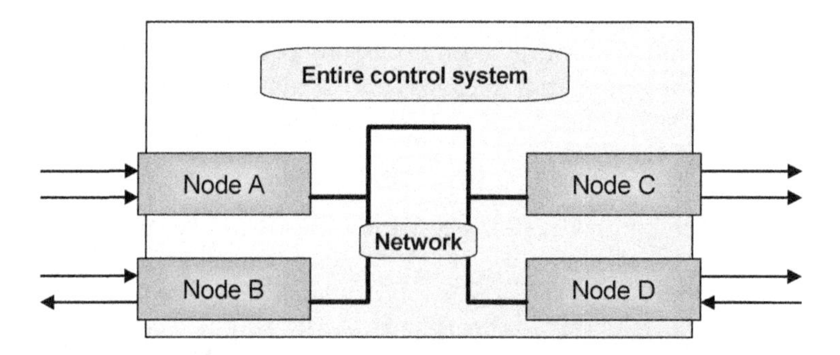

Figure 1.3 Inputs and Outputs - Embedded

However, in many embedded networks the scenario becomes more complex, as shown in Figure 1.3. Often embedded networks use distributed control, meaning there is no master and each node has more intelligence to decide on its own what to do with the data. As a result the data flow is more flexible; an input node sends its data directly to an output node which by itself decides when and how to switch its outputs. What can be confusing to beginners in embedded networking is that at some point the message sent as a result of an input "mysteriously" becomes an output.

As the same figure shows, when discussing embedded networks one should also avoid the terms "input node" or "output node" because many nodes (such as those illustrated by nodes B and D) might have both inputs and outputs.

In general, it is best to refer to the terms "inputs" and "outputs" only as long as these terms directly correspond to the inputs and outputs of the application, not of the network. For any data in transition on the network simply refer generically to "network variables" or "process data variables." In this book we will primarily use the term "process data variables" because in CANopen the term "network variables" has a specific meaning and should not be used in a general sense.

1.1.3.4 Master/Slave

Talking about masters and slaves in a network implies that the master has some sort of control function over the slaves. Typically this involves scanning the network and detecting the insertion or removal of slave nodes and the configuration of the nodes (informing them about communication channels and methods to use). Functionality may also include shutdown and/or reset of single nodes or the entire network.

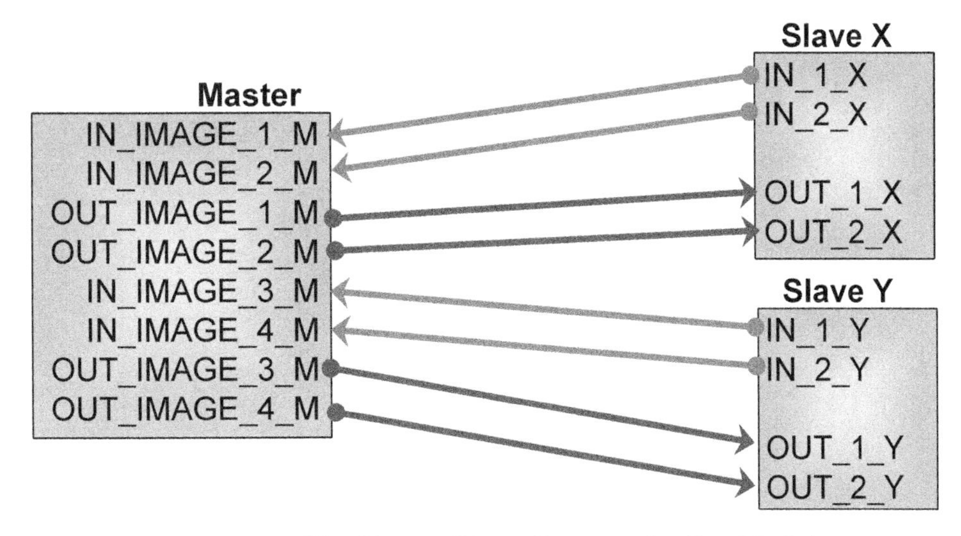

Figure 1.4 The Master/Slave Communication Model

A master/slave communication method as shown in Figure 1.4 refers to master-driven communication. In such an environment the slaves cannot typically communicate by themselves or with each other. Only the master may initiate communication and slaves only respond when they receive such a request from the master.

In comparison, Figure 1.5 shows a direct, master-less communication model, where all devices can directly exchange information without the requirement to route messages via a master. Obviously such a direct communication model is much more efficient, as it has less communication overhead. A single message is sufficient to send data from one device to another. In a master/slave communication model two messages would be required – one from the input to the master and one back from the master to the output.

A "flying master" is a dormant master that can become active and take over "on-the-fly" to be the new master. There are different methods for determining when and how such a dormant master can become active. One possibility is using a negotiation phase where a node on the network asks the existing master for permission to become the new master. Another method is that the dormant master monitors all communication on a network and recognizes when the existing master fails. In that case it automatically becomes active and replaces the master that failed.

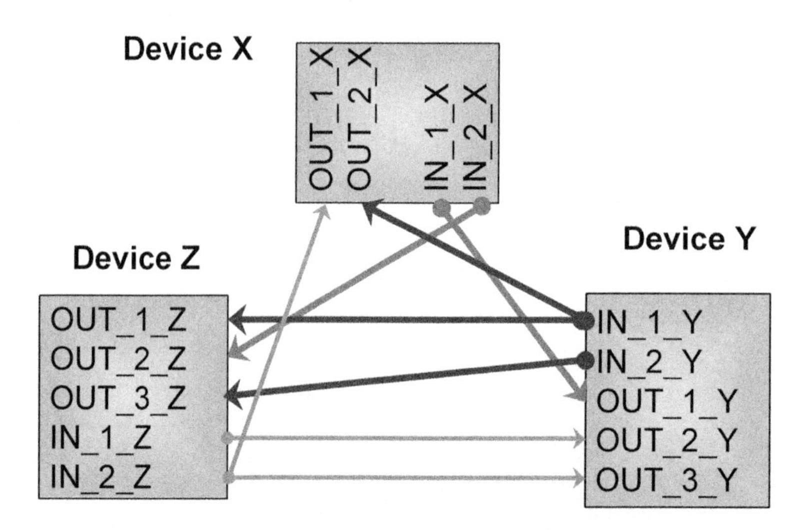

Figure 1.5 Direct Communication Model

1.1.3.5 Client/Server

In a client/server environment the server provides "services" to the network. A typical service could be serving data access points (inputs and outputs) to the network. A client is a network node making use of these services. Whether a network module becomes a server or client is completely unrelated to its status as a master or slave. Both masters and slaves can implement client and server functionality.

1.1.3.6 Producer, Consumer

A producer transmits data to the network and a consumer receives data from the network. For a specific set of data there can only be one producer and there is at least one, but possibly multiple consumers for that data.

1.1.3.7 Point-to-point, Multicast, Broadcast

A computer network typically supports multiple communication channels and methods. A point-to-point communication involves just two nodes on the network communicating directly with each other. All other nodes on the network either do not see this channel or ignore it, so they are not affected by the communication going on between the other nodes.

A multicast is the transmission of one message to multiple nodes. Typically the message is not duplicated to achieve a multicast, rather the consumers are configured to simultaneously receive the single message transmitted.

A broadcast is the transmission of one message to *all* nodes connected to the network. A master typically uses broadcasts to issue network-wide commands (commands affecting every node) or to signal an emergency.

1.1.3.8 Message Triggering: Polling, Time Driven, Event Driven, Change-of-State (COS) and Time Triggered

The overall performance of a network in terms of achievable data bandwidth and latency times often depends on the message triggering method chosen. So when and how does a message with process data get transmitted? The following is just a summary of some of the basic methods. These methods are covered in more detail in later chapters, organized according to where they are used in a CANopen system.

The most traditional method coming from pure master/slave environments is polling. With this method a master polls the inputs as required by the control algorithm in the master. Because an additional polling message is required, the overhead is fairly large, decreasing available bandwidth.

In time driven communication, the producers transmit messages automatically on a fixed time basis, for example every 50 milliseconds. This method makes the bandwidth requirements and worst-case delay times very predictable.

Time driven communication can be divided into methods using a local or a global timer. When local timers are used, each node has its own timer and individually transmits the message(s) upon timer expiration. Because the local timers are unsynchronized, the timing relationship between nodes is unspecified. On the other hand, if a global timer is used all transmissions are synchronized since all nodes will be using the same timer reference.

Event driven on a change-of-state allows for the fastest possible reaction time, because data gets transmitted as soon as it changes. Bandwidth usage is optimized, because data does not need to be transmitted if there are no changes. Unfortunately, this method is the least predictable, since many input changes in a short time will create message bursts on the network.

In cases where an input changes constantly, such a trigger method would cause continuous network traffic. To prevent this, CANopen defines an "inhibit time" during which a message may not be re-transmitted. When transmitting a message, a node needs to wait for the inhibit time to expire before it may re-transmit the message with the same variables.

Time triggered communication provides synchronization signals with time windows. A producer may transmit its message in a time window that starts a certain number of microseconds after the synchronization signal and has a specified length of several microseconds. This method allows for the reserving of bandwidth in the form of time slots for certain communication channels.

It should be noted that in CANopen all of these trigger mechanisms are available and combinable (except for time triggered communication). More detailed examples are described in Chapter 2, Section 2.5.

Let's take a look at an example to help clarify the terminology covered so far. The network in Figure 1.6 has one master and three slaves. The slaves provide their configuration data as a service to the network. They implement server functionality to do so – they serve that data to the network. The master becomes a client and requests the "health" data from the clients using a point-to-point communication channel.

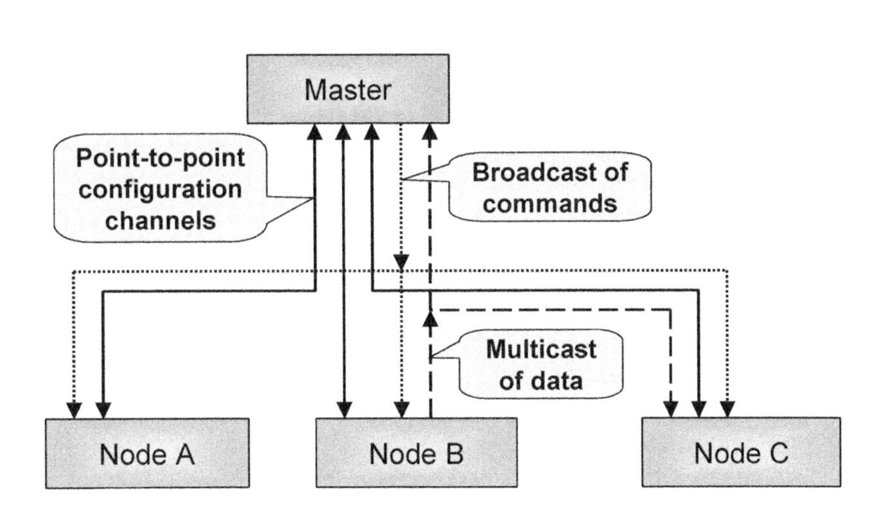

Figure 1.6 Sample Network Layout

The master uses a broadcast channel to inform all slaves with one message to switch into a specific operation state (run/operate or stop).

Node B is an input node that produces multicast messages of the input data on an event driven basis to the master and Node C – which in turn become consumers.

1.1.3.9 Real-time, Latency

Many embedded control systems, especially those used in industrial automation applications, have real-time requirements. This means that data needs to be processed immediately, in real-time, within a specified time slot that may not exceed a predefined limit. There is no general specification on how fast a control system needs to be to qualify as "real-time system" – it just needs to be fast enough to be able to handle the requirements of the application.

If the real-time requirements are such that an input needs to influence an output within 10 milliseconds, then the control system must guarantee that a change to the input affects the output accordingly in 10 milliseconds or less.

If a network is involved, transmitting messages may involve latency times – the time a message can be delayed by not having immediate access to the network. This may be because there is another message currently in progress on the network that cannot be interrupted.

As an example, let's see what individual times make up the total maximum time for an I/O cycle – that is, an input read being transmitted via a network and switching an output somewhere else on the network.

Input-scan cycle time:
Any microcontroller implementing an input typically scans the inputs at a fixed rate, such as every 100 microseconds. The worst case would be that we just did a scan, the input changes, but it takes us the entire scan cycle time (100 microseconds) to re-check the input and recognize the change.

Input filter/debounce time:
In order to ensure that an input signal is stable and not just a disturbance on the input line, many systems require that an input signal is detected continously for a specified time period. This time period is application specific.

Software processing time of input module:
The software in the input module reads the input signal and transfers it to the peripheral handling the network. Depending on other interrupts or tasks running on the microcontroller, the execution time for this piece of software can vary. In a real-time system it must be clearly determinable what the maximum runtime is.

Network latency:
The network message that the input module is trying to send might not be immediately sent if other network traffic is currently ongoing. For real-time systems it must be clearly determinable what the maximum latency is.

Network transmit time:
The time it actually takes to send the data via the network.

Software processing time of output module:
The software in the output module detects the receipt of a network message containing new process data and transfers the data to the appropriate output. Depending on other interrupts or tasks running on the microcontroller, the execution time for this piece of software can vary. In a real-time system it must be clearly determinable what the maximum runtime is.

1.1.3.10 Physical Stuff – Signals, Wires, Speeds and Network Layout

Before a communication channel between two or more computerized systems can be established, some basic physical decisions have to be made. What is the physical transmission media chosen? Wire, radio signals or something else? And if wire, what kind? And how will the physical signal look on it? And what will be the speed or the maximum bandwidth?

For embedded and industrial communications, wire is the first choice when it comes to the physical media. Most common is a "regular" wire, meaning no special demands on impedance, resistance or conductance are made. For noisy environments, with a lot of Electro Magnetic Interference (EMI), twisted and shielded wires are preferable. Many embedded networks try to use readily available wiring like Ethernet cables, phone cables, or serial cables as used on PC COM ports.

The limiting speed factor for embedded networks is that all connected microcontrollers need to be able to deal with the speed used. A network running at 1Mbps can transmit one bit per microsecond. An 8-bit microcontroller that only executes one instruction per microsecond needs several instructions to transfer a byte from one location to another, and could just about keep up with the communication of a 1Mbps network.

> How much CPU performance is required to handle a network operating at the highest speed rates and 100% busload?
>
> Just to give a quick example: "highest speed rates" means that the length of a single message is roughly between 50 and 150 microseconds. So the worst case for the receiver is that a message comes in and needs to be processed every 50 microseconds. If the receiving microcontroller cannot keep up, messages might potentially get lost.
>
> Later in Chapter 5, Section 5.3 we will see how sophisticated implementations of CAN interfaces help to keep down the workload for the microcontroller. By offering hardware filtering mechanisms these implementations can be configured to ignore messages that are of no interest to the local microcontroller. The microcontroller only needs to react if a message comes in that is meant to be received by the local microcontroller.

The network layout refers to the physical connection of the nodes to the networks. Common layouts are stars and buses. A star uses a central hub and all nodes connected to the network are connected to that hub. A bus is a line and nodes may be connected anywhere on that line. Some buses allow junctions or drop lines (lines splitting off from the bus), some do not. For more details on the physical layout, see Section 5.2.6.

1.1.3.11 ISO 7-Layer Reference Model

The standard network communication model is the ISO layer model that defines 7 layers from the physical media up to the application interface [ISO7498]. Most on-chip communication interfaces usually only implement layer 1 (Physical Layer) functionality. Some, like CAN, also offer partial layer 2 functionality (Data Link Layer). Functionality from the layers above is usually implemented in software only. Protocol standards that implement these layers or parts thereof are referred to as "higher-layer protocols."

It should be noted that not all layers are implemented for embedded networking applications. Just to give an example, it would not really make much sense to add overhead for long-distant routing of messages if the network does not have any "long distance" functions.

Also, traditional 7-layer implementations would require an interface between any two layers next to each other, resulting in an overhead that is unacceptable for embedded applications. That is why higher-layer CAN protocols only implement selected functionality from the higher layers, to minimize the overhead.

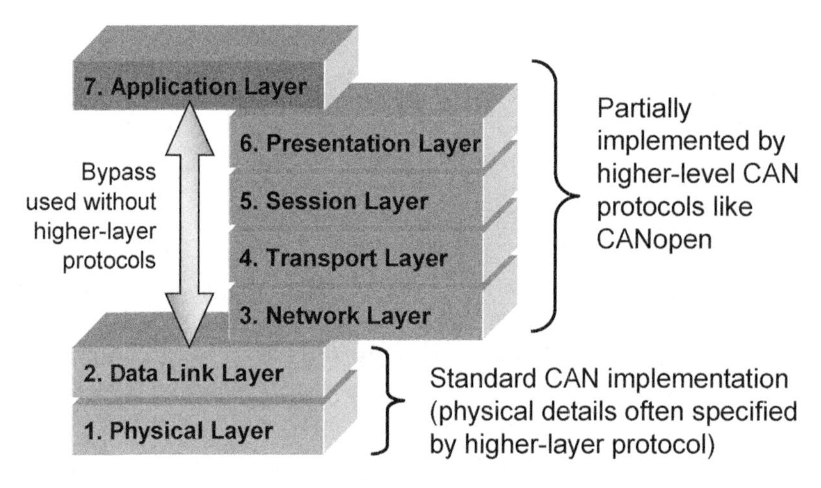

Figure 1.7 The ISO 7-layer Reference Model

The ISO 7-layer network reference model is used to classify the functionality provided by a communication network system. The model expects a clear separation between these layers with defined interfaces between them, to achieve interchangeability. For more information on this reference model see [Comer00].

As noted above, this type of implementation is not always suitable in embedded systems, as the overhead of all these interfaces would be too big to implement efficient communication systems for lower-end microcontrollers with limited resources.

With CAN, most parts of the physical and data link layers are implemented in hardware and there is no common, standardized software interface. Some applications put their own application layer directly on top of the data link layer. However, with CANopen at least parts of the other layers are implemented.

1. Physical Layer:

   - Describes the physical interconnection between network nodes
     (CANopen: specifies usage of ISO 11898, high-speed)

   - Includes electrical characteristics of signals used
     (CANopen: chosen transceiver uses differential signal)

   - Defines "bit-level" communication
     (CANopen: bit generation, synchronization)

2. Data Link Layer:

   - Bits are combined into frames
     (CANopen: CAN data frames)

   - Includes error detection via checksums
     (CANopen: provided by CAN)

   - Defines set of acknowledgements to determine successful transmission
     (CANopen: provided by CAN)

   - Enables successful point-to-point communication to the next bridge or gateway, but not beyond
     (CAN/CANopen: not provided)

3. Network Layer:

- Includes concepts of destination addressing and routing (CANopen: SDO channels)

- Provides interaction functionality between a host and the network (CANopen: configuration via SDO)

- Uses fragmentation to allow transmission of messages larger than allowed with frames
(CANopen: segmented/fragmented transfer supported)

- Able to detect and respond to network bandwidth limitations
(CANopen: not provided)

4. Transport Layer:

   - Provide end-to-end reliability: communication between source and destination hosts
   (CANopen: partially provided by NMT services, see Chapter 2, Section 2.6).

   - Double-checks that no switch, bridge or gateway in between end-to-end communication has failures
   (CANopen: no long-distance routing supported)

5. Session Layer:

   - Allows different hosts on the network to begin and end communication sessions
   (CANopen: not typically used)

   - Token management: Only the side holding a token may perform critical functions like a write access to a shared data base record
   (CANopen: SDO channel management)

   - Synchronization: Can be used for large data transfers – supports resume of an interrupted transfer
   (CANopen: SDO block transfer mode available with abort, but no resume)

6. Presentation Layer:

   - Handles data representation and encodes data in a standardized way
   (CANopen: Object Dictionary, defined data types)

- Data compression
  (CANopen: not supported)
- Encrypting/Decrypting
  (CANopen: not supported)

7. Application Layer:
   - Application programs making use of the network

1.2 Code Requirements for Embedded Systems

Objective

CANopen is mostly used in embedded systems. For those of you who come from a pure PC programming environment, we will point out a few things that you should be aware of when jumping into the programming of embedded systems. If on the other hand you don't need to implement CANopen nodes yourself and you are just integrating or configuring CANopen networks, you may want to continue your reading with Section 1.4.

Those of you with multiple years of experience in the "embedded field" by either designing and/or programming microcontrollers might be tempted to skip this section. However, we would recommend that you at least glance at it, as we will specifically point out the impact that typical limitations of embedded systems have on CANopen implementations.

We address topics such as limited resources (memory and CPU performance), limited debugging environments, typically available communication channels and real-time requirements.

If you are a newcomer to embedded systems, you should consider additional literature such as [Barr99], [Berger01] or [Ganssle00].

The field of "embedded systems" can be divided into two main categories. On one hand we have the high-volume electronics typically used in many consumer products or other every-day products, as well as many products from the sensor and actuator level of the automation pyramid shown in the previous section. On the other hand we have the lower-volume specialty electronics that run some very specific control tasks, often situated in the Controller or Process Control levels of the automation pyramid. In very rough terms, high-volume refers to systems used in quantities of hundreds of thousands whereas low-volume in this context indicates a maximum of a few thousand.

If CANopen is used in embedded systems it is important to know which category it is in. In general, the low-volume applications tend to be less price sensitive and can afford to use microprocessors or microcontrollers with more horsepower and more memory, facilitating the implementation of CANopen. Commercial CANopen source code implementations can be purchased and integrated – typically without running into any performance or memory requirement issues.

However, on high-volume embedded systems price is a very important factor and there will be a certain limit on the resources (both in CPU processing time and memory usage) that can be made available to implement CANopen. Depending on the CANopen functionality required by an application, it might be impossible to implement full-blown CANopen on a lower-end 8-bit microcontroller.

The following is a list of resource constraints typical for embedded systems and a summary of how these constraints affect a desired CANopen implementation.

1.2.1 CPU/MCU Performance

CANopen is very flexible, so the amount of CPU processing time required for handling the communication itself greatly depends on the CANopen functionality implemented. CANopen can be handled by 8-bit microcontrollers running at speeds as slow as executing just one or a few assembly instructions per microsecond. However, there are typically some constraints one might run into. For example, the maximum bus speed supported might be slower than 1Mbps, or, if the 1Mbps rate is supported, it may be that a device cannot handle a maximum 100% busload.

In addition, Section 5.3 shows how different CAN controller implementations offered by different chip manufacturers impact the MCU performance required to handle the CANopen communication. Some CAN controllers have advanced filtering and/or buffering techniques, greatly reducing the burden on the MCU.

Systems using an 8-bit microcontroller unit would either need to be prepared to sacrifice a significant share of the MCU processing time for handling the CANopen communication, or sacrifice CANopen performance. There might be bursts where some 50% or more of the MCU time needs to be dedicated to the CANopen communication.

Obviously this would not leave enough resources remaining for a demanding application such as a multi-phase motor control. However, it is more than adequate for simple sensors (such as temperature sensors).

1.2.2 Real-Time Requirements

Another performance factor to consider is the real-time behavior, defined as the guaranteed response time to an event. For example, can it be guaranteed that once a CAN message is received with new output data, that this data will actually be applied to the output pins within a certain time limit? Applications with high real-time demands may require that this time be a fraction of a millisecond.

8-bit microcontrollers that are based on commercial, portable CANopen source code might have a tough time guaranteeing such a value unless processor and possibly application-specific optimizations are made to the code.

If an application has specific real-time demands it may be necessary to either use a more powerful microcontroller or to hand-optimize the CANopen code towards the application. The drawback is that after such optimization it is much tougher to port the code to different microcontroller architectures.

1.2.3 Code Memory Space

The code memory size required for a CANopen slave protocol stack varies greatly. Not only does C source code compile very differently on various microcontroller architectures, the code size varies even more depending on which CANopen features are enabled or disabled. Most commercial source codes allow code segments to be included or excluded from the program via C "#define" statements. If, for example, the optimized block transfer routines are not required, there is no need to actually include and implement the code for this function.

On an 80C51 microcontroller code sizes for CANopen can vary from 2kbyte for minimal bootloader functionality (not truly implementing a full CANopen node) versus 4kbyte-5kbyte of code for minimal CANopen implementations (like MicroCANopen, see Section 6.3) all the way up to 25kbyte-45kbyte for a full-blown CANopen slave node with all the bells and whistles.

CANopen masters or managers vary even more in functionality and may use considerably more code memory.

Although the overall situation is similar on 16-bit and 32-bit microcontrollers, they typically have more overall code space available so saving a few kilobytes of code is not as crucial as on an 8-bit device.

1.2.4 Data Memory Space

In general, the data memory requirements depend on factors similar to the code memory requirements. In keeping with the example from above, a correct implementation of the block transfer mode requires a RAM buffer for all the data received via the "block transfer." Again the requirements have a wide range from 100-200 bytes for minimal implementations like MicroCANopen and some 500 bytes to 1kbyte for full-blown CANopen implementations. This is without the process variables themselves – so all variables transmitted or received by a node typically need additional RAM space. Nodes that need to receive or monitor *all* process variables may need several hundred bytes of additional RAM.

1.2.5 Non-volatile Data Storage

If a CANopen node can be re-configured during operation and such configurations need to be stored or reloaded after start-up, non-volatile storage memory such as an EEPROM is required. Typically three sets of configuration data are required – two in EEPROM and one in RAM. The manufacturer default configuration and the last saved configuration are held in EEPROM, while the current configuration is held in RAM.

1.3 Communication Requirements for Embedded Networking

> Objective
>
> The previous section examined code requirements. In this section we will outline the basic communication requirements.
>
> A question often asked by novices is *Why do we need anything besides TCP/IP anyway? That is the standard in networking, so why not use it in embedded systems, too?* This chapter will answer these questions and outline the requirements desired by many embedded networking applications.

1.3.1 Higher-Layer Protocol

While many communication/networking technologies are available on-chip with even the lowest-priced microcontrollers, they lack a dominant higher-layer protocol standard. All the serial interfaces typically provided on-chip with many microcontrollers only includes some sort of layer 2 (Data Link layer) interface. That means that some functionality is provided to transmit and receive data, but it is not defined when and how messages go over the network and what kind of data they contain. As soon as someone starts to specify things like data types (includes bit and byte order, for example, to ensure that everybody knows if the hi-byte or low-byte comes first in a word) or message identifiers to be used for specific services (either to recognize specific variables or configuration settings), he/she defines a higher-layer protocol.

Many applications still use proprietary higher layer protocols. Typical pitfalls with proprietary protocols are:

- They must be very well documented, otherwise they are only usable by the people who invented them
- New team members have no other source for learning than the in-house documentation and possibly in-house cross training
- No third party, off-the-shelf development and test tools are available for the protocol; they must be developed in-house
- No access to plug-and-play modules of third parties

For embedded networking applications a standardized higher-layer protocol is desirable to avoid the pitfalls listed above. However, one should pay careful attention when choosing a protocol for embedded networking to ensure the requirements detailed in the following sections are met.

1.3.2 Price, Performance, Resources

As outlined in the previous sections, many embedded applications are price-sensitive and thus cannot incorporate the hardware and software required to handle TCP/IP in every network node. There are indeed implementations of small embedded TCP/IP nodes such as embedded webservers, however, these still require a lot of resources, not only in CPU performance but also in memory for storing the data files actually "served" by the web server.

As of today it is not yet imaginable that such implementations could become so affordable that we could put one in every light bulb. However, many long-life, low-power lighting technologies already use microcontrollers today. So utilizing some other, more affordable networking technology will most likely happen first.

In addition to cost there are also technical considerations. TCP/IP is not very suitable for control purposes simply because it was not designed for that purpose. For example, (and this becomes most visible in the actual message definitions) in Ethernet single messages can have up to 1500 bytes of data and messages have an overhead of about 24 bytes (preamble, addresses, type info, checksum). The TCP/IP layers add additional overhead resulting in even more bytes wrapped around the data, potentially allowing messages of up to 64kbytes. If a node only implements a simple analog sensor (such as temperature, pressure, speed, distance, or similar) it typically only has one variable of 8 or 16 bits to report. If the node used TCP/IP on Ethernet as the communication network, every data word transmitted would result in some 50 or more bytes of overhead being transmitted with it. And this does not include any overhead one might have for establishing a communication channel between two nodes.

To summarize, a "usable" embedded networking technology must work on some of the lowest priced microcontrollers. These are typically 8-bit devices with just a few kbytes of code space and a few hundred bytes of RAM. The technology must use one of the existing communication channels available on-chip (like UART, I^2C, CAN or others) and preferably the technology should not require a lot of code overhead for handling the communication. That overhead, however, depends on the higher-layer protocol being used.

1.3.3 Definition of Data Types and Process Variables

A higher-layer protocol usable for embedded networking would need to specify ways to recognize variables. There must be some methods in place to define data types (such as signed and unsigned integers of different lengths) and identify the variables themselves. If a temperature sensor transmits a temperature, the receiving party or parties need to be able to recognize that this is the temperature value.

It would also be desirable to be able to directly request a specific variable using a monitoring or analysis tool. If a node has several variables (maybe an entire array of temperatures), it would be nice to be able to request a specific variable: "Please send the information for temperature sensor number 3 now."

This should work in both directions - both read and write accesses to variables should be available to masters or monitoring/analysis tools.

1.3.4 Exchanging Process Variables

If a network technology is used that supports multi-master (any node can send a message at any time, collisions are resolved) and multicast or broadcast (a message transmitted is received simultaneously by a group of nodes or all nodes), then it should be possible to take advantage of these features.

CAN supports these features and thus it is possible to set nodes to individually decide when to transmit a message, for example by using change-of-state or event time trigger mechanisms.

It should also be possible to directly link variables between devices instead of only offering the master-slave communication model. So if a process variable produced by one node is required by several other nodes, it should be possible to configure all the receiving nodes to directly consume that variable whenever transmitted over the network. Without such a feature the interference or translation of a master is required.

1.3.5 Configuration of Network Devices

Preferably, network nodes are at least in part (re-)configurable via the network itself. A master or configuration tool should be able to read and/or set network parameters in individual nodes that define the communication behavior of that node.

This could include things such as how often to send a heartbeat message or which process variables are transmitted from where, to where, in which messages and when (triggering mechanisms).

1.3.6 Off-the-Shelf, Plug-and-Play

One of the bigger challenges is the demand for off-the-shelf, plug-and-play support. For a network technology this means that nodes (like I/O modules, sensors, actuators, etc.) are available from several providers and are interchangeable.

System designers that build a network can use these components and integrate them along with their own network nodes. This way only nodes with specific requirements would need to be developed from scratch. Generic I/O nodes would not need to be (re-)developed, but could be acquired from third party providers instead.

1.4 Introduction to CANopen from the Application Level

Objective

In the previous sections we have examined general requirements for embedded networks. In this section we would like to introduce CANopen and point out how many of the embedded networking requirements are met by CANopen.

This section is an introduction to the primary functionality provided by CANopen and is intended to give students a quick start into the main ideas of CANopen.

The chapters following this section will repeat some of the basic information presented here and add technical details that are missing in this first overview of CANopen functionality.

1.4.1 The Object Dictionary Concept

The core of any CANopen node is the Object Dictionary (OD), a lookup table with a 16-bit Index and an 8-bit Subindex. This allows for up to 256 Subentries at each Index.

Each entry can hold one variable of any type (including a complex structure) and length. In the following sections the terms Index, Subindex and Subentry will be used when describing such Object Dictionary entries.

All process and communication related information/data is stored as entries in pre-defined locations of the Object Dictionary. Unused entries do not need to be implemented.

The Object Dictionary not only provides a way to associate variables with an Index and Subindex value, it also specifies a data type definition table. The entries starting at Index 1 are exclusively used to specify data types. Table 1.1 shows the first seven entries in the Object Dictionary defining some commonly used data types. The complete listing of pre-defined data types is given in Chapter 2, Section 2.2.3. In addition, CANopen also supports application specific data types that can be added to the list of supported data types.

Index	Data Type
1	BOOLEAN
2	INTEGER8
3	INTEGER16
4	INTEGER32
5	UNSIGNED8
6	UNSIGNED16
7	UNSIGNED32

Table 1.1 Object Dictionary Entries Starting at Index 1 Define Data Types

It should be noted that the entries mentioned above are only used to define data types, not to store any variables. The Object Dictionary entries beyond 1000h are used for variable storage; if an entry is specified to be of type "UNSIGNED16" then an alternate description of the data type (for example, used in electronically readable specifications) is used to indicate it is of data type 6.

As specified, the Object Dictionary satisfies the basic networking requirement of being able to define data types and place variables into the network nodes. If a specification says that a node must have a variable called "X-Position" which is located at

Index 2000h, Subindex 0 and its data type is 4, then according to the Object Dictionary the data type is INTEGER32, an integer value of 32 bits.

Index	SubIdx	Type	Description
1000h	0	UNSIGNED32	Device Type Information
1001h	0	UNSIGNED8	Error Register
1017h	0	UNSIGNED16	Heartbeat Time
1018h			Identity Object
	0	UNSIGNED8	= 4 (Number of sub-index entries)
	1	UNSIGNED32	Vendor ID
	2	UNSIGNED32	Product Code
	3	UNSIGNED32	Revision Number
	4	UNSIGNED32	Serial Number

Figure 1.8 Mandatory Object Dictionary Entries Supported by all CANopen Nodes

Figure 1.8 lists the mandatory Object Dictionary entries that every CANopen node must implement to be CANopen compliant. Primarily, these provide the device type information that gives an indication of which device profile a device belongs to (if any), an error register, and an identifier record. The heartbeat is a low-priority status message sent by a node on a periodic basis. The heartbeat time is listed here because every node must support either the heartbeat or node guarding mechanism; today heartbeat is the recommended, preferred method.

1.4.2 Device Profiles

Although the Object Dictionary concept allows for structuring the data that needs to be communicated, there is still something missing: Which entry in the dictionary is used for what? The dictionary is far too big to allow the master to take "wild guesses" and simply try to access certain areas of the dictionary to see if they are supported.

The solution is simple. First of all, there are a few mandatory entries that all CANopen nodes must support. These include the identity object with which a node can identify itself, and an error object to report a potential error state. In addition, optional entries are specified by the CANopen specification. The Device profiles are add-on specifications that describe all the communication parameters and Object Dictionary entries that are supported by a certain type of CANopen module. Such profiles are available for generic I/O modules, encoders and other devices.

A master or configuration tool can read-access the identity object of any slave node using a Service Data Object or SDO (a messaging protocol – more about this shortly). As a reply, it receives an SDO with the information about which device profile a module conforms to. Assuming the master knows which object entries are defined for a particular device profile, it now knows which Object Dictionary entries are supported and can access them directly.

There may be instances where an application requires the implementation of non-standardized, manufacturer-specific Object Dictionary entries. This is not a problem, because CANopen is truly "open." Additional entries that disable or enable a certain functionality that is not covered by one of the existing device profiles can be implemented in any device, as long as they conform to the structural layout of the Object Dictionary.

1.4.3 Electronic Data Sheets

Electronic Data Sheets (EDS) offer a standardized way of specifying supported Object Dictionary entries. Any manufacturer of a CANopen module delivers such a file with the module, which in layout is similar to the ".ini" files used with Microsoft Windows operating systems. (Note: a future standard for EDS files based on XML is currently in development.)

An example of an Object Dictionary entry in an EDS file is:

```
[1000]
ParameterName=DeviceType
ObjectType=0x07
DataType=0x0007
AccessType=ro
DefaultValue=0x00030191
PDOMapping=0
```

The example above shows the EDS definition of the Object Dictionary entry [1000h,00h]. The data type is 7 (UNSIGNED32, see Table 1.1).

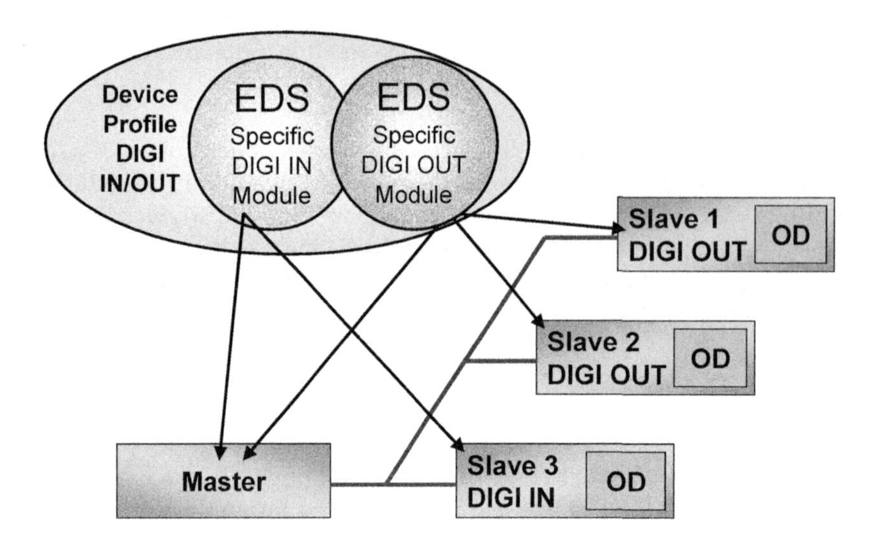

Figure 1.9 Electronic Data Sheets (EDS) Specify the Contents of Object Dictionaries

A CANopen master or configuration tool running on a PC with a CAN card can directly load the EDS into its set of recognized devices. Once a device is found on the network, the master or configuration tool will try to find the matching EDS. Once found, all supported Object Dictionary entries are known by the master/configuration tool.

Figure 1.9 shows the relationship between Device Profiles and Electronic Data Sheets. The Device Profile specifies the minimum entries that need to be supported by a device conforming to the profile. However, the EDS might only specify objects that are specific to a certain manufacturer or sub-type of module.

Device Profiles and Electronic Data Sheets are the basic functionality needed to meet the requirement for "off-the-shelf" availability of network devices. From the communication point of view, any two nodes that conform to the same EDS are interchangeable - their Object Dictionaries are identical and they have the same communication behavior.

1.4.4 Accessing the Object Dictionary: SDOs

The next requirement is that of a direct communication channel. A master or configuration tool needs to be able to read and/or write the Object Dictionary entries of all the nodes connected to the network.

CANopen supports such a basic client/server communication method by implementing a point-to-point communication mode that allows for the issuing of read or write requests to the node's Object Dictionary. Messages that contain requests or answers to/from the Object Dictionary are called Service Data Objects (SDO).

It should be noted that by default only one node in the system has the right to actively initiate this SDO communication mode. Typically this is some sort of master/manager. However, there are ways for other nodes such as configuration or analysis tools to request the use of SDO communication channels.

The default SDO communication is a master-driven (client) request/response communication. The master/manager "owns" all the SDO communication channels and has one channel available to each node in the system. Only the node that owns a channel may send an SDO read or write request to the node (and Object Dictionary in it) and the node addressed must reply with an SDO response either confirming the write access or replying to the read-request (server, because the node "serves" its Object Dictionary data to the network).

It should also be noted that SDOs support something called "segmented transfer" that allows Object Dictionary entries of any size to be transmitted. If the content does not fit into a single message, it is automatically segmented and distributed via multiple messages.

The Service Data Object methodology allows master-driven read/write access to all Object Dictionary entries of all nodes connected to the network. Strictly speaking, this functionality by itself would already allow simple master-driven network systems. As both process and configuration data are part of the Object Dictionary, the process data could be updated using SDO transfers.

However, for a number of reasons this would not be a very efficient implementation. First, it only implements a polling scheme where the master must handle *all* inputs and outputs. Second, it also adds a lot of message overhead. To get an input to an output, four messages have to be transmitted via the network:

1. Master sends SDO read request to input node.
2. Input node replies with SDO response and the data.
3. Master sends SDO write request to output node.
4. Output node confirms with an SDO response.

Third, by definition SDOs always have a message length of 8 bytes, even if an SDO only contains one data byte or a simple acknowledgement without process data.

In summary, the Service Data Object ensures a basic access method to any entry in the entire Object Dictionary of any node. However, for pure process data communication a more efficient methodology is required.

Because the configuration data is available via SDO accesses, SDOs fulfill the requirement for "plug-and-play." A system integrator who needs a specific I/O node, such as a rotary encoder, can choose any product conforming to the Device Profile for encoders. The system integrator or technician can then use CANopen configuration or master software to configure the node to perform the communication actions as demanded by the specific application.

1.4.5 Increased Performance with PDOs

For most applications, the SDO communication is not efficient enough to handle the exchange of real process data; the overhead is just too big and the message triggering methods are too limited (master-driven polling only).

Because CAN supports the multi-master communication concept (any node can send a message at any time and collisions are resolved by message priority), a more direct communication method is required to allow for more efficient, higher-priority access to process data.

The Process Data Object (PDO) implements an optimized solution for placing multiple process data variables from the Object Dictionary into a single CAN message of up to 8 bytes.

1.4.5.1 PDO Mapping

A PDO is like a "shortcut" to several process data variables in the Object Dictionary. Via a process called PDO mapping (all implemented through Object Dictionary entries), any dictionary entry can be mapped to data in a PDO, the only limit being that in total a PDO cannot contain more than 8 bytes.

Index	SubIdx	Type	Description
1A00h			1st Transmit PDO - Mapping
	0	UNSIGNED8	= 4 (Number of used map entries)
	1	**UNSIGNED32**	**= 6000 01 08h (Idx 6000h, SubIdx 1, 8 bit)**
	2	**UNSIGNED32**	**= 6000 02 08h (Idx 6000h, SubIdx 2, 8 bit)**
	3	**UNSIGNED32**	**= 6401 01 10h (Idx 6401h, SubIdx 1, 16 bit)**
	4	**UNSIGNED32**	**= 6401 02 10h (Idx 6401h, SubIdx 2, 16 bit)**
6000h			Process data, digital inputs
	0	UNSIGNED8	= 2 (Number of sub-index entries)
	1	**UNSIGNED8**	**8-bit digital input**
	2	**UNSIGNED8**	**8-bit digital input**
6401h			Process data, analog inputs
	0	UNSIGNED8	= 2 (Number of sub-index entries)
	1	**UNSIGNED16**	**16-bit analog input**
	2	**UNSIGNED16**	**16-bit analog input**

TPDO1	D IN 1	D IN 2	A IN 1		A IN 2		Unused	
	Byte 1	Byte 2	Byte 3	Byte 4	Byte 5	Byte 6	Byte 7	Byte 8

Figure 1.10 Process Data Object (PDO) Mapping Example

Consider the PDO mapping example in Figure 1.10. A CANopen input node supports two digital inputs of 8 bits each and two analog inputs of 16 bits each. In conformance with the Device Profile for Generic I/O modules, Object Dictionary entries at Index 6000h store the two digital inputs of 8 bits each, and entries at Index 6401h store the two analog inputs as two words.

The Object Dictionary entries at Index 1A00h specify the PDO mapping, indicating which bits of which Object Dictionary entries are used in the Transmit PDO 1 (TPDO1), filling the TPDO bit-by-bit. Note that this mapping can really be done on a bit-level. Each entry starts using the first available free bit in the PDO and occupies as many bits as it requires.

The second Subentry (Subindex 1) at Index 1A00h maps object 6000h, Subindex 1, 8 bits to the first bits of the TPDO1. The next Subentry (Subindex 2) at Index 1A00h maps object 6000h, Subindex 2, 8 bits to the next free bits of the TPDO1, and so on. In this example the remaining bits of TPDO1 (data bytes 7-8) remain unmapped and unused.

Which PDOs are pre-defined for specific nodes along with their default mapping is specified in the Device Profile and the Electronic Data Sheet.

1.4.5.2 PDO Linking

When it comes to the communication partners involved, PDOs have a default arrangement similar to SDOs. The default state is that the master is the only node that receives Transmit Process Data Objects (TPDO), and only the master may send Receive Process Data Objects (RPDO) to the slaves. In other words, it ensures that a pre-defined connection is usable by default, since unique CAN message identifiers are assigned to each supported PDO – one unique ID for each TPDO and one for each RPDO. In CANopen terms, the COB ID is the Connection Object Identifier that contains the CAN message ID and some additional configuration bits, such as a bit to enable and disable the PDO.

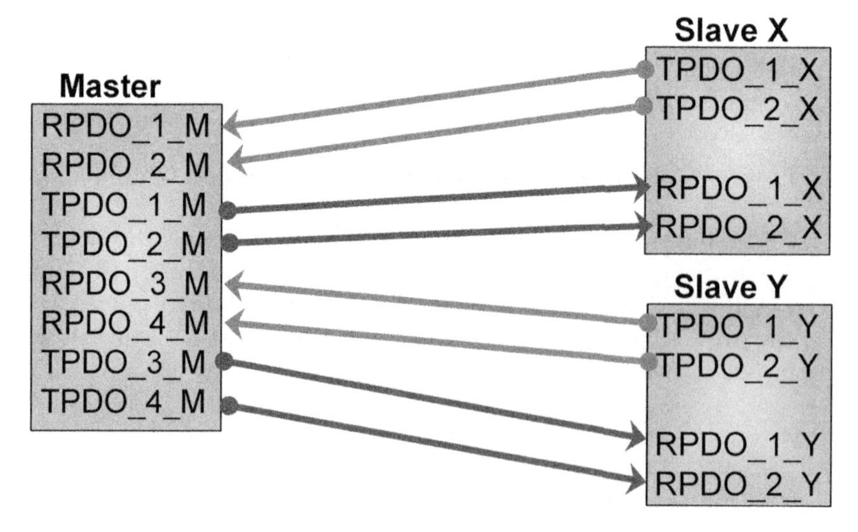

Figure 1.11 Default PDO Linking - Master/Slave Model

During the initialization and configuration cycle, the PDO linking can be changed. A master could inform one or multiple output modules that they should directly listen to a specific TPDO of an input module. Again, a TPDO correlates to a unique COB ID, a CAN message identifier. So in short, a node is informed as to which message frames it should listen to and which ones it can ignore.

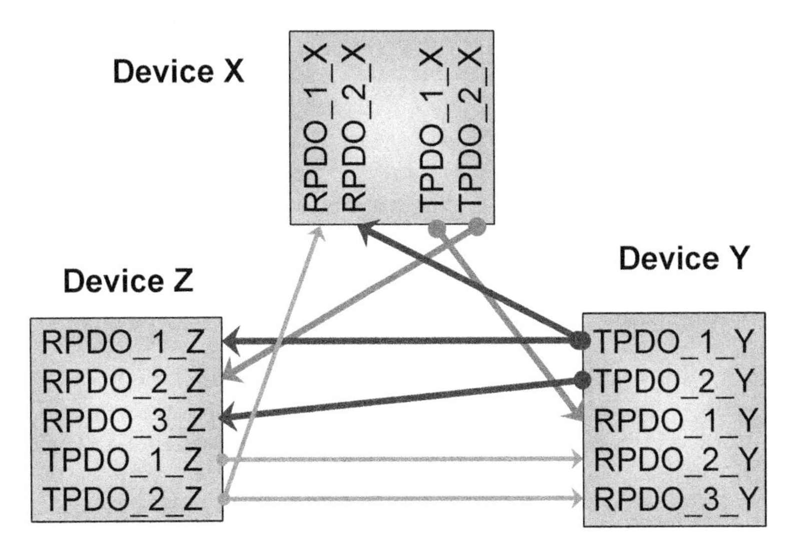

Figure 1.12 Optimized, Direct PDO Linking

Once these new linking settings are done and the network enters into the operational mode, the master would not need to get involved in the process data communication and could focus on other things like network management.

1.4.5.3 PDO Triggering

Now that a "shortcut" is available that allows several Object Dictionary entries to be packed into one message, what are the options for triggering a PDO? CANopen supports a total of four trigger modes:

1. **Event driven**: If the input device recognizes a change-of-state (COS) on any of its inputs, it updates the data in the Object Dictionary and the PDO and then transmits the PDO. This mode allows for some of the fastest response times.

2. **Time driven**: A PDO can be configured to be transmitted on a fixed time basis, for instance every 50 milliseconds. This mode helps to make the total busload more predictable.

3. **Individual Polling**: Using a regular CAN feature, the remote request frame, a PDO is transmitted only if the data is specifically requested by another node. Note: Using this feature in new designs is not recommended, as the specific implementation of remote requests varies between different CAN controllers!

4. **Synchronized**: A special mode allowing for a synchronized polling as required by many motion control applications.

These trigger modes are explained in detail in Chapter 2, Section 2.5.

The PDOs allow for the implementation of very efficient and flexible communication models. Being able to put multiple variables into a single message and sending them directly from one node to another (or a group of others) is a network service that is rarely available in more traditional industrial automation networks.

1.4.6 Network Management (NMT)

CANopen allows for a Network Management Master to watch over all nodes to see if they are operating within their parameters. Upon failure of a node or the reception of certain alarm/emergency messages it can initiate the appropriate recovery or shut-down procedures.

There are different options as to how this supervising of nodes is implemented. The latest version of CANopen recommends the usage of heartbeat messages. This allows nodes to supervise each other, even without a Network Management Master (if necessary).

The idea is that each node emits a regular heartbeat message as long as it is alive and operating within its parameters. If all nodes produce such a heartbeat, every node can monitor all the heartbeats of its communication partners. This is especially helpful in COS (change-of-state) systems, where data messages can occur very sporadically and might not be transmitted for a long time. Using the heartbeat protocol, all nodes at least know that their communication partners are operational, even if they do not receive PDOs with new data from them.

2 The CANopen Standard

*"The most important thing in a programming language is the name.
A language will not succeed without a good name.
I have recently invented a very good name
and now I am looking for a suitable language."*
Donald Knuth

This chapter focuses on the technical side of CANopen as specified by CiA DS301 V4.02 (CAN in Automation Draft Standard, www.can-cia.org) [CiADS301] and EN50325-4 (Cenelec European Committee for Electrotechnical Standardization) and also introduced by [Farsi99]. Besides DS301, there are many additional CANopen related standards published by the CiA. These include several frameworks, device profiles and application profiles. An overview of these standards follows in the next chapter.

CANopen is "open" in three ways: First, CANopen is "open" because the technology is laid open and by itself does not require payment of any license fees.

Second, CANopen enables a network designer to combine both CANopen compliant and proprietary CAN nodes into one network. It just needs to be ensured that the CANopen nodes and the proprietary CAN nodes do not interfere with the CAN message identifiers used by each other.

Third, CANopen can easily be extended or customized towards a specific application. CANopen consists of a small set of mandatory functionality and a huge set of optional functionality. Only the mandatory functionality must be implemented in each node to be CANopen compliant. The system designer may pick from the pool of optional functionality exactly those functions needed for a particular application. In addition, CANopen is expandable and tolerates future functionality, even allowing manufacturers to implement functionality that is not yet available in the CANopen drafts or standards.

When explaining network protocol standards, tutors and authors face the "structuring challenge." Do we explain the protocol stack bottom-up (starting at lowest level working upwards) or top-down (starting at highest level working downwards)? The benefit of a top-down approach is that one starts directly at the level to which the application interfaces. So it is up to each individual student or reader to follow along to her/his desired level of detail. In general, we stay with the top-down approach and only deviate from it if we think it helps to better understand the concept under discussion.

2.1 Using Identifiers and Objects

Objective

Newcomers to CANopen easily confuse some of the terms used. Often this is due to the confusing and sometimes conflicting usage of the words "identifier" and "object."

This section is intended to make you aware of this, and to clarify usage.

One of the challenges for newcomers to any network technology is to catch up on the terminology used and understand the different terms and abbreviations. In CANopen the most often used words are "identifiers" and "objects." However, there are many kinds of identifiers and objects in CANopen, so it is extremely important to recognize the differences.

There are 3 different kinds of "identifiers" within CANopen: the Node ID, the Object Dictionary Indexes, and the COB ID or CAN ID.

- The Node ID is used to identify a specific CANopen node. The allowed range for Node IDs is from 1 to 127.
- The Object Dictionary's Index and Subindex (16-bit and 8-bit "identifier") are used to identify a specific variable (which can be process data or configuration data) within a node. The Object Dictionary is described in detail in the next section.
- The COB ID is the "Connection Object ID" and primarily identifies a specific message on the network. This ID directly corresponds to the CAN message ID. In a CANopen system, a COB ID is unique and used for one specific communication channel (from one node to one or more other nodes). In addition, the COB ID may include control bits, such as an enable/disable bit.

CANopen also uses multiple definitions involving the term "object." There are objects such as Object Dictionaries (OD), Process Data Objects (PDO), Service Data Objects (SDO) and Connection Objects (COB), among others.

- The Object Dictionary's primary function is to store variables and constants, both process data and configuration data, in some sort of look-up table. The Object Dictionary is described in detail in the next section.
- Process Data Objects are messages (or frames) that contain process data.
- Service Data Objects are messages (or frames) that contain service/configuration data.
- COB IDs are used whenever a message (or frame) needs to be assigned to implement a service. For example, each SDO requires the assignment of two COB IDs: one for the client sending requests to the server and one for the server sending responses back to the client.

2.2 The CANopen Object Dictionary

Objective

This section introduces one of the core aspects of CANopen, the Object Dictionary.

The Object Dictionary was briefly discussed in Chapter 1, however this section will provide in depth information on how the Object Dictionary is organized, how to read the Object Dictionary entries in the CANopen specifications, the various types of entries, what the Object Dictionary contains and how to make accessing the Object Dictionary easy.

2.2.1 What is the Object Dictionary?

The Object Dictionary (called "OD" for short) is like a table that holds all network-accessible data, and each CANopen node must implement its own Object Dictionary.

The Object Dictionary contains a description of the CANopen configuration and functionality of the node it is stored in, and may be read and written to by other CANopen nodes. In addition, the Object Dictionary is used for storing application specific information that is used by the node in which it is stored. This information can also be used by other nodes on the network.

By writing data to the entries in the Object Dictionary of a node (and sometimes by reading from them), the node can be instructed to perform an operation of some kind, for example sampling current temperature or G-forces and making the sampled data present in the Object Dictionary to be read by others.

By reading the entries in the Object Dictionary of a node, other nodes may find out some information about what the node does and how it operates. Whether complete descriptive information or only minimal information about the node is present in the Object Dictionary can vary from application to application depending on the requirements of the network design. However, some information in the Object Dictionary is mandatory and must be present. Which information is mandatory often depends on which CANopen features are implemented by the node.

Object Dictionary Example

Suppose a CANopen network implements a system that precisely controls motors or some other type of precision actuator (perhaps for a robot arm) and that the performance of the motors varies with temperature.

In implementing the CANopen network a node may be responsible for controlling the motors or actuators. The Object Dictionary of that node can contain the current position of each of the motors, allowing it to be read by any other node on the network. In addition, other nodes on the network are able to write new positions to the Object Dictionary of the node, thereby causing the node to make the motors move as required.

A second node on the network contains a temperature sensor and knowledge of how temperature affects the motors. During power-up of the network, a network master stores initial calibration data in the node by writing to the node's Object Dictionary. This calibration data is then made available to other nodes on the network by reading the Object Dictionary. As the temperature changes, the node modifies the calibration data stored in it, and therefore the data available to other nodes reading the Object Dictionary.

2.2.2 Object Dictionary Organization and Contents

The Object Dictionary is organized as a collection of entries, rather like a table. Each entry has a number called an Index, which is used to access the entry. The Index is 16 bits in size giving a maximum of 65,536 entries. Each entry in the Object Dictionary may have up to 256 Subentries, referenced using an 8-bit value called the Subindex. Each entry has at least one Subentry.

Not all entries in the Object Dictionary are implemented or used, creating gaps in the table. For example, the entries with Indexes 0000h - 09FFh are often not implemented, but the entry with Index 1000h is always implemented.

It is common practice to use hexadecimal (base 16) when referring to Object Dictionary Indexes and Subindexes. The CANopen specifications use hexadecimal notation for these values.

For entries that store only one value, there is only one Subentry at Subindex 00h. Entries that store more than one value must have a Subentry for each value, and store the number of the highest Subentry at Subindex 00h.

Object Dictionary Examples

Index 2000h stores a single 8-bit value:

Index 2000h, Subindex 00h = 8-bit value

Index 2001h stores two 8-bit values:

Index 2001h, Subindex 00h = 2
Index 2001h, Subindex 01h = first 8-bit value
Index 2001h, Subindex 02h = second 8-bit value

When referring to an Object Dictionary entry with only one Subindex the Subindex is omitted from the description. For example, consider the phrase "reading Index 1000h." The lack of a Subindex implies that the entry has only one Subindex, numbered 00h.

The Object Dictionary contains several different types of data. The data may be stored in standardized and custom data types (integers, strings, etc.) and the descriptions of the data types used are also stored in the Object Dictionary. In addition, the Object Dictionary stores the configuration information for the CANopen communications used by the node, any manufacturer specific information, and various data for device profiles.

The 65,536 possible Indexes are divided up into sections structuring the Object Dictionary.

Index Range	Description
0000h	Reserved
0001h – 0FFFh	Data Types
1000h – 1FFFh	Communication Entries
2000h – 5FFFh	Manufacturer Specific

Table 2.1 Object Dictionary Organization

Index Range	Description
6000h – 9FFFh	Device Profile Parameters
A000h – FFFFh	Reserved

Table 2.1 (Continued) Object Dictionary Organization

2.2.3 Data Types

The Object Dictionary entries in the data type section (0001h to 0FFFh) do not store any variables; they are only used for the definition of data types. If physically implemented in a node, reading these entries returns the data size of that data type in bytes, or an error if the data type is not used in the node. This mechanism allows a configuration tool to read the data types section to determine which data types are actually used in the node.

The Object Dictionary may store both standard/pre-defined and manufacturer defined data types. In addition, the CANopen specification defines two basic classes of data types, Standard and Complex. To organize the range of Indexes used for defining the data types, the data types section of the Object Dictionary is further divided into the sections shown in Table 2.2.

Index Range	Description
0001h – 001Fh	Standard Data Types
0020h – 0023h	Pre-defined Complex Data Types
0024h – 003Fh	Reserved
0040h – 005Fh	Manufacturer Complex Data Types
0060h – 007Fh	Device Profile Standard Data Types
0080h – 009Fh	Device Profile Complex Data Types
00A0h – 025Fh	Multiple Device Modules Data Types
0260h – 0FFFh	Reserved

Table 2.2 Data Type Storage in the Object Dictionary

The Multiple Device Modules data types stores both standard and complex data types when more than one device profile is used.

When implementing a CANopen node it is possible to define custom complex data types in the Manufacturer Complex Data Types section of the Object Dictionary.

2.2.3.1 Standard Data Types

Table 2.3 lists the Standard Data Types, their descriptions and the Object Dictionary locations where they are defined.

Standard Data Type	Description	Stored in OD Index
BOOLEAN	Single bit value 0 or 1 indicating false or true	0001h
INTEGER8	8-bit signed integer	0002h
INTEGER16	16-bit signed integer	0003h
INTEGER24	24-bit signed integer	0010h
INTEGER32	32-bit signed integer	0004h
INTEGER40	40-bit signed integer	0012h
INTEGER48	48-bit signed integer	0013h
INTEGER56	56-bit signed integer	0014h
INTEGER64	64-bit signed integer	0015h
UNSIGNED8	8-bit unsigned integer	0005h
UNSIGNED16	16-bit unsigned integer	0006h
UNSIGNED24	24-bit unsigned integer	0016h
UNSIGNED32	32-bit unsigned integer	0007h
UNSIGNED40	40-bit unsigned integer	0018h
UNSIGNED48	48-bit unsigned integer	0019h
UNSIGNED56	56-bit unsigned integer	001Ah
UNSIGNED64	64-bit unsigned integer	001Bh
REAL32	32-bit single precision floating point number	0008h
REAL64	64-bit double precision floating point number	0011h

Table 2.3 Standard Data Types

Standard Data Type	Description	Stored in OD Index
VISIBLE_STRING	A text string containing printable ASCII characters	0009h
OCTET_STRING	An array of 8-bit unsigned integers	000Ah
UNICODE_STRING	An array of 16-bit unsigned integers	000Bh
TIME_OF_DAY	48-bit value representing days since January 1, 1984 and milliseconds since midnight	000Ch
TIME_DIFFERENCE	48-bit value representing a number of days and milliseconds since midnight	000Dh
DOMAIN	Block of data	000Fh

Table 2.3 (Continued) Standard Data Types

Each Subindex in the Object Dictionary uses one of the Standard Data Types listed in this section.

Often a shorthand notation is used to refer generically to some of the data types:

INTEGERx	a signed integer stored using x bits
REALx	a floating point value stored using x bits
UNSIGNEDx	an unsigned integer stored using x bits
VISIBLE_STRINGx	a string containing x characters
OCTET_STRINGx	a string containing x bytes

Table 2.4 Data Type Shorthand Notation

The DOMAIN type is a block of application specific data that can be any length desired. This provides an open-ended and flexible data type that is often used for various purposes, from chunks of configuration data to node firmware. Detailed descriptions of the other data types may be found in the CANopen specification [CiADS301].

Data types consisting of multiple bytes are transferred using little-endian format, which specifies that the least significant byte of the value is stored or transferred first, and the most significant byte is stored or transferred last.

> Standard Data Type Example
>
> Suppose we wish to store a current temperature value in the Object Dictionary. We could do this by using a REAL32 at Object Dictionary entry 2000h. Recall that each Object Dictionary entry must have at least one Subentry that uses Subindex 00h. Therefore at Index 2000h, Subindex 00h the data type will be REAL32.

A CANopen node can also allow the reading of the Object Dictionary entries that define the standard data types. When read, they return the bit size of the type. For example, the type UNSIGNED16 is defined at Object Dictionary entry 0006h. When entry 0006h is read it can return the value 16.

2.2.3.2 Complex Data Types

Complex Data Types are types that contain one or more of the standard data types grouped together, allowing sets of data to be constructed. This is analogous to structures in the C programming language.

Complex Data Types are really a shorthand or simplification for describing Object Dictionary entries that use different types for each of their Subentries, and are useful when a specific collection of data types are to be used frequently.

> Complex Data Type Example
>
> Let's suppose we want to store the details of an error message. We would need to know the error number and the text for the error message. To do this we could define a complex data type called ERROR_MESSAGE defined as:
>
> UNSIGNED16 - Error Number
> VISIBLE_STRING - Error Text
>
> Once the type is defined we could use it in the Object Dictionary. For example, we could say that Object Dictionary entries 2000h – 200Fh have the type ERROR_MESSAGE in order to create a place to store 16 error messages. Taking a closer look at Object Dictionary Entry 2000h, it would look like the following:
>
> Index 2000h:
> Subindex 00h - stores the value 2 indicating highest Subindex of 2
> Subindex 01h - has the type UNSIGNED16
> Subindex 02h - has the type VISIBLE_STRING

There are four pre-defined complex data types defined in the CANopen specification. The types are shown in Table 2.5.

Data Type	Description	Stored in OD Index
PDO_COMMUNICATION_PARAMETER	Record to hold the communication parameters used for a PDO	0020h
PDO_MAPPING	Record to hold the mapping parameters used for a PDO	0021h
SDO_PARAMETER	Record to hold the communiucation parameters used for a SDO	0022h
IDENTITY	Record to hold identity information, such as vendor ID and product ID	0023h

Table 2.5 Predefined Complex Data Types

Taking a closer look at the PDO_COMMUNICATION_PARAMETER complex data type in the CANopen specification reveals it is defined as follows:

Index	Subindex	Name	Type
0020h	00h	Number of highest Subindex	UNSIGNED8
	01h	COB ID	UNSIGNED32
	02h	Transmission Type	UNSIGNED8
	03h	Inhibit Time	UNSIGNED16
	04h	Reserved	UNSIGNED8
	05h	Event Timer	UNSIGNED16

Table 2.6 Complex Data Type Example

The type is made up of a collection of 8-bit, 16-bit and 32-bit values. Note that Subindex 00h always has the type UNSIGNED8 when there is more than one Subindex.

A CANopen node can allow the reading of the Object Dictionary entries that define the complex data types. When read, they return the Object Dictionary Index for the data type encoded as an UNSIGNED8.

Reading Complex Data Type Example

Suppose the Error Message type given earlier was defined in the manufacturer specific complex data type area at Index 0040h. Reading each of the three Sub-indexes would return the following values:

Index 0040h:
Subindex 00h - returns 2 for highest number of Subindex
Subindex 01h - returns 06h for UNSIGNED16
Subindex 02h - returns 09h for VISIBLE_STRING

2.2.4 Communication Entries

The communication entries in the Object Dictionary describe most of the aspects of the CANopen communications used by the node. Many of the entries are or can be made writeable, allowing configuration of a node by other nodes on the network. The entries occupy the Index range 1000h – 1FFFh in the Object Dictionary.

Table 2.7 gives an overview of all the communication entries. Following the table, the mandatory entries are described to give some examples for available entries. Mandatory entries are those that must be implemented in a node in order to be CANopen compliant. An additional listing can be found in the reference section and in the CANopen standard [CiADS301].

Index	Name
1000h	Device Type
1001h	Error Register
1002h	Manufacturer Status Register
1003h	Pre-defined Error Field
1005h	COB ID SYNC
1006h	Communication Cycle Period
1007h	Synchronous Window Length

Table 2.7 Communication Entry Overview

Index	Name
1008h	Manufacturer Device Name
1009h	Manufacturer Hardware Version
100Ah	Manufacturer Software Version
100Ch	Guard Time
100Dh	Life Time Factor
1010h	Store Parameters
1011h	Restore Default Parameters
1012h	COB ID Time
1013h	High Resolution Time Stamp
1014h	COB ID EMCY
1015h	Inhibit Time EMCY
1016h	Consumer Heartbeat Time
1017h	Producer Heartbeat Time
1018h	Identity Object
1200h – 127Fh	Server SDO Parameters
1280h – 12FFh	Client SDO Parameters
1400h – 15FFh	RxPDO Communication Parameters
1600h – 17FFh	RxPDO Mapping Parameters
1800h – 19FFh	TxPDO Communication Parameters
1A00h – 1BFFh	TxPDO Mapping Parameters

Table 2.7 (Continued) Communication Entry Overview

2.2.5 Mandatory Entries

2.2.5.1 Device Type (1000h)

The Device Type is a 32-bit value that describes in a limited way some of the capabilities of the node. For example, it can describe if the node is a digital input/output module, and if so, whether inputs and/or outputs are implemented.

2.2.5.2 Error Register (1001h)

The Error Register is an 8-bit value that can indicate if various generic errors have occurred in the node, for example, current error, temperature error, communication error, etc. The only bit that must be implemented is the generic error bit. There is a manufacturer specific bit available to indicate an application specific error. This byte is also transmitted in Emergency Objects.

2.2.5.3 Guard Time (100Ch)

Nodes must support either heartbeats or node guarding. Both mechanisms are discussed later in this chapter. To summarize, these mechanisms allow nodes to determine if a specific node is alive and well and able to communicate to the network, along with the node's current state. The Guard Time is a 16-bit value that specifies how frequently the node guarding request is transmitted by the master or must be received by the node. This entry must be implemented if heartbeats are not used.

2.2.5.4 Life Time Factor (100Dh)

The Life Time Factor is an 8-bit value that works with the Guard Time. It specifies how many multiples of the Guard Time must pass without transmission from the master or reception of a response from a slave before an error condition is generated. This entry must be implemented if heartbeats are not used.

2.2.5.5 Producer Heartbeat Time (1017h)

If the node is not using node guarding then it must implement heartbeats. This entry specifies how often the node should transmit heartbeat messages. It can be set to zero, however, to disable heartbeat transmission. This entry must be implemented if node guarding is not used.

2.2.5.6 Identity Object (1018h)

The Identity Object provides identifying information about the node. It must contain at a minimum the CAN In Automation assigned Vendor ID, which is unique to a particular vendor. It may also contain a product code to identify the product the node is in, a revision number and a serial number.

2.2.6 Manufacturer Specific Entries

This section of the Object Dictionary, using Indexes from 2000h to 5FFFh is left completely open by the CANopen specification for application specific use. Whenever the application requires storage of data or configuration of operations that are outside of any CANopen standard (including frameworks, device profiles and other standards), they are located in this section of the Object Dictionary.

> Manufacturer Specific Entry Example
>
> Suppose our node featured a real time clock. We might want to make the current time available in an Object Dictionary entry so other nodes on the network can read it. We could achieve this by defining the following Object Dictionary entry in the Manufacturer Specific section:
>
> Index 2000h:
> Subindex 00h - 3 (UNSIGNED8)
> Subindex 01h - Hours (UNSIGNED16)
> Subindex 02h - Minutes (UNSIGNED8)
> Subindex 03h - Seconds (UNSIGNED8)

2.2.7 Device Profile Parameters

The CANopen specification [CiADS301] provides a variety of communication services. Once a specific node is implemented, the designer of the node (or the network where it will be used) has to specify which of these communication services are used and how. A Device Profile specifies the process data variables a node knows and the default configuration and communication settings. There are proprietary profiles, as well as CiA standardized Device Profiles and Application Profiles. For more information on these, see Chapter 3.

CiA Device Profiles standardize specific types of nodes, for example a generic Input/Output module. Specifications are published for various device types and, in order to implement them, they use Object Dictionary entries located in the Device Profile Parameters section.

Device Profile Example

In the Device Profile CiA DS 401 Generic I/O [CiADS401] the Object Dictionary entry 6000h allows up to 2032 digital inputs to be read, 8-bits at a time.

Index 6000h:
Subindex 00h - 1 – 254(UNSIGNED8)
Subindex 01h - Read inputs 1 – 8(UNSIGNED8)
Subindex 02h - Read inputs 9 – 16(UNSIGNED8)
--
--

2.2.8 Reading the CANopen Specification

The CANopen specification [CiADS301] can be hard to follow but, like most things, once you have stared at it for long enough it starts to make sense. This is especially true if you figure out which of the many standards, frameworks and device profiles available are relevant for your application.

This section aims to give you a jump-start on understanding the Object Dictionary descriptions contained in the specification.

The following headings are used in the specifications to describe Object Dictionary entries. Their names are sometimes used inconsistently so both versions (where applicable) are listed below.

The bullets after the table describe each of the headings further.

Heading	Description
Index	the Object Dictionary Index
Object or Object Code	the Object type
Name	the name of the entry
Type or Data Type	the data type

Table 2.8 Specification Headings

Heading	Description
Acc. or Access Attributes	read and write attributes
M/O or Category	indicates if the entry is mandatory or optional

Table 2.8 (Continued) Specification Headings

- Index has been covered previously.
- The Object or Object Code is used to indicate the type of the object as a whole.

Object Code	Description
NULL	No data fields
DOMAIN	A large variable amount of data
DEFTYPE	Defines a standard data type
DEFSTRUCT	Defines a complex data type
VAR	A single value
ARRAY	An entry with more than one Subindex, with each Subindex (except 00h) having the same data type
RECORD	An entry with more than one Subindex, with each Subindex (except 00h) having differing data types

Table 2.9 Object Codes

- The Object Codes are not stored in the Object Dictionary, and therefore cannot be read from the Object Dictionary. By reading the number of Subindexes and knowing the type of each Subindex (which is necessary for using the data read), along with the Index of the entry (is it a data type declaration or not?), the Object Code information is largely redundant and can be ignored. Note, however, that it is present in the Electronic Data Sheets and Device Configuration Files which are explained in the following section.
- Name and Data type have been covered previously.

- The access attributes are straightforward and indicate whether an entry can be read, written or both.

Attribute	Description
RW	Read and write access
WO	Write only
RO	Read only
CONST	Read only, Data is constant

Table 2.10 Access Attributes

- The M/O or Category sections indicate if a specific entry or Subindex needs to be implemented or not for CANopen conformance.

Category	Description
Mandatory	Must be implemented
Optional	May be implemented if desired
Conditional	Must be implemented if certain other entries or features are implemented

Table 2.11 Categories

2.3 The Electronic Data Sheets (EDS) and Device Configuration Files (DCF)

Objective

EDS and DCF file formats are used in CANopen to describe the Object Dictionary implemented in a specific node. In this section we point out how these files are generated, maintained and used.

In order to provide CANopen software tools such as monitors, analyzers and configuration tools with a way to recognize which Object Dictionary entries are available in CANopen nodes, an electronically readable file format is required. CANopen specifies such a format called Electronic Data Sheet (EDS). An EDS is the electronically readable version of an Object Dictionary specification.

2.3.1 EDS Format and Editing

The format of the EDS is specified in [CiADSP306]. It is similar to that of Microsoft Windows ".ini" files and a regular ASCII-editor could be used to read and/or modify it. However, in order to be compliant with the standard, entries must not only have the appropriate parameters but several entries must also be cross-referenced. Thus trying to edit and maintain an EDS with an ASCII-editor, although possible, is not really practical.

Excerpts from a typical EDS file:

```
[1018]
ParameterName=Identity object
ObjectType=0x9
SubNumber=3

[1018sub0]
ParameterName=Number of entries
ObjectType=0x7
DataType=0x0005
AccessType=ro
DefaultValue=3
PDOMapping=0
LowLimit=1
HighLimit=4

[1018sub1]
ParameterName=Vender ID
ObjectType=0x7
DataType=0x0007
AccessType=ro
DefaultValue=0x0400005A
PDOMapping=0

[1018sub2]
ParameterName=Product code
ObjectType=0x7
DataType=0x0007
AccessType=ro
DefaultValue=0x03
PDOMapping=0

[1018sub3]
ParameterName=Revision number
```

```
ObjectType=0x7
DataType=0x0007
AccessType=ro
DefaultValue=0x0000002F
PDOMapping=0
```

There are several commercial software tools available that support the generation and maintenance of EDS files. These tools take adding or removing dictionary entries to the drag & drop level; all standard Object Dictionary entries are pre-defined and can be added to a new EDS with a few mouse clicks. As an example, Figure 2.1 shows how such an editor displays the Identity Object entry (1018h).

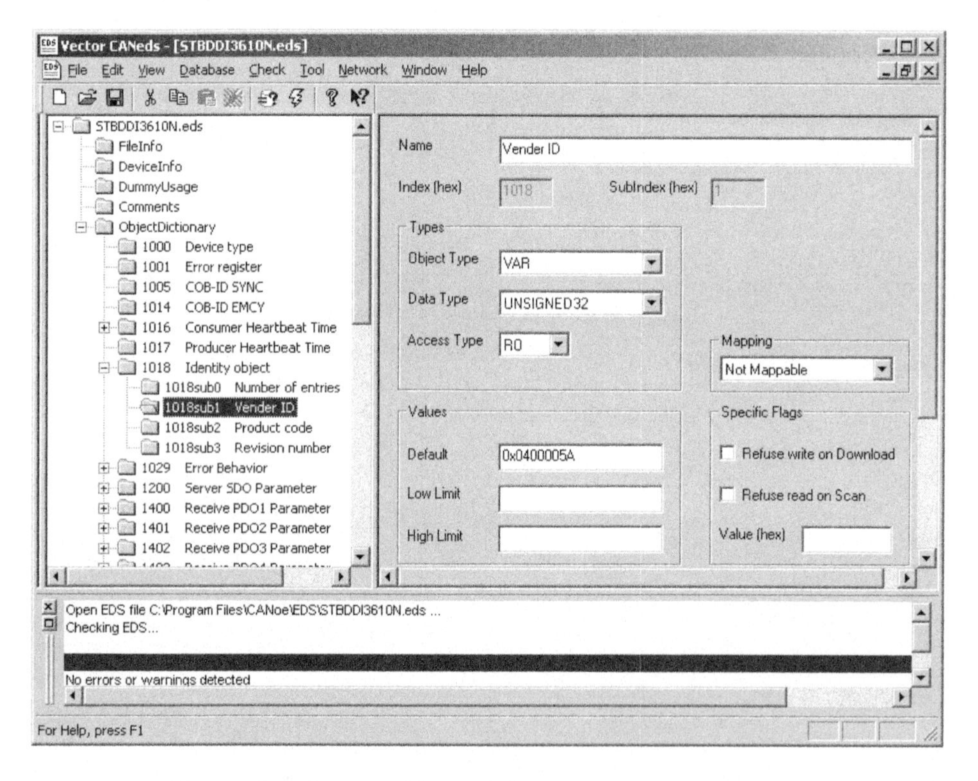

Figure 2.1 Screen Shot of Vector's CANeds Editor

For those of you who would still like to use an ASCII editor we recommend starting with one of the examples published at www.CANopen.us. When editing EDS files "manually" with an ASCII editor you might want to double-check after any changes to see if it still conforms to the standard. The web page listed above also has a link to a free EDS checker tool offered by Vector. This tool checks to see if an EDS conforms to the standard [CiADSP306] and displays appropriate error messages if not.

2.3.2 EDS Usage

There are several tools available that can work with EDS files. High-end CAN monitors and analyzers or CANopen configuration tools can extract symbolic information from these files and use them in their displays. A monitor or analyzer with this feature can listen to CANopen traffic on the network and associate the symbols of these files with the messages seen on the network. So if there is a process data variable defined in an EDS that is called "Boiler Temperature" and that value is transmitted over the CANopen network, these tools can directly make the symbolic link and display the text "Boiler Temperature" along with the current value transmitted.

Other tools that work with EDS files are high-end CANopen masters. Such a CANopen master is typically used in a system to receive all inputs, run some control algorithm and then transmit all outputs. A CANopen master that can read EDS files can use the symbolic names from the EDS file in the control algorithm. So in the case of the example above a variable called "Boiler Temperature" is available to be used by the control algorithm.

Another tool which utilizes the EDS file is the CANopen Conformance Test. The CANopen Conformance Test is available through National Instruments and is used by the CiA to test if a CANopen device is CANopen compliant. This test not only checks for CANopen conformance in general, it also tests if a node implements all the Object Dictionary entries specified in its EDS file.

Engineers working on applications that do not require 100% CANopen conformance might be tempted to skip writing an EDS file for their node(s). As consultants with practical experience in the field, we strongly recommend that you create EDS files even in those cases where CANopen compliance is not required.

Any CANopen network design will eventually reach the state where two or more nodes will communicate with each other. On that first contact, there is a good chance that the communication will not happen precisely as the system designer(s) had in mind. So then the next question is: *Which of all the nodes connected is the one that is doing something wrong?*

If the only basis for each node's implementation is a written specification, the debugging process to follow will always be locked to "manual" mode; read the specification for all nodes suspected of ill behavior, interpret it (and there is always room for interpretation) and double-check what the nodes are actually doing.

However, if EDS files are used, the debugging process can eventually be automated. There is only minimal room for "interpretation variance" and there are several tools (as listed above) that directly work with EDS files to simplify the debugging process. Even the CANopen conformance test can be useful for this scenario: it can be used to confirm that the features that should be CANopen compliant are indeed CANopen compliant.

2.3.3 DCF Format and Usage

The format of the Device Configuration File is almost identical to the EDS. However, the usage is very different and justifies giving it a separate name and not just referring to it as some sort of "EDS variant."

The EDS defines the format of an Object Dictionary that may apply to multiple nodes. The idea of a DCF is to store the configuration parameters of a specific node.This can include minimum, maximum and default values for each entry. The DCF stores a specific setting, the current value that an Object Dictionary entry has or should have. The idea is that a CANopen configuration tool or master can use the EDS to find out which entries are accessible in a node, and they can use a DCF to store (or retrieve) the values that a node has in these Object Dictionary entries. Thus it becomes possible to save and restore all settings of a node: to save current settings a tool/master would read all Object Dictionary entries of a node and store the values read in a DCF. A restore of a node can be performed by reading the values from the DCF for the node and writing them to the Object Dictionary entries.

> As an example, a system could feature several digital I/O modules that are *all* implemented in accordance with an EDS for "generic I/O." However, during operation some of these nodes might be configured to be exclusively inputs and others to be exclusively outputs. The specific configuration of each *individual* node is stored in its own Device Configuration File (DCF).

2.4 Accessing the CANopen Object Dictionary (OD) with Service Data Objects (SDO)

> Objective
>
> Now that we have a method for defining the data in a node (the Object Dictionary) that can be shared via the network, we need a method to access it. This section explains the Service Data Objects (SDO) – the method used to implement generic access to the Object Dictionary of a node by using request and response messages.

2.4.1 Client and Server

Each CANopen node not only implements its own Object Dictionary, it also implements a server that handles read and write requests to its Object Dictionary. So a master or configuration tool acts as a client to that server and can send read or write requests to it. As an example, a configuration tool could send a request: "Node Number 5, I need to know what you have at Index 1000h, Subindex 00h." Node 5 would recognize the request and reply with a response: "Whoever requested it, here is the data that I have in my Object Dictionary at Index 1000h, Subindex 00h."

2.4.2 Message Identifiers Used for SDOs

As discussed earlier, CANopen uses unique message identifiers – one message ID is only used for one purpose in an entire CANopen network (this is a requirement of the CAN arbitration feature that is explained in Section 5.2.8). There are some exceptions to this rule that are primarily used for specific configuration services during initialization, test or debugging.

In order to implement a point-to-point communication channel two such message identifiers need to be reserved; one to send requests to a specific node and one for responses sent by that node. Figure 2.2 shows the message identifiers that are used by default. The message identifier that is used to send a request to a specific node is calculated by adding the Node ID of that node (1-127) to a base address of 600h. Thus addresses 601h to 67Fh are used to provide 127 channels from one client to up to 127 servers. The message identifier that is used to send a response from each node back to the client who sent the request is calculated by adding the Node ID of the node (1-127) to a base address of 580h. Thus addresses 581h to 5FFh are used to provide 127 channels from as many as 127 servers back to the client.

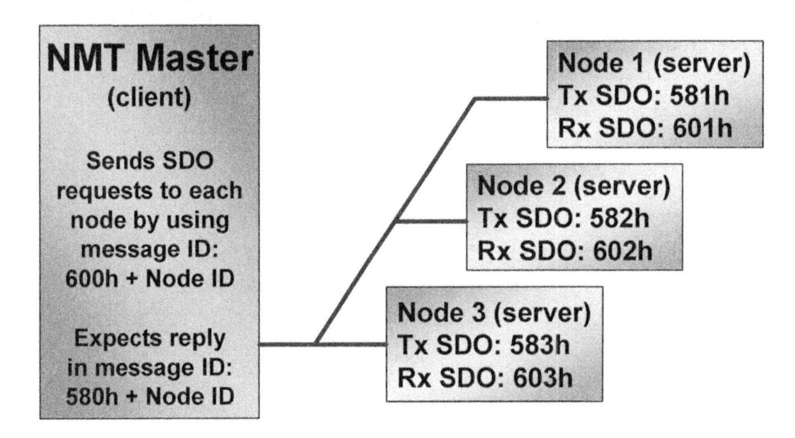

Figure 2.2 Default Message IDs for SDO Communication

It should be noted that the default scheme used for assigning the message identifiers only allows one client to be on the network. Because message IDs must be unique, no two devices have the right to send SDO requests to the same node at the same time. The entire SDO communication was designed around the idea that only one node in the system needs the power to access each and every Object Dictionary entry in each and every node. This is either a configuration tool or some sort of master responsible for configuration.

Advanced Features

Where required, CANopen optionally allows the implementation of either a method to perform SDO channel sharing or a method that provides additional SDO channels. The latter allows a single node to implement multiple SDO servers. So besides using just one message identifier pair to reserve a request and a response channel for the SDO, such nodes would reserve additional message identifier pairs for each additional SDO server implemented. A node with two servers could provide both a master and a configuration tool access to its Object Dictionary at the same time. There is no default scheme regarding which message identifiers should be used for such additional channels.

The other method for allowing multiple clients is to implement SDO channel sharing. Instead of having the servers implement multiple SDO channels the clients implement a method of sharing the existing channels. This is implemented via a so called SDO Manager that is responsible for all SDO channels and by default is the only client that may use any of the channels. Other clients that would like to use a specific SDO channel (such as a configuration tool only connected to the network for maintenance) have to request the channel from the SDO Manager and may only use it after the SDO Manager assigns it to them. This method is described in detail in [CiADS302].

2.4.3 SDO Message Contents

Every SDO request and response message contains 8 bytes of data of which the first byte is a so-called "specifier." The bits in it primarily specify whether this message contains a read, write or abort (error indication). Other bits are used to indicate if this is an "expedited transfer" where all data exchanged is part of this message, or a "segmented transfer" where the data does not fit into one message and multiple messages are used. The optional "block transfer" is optimized for the transfer of large data blocks and is described in Section 2.8.5.

Typically bytes 2 to 4 contain the "multiplexor" – the combination of 16-bit Index and 8-bit Subindex identifying the Object Dictionary entry that is accessed with this SDO. The byte order for the multiplexor is as follows: low byte of 16-bit Index, high byte of 16-bit Index and 8-bit Subindex.

The remaining bytes (5 to 8) are used to transmit data where applicable. If the data transferred is 4 bytes or less it is typically part of the message (expedited), otherwise it follows in additional messages (segmented).

In the case of expedited transfers the number of bytes used for data is indicated by additional bits in the specifier.

In the case of a segmented transfer the first SDO request and SDO response do not contain data, but an indication of how many bytes will need to be transferred in total. Each segment transmitted after that also contains the specifier byte and up to 7 data bytes. The specifier contains bits that specify if this is the last segment and if it is, how many of bytes in the current message are data bytes that belong to the transfer.

At any time during a transfer, any of the two communication partners may abort the communication by sending an SDO Abort message.

The detailed message contents of all SDO messages is explained in Section 2.8.

2.4.4 SDO Download vs. Upload

Per [CiADS301] an "SDO Download" implements a write access to the Object Dictionary of a node and an "SDO Upload" implements a read access.

In the authors' experience these terms are easily confused, and they are only listed here for completeness. In the following we will use the terms "SDO Read Access" and "SDO Write Access." These are easily understandable, especially as they correspond to the access type field available for all Object Dictionary entries, where possible values include read-only, read-write and write-only.

2.4.5 SDO Usage Limitation

The SDO Read Access and SDO Write Access as explained in this section provide a mechanism for generic read and write access to the Object Dictionary of each node on the network. Because all configuration and process data of a node is part of its Object Dictionary, the SDO transfers can be used to access the process data and it would be possible to implement a communication system entirely based on SDO communication. However, that was never the intention of the SDO communication (recall that SD stands for "Service Data") and thus this communication mode is not very efficient.

For real-world implementations a leaner, more efficient communication method is desirable in order to minimize the communication overhead and to make best usage of the available bandwidth. In CANopen, this lean and efficient communication method is provided by the Process Data Objects (PDOs).

2.5 Handling Process Data with Process Data Objects (PDO)

Objective

So far we have established communication channels that allow a master or configuration tool to get access to all the Object Dictionary entries in a node. However, this is not an efficient communication model for sending process data.

The PDOs provide a far more sophisticated service for process data. The PDO communication services are explained in this section.

CANopen is primarily intended to run on CAN; a message oriented communication system capable of transmitting up to 8 bytes in a single message – in other words, it is *optimized* for CAN. It should be noted that although optimized for CAN it can still run on other, completely different network technologies such as I2C or Ethernet.

The obvious demands resulting from the CAN functionality are that it must be possible for nodes to transmit their data whenever they want to (and not be required to wait for another node to poll them) and to place multiple process data variables into a single CAN message.

All these demands are fulfilled by CANopen's Process Data Objects, or PDOs.

In CANopen we distinguish between Transmit Process Data Objects (TPDOs) and Receive Process Data Objects (RPDOs). When looking at single nodes this terminology indicates if a PDO is produced or consumed by this node. So for each PDO in a system there is exactly one node producing it, and for that node this PDO is a TPDO. There is also at least one node (but perhaps multiple nodes) that receive and consume the PDO. For all nodes consuming it, the PDO is an RPDO. This is illustrated in Figure 2.3.

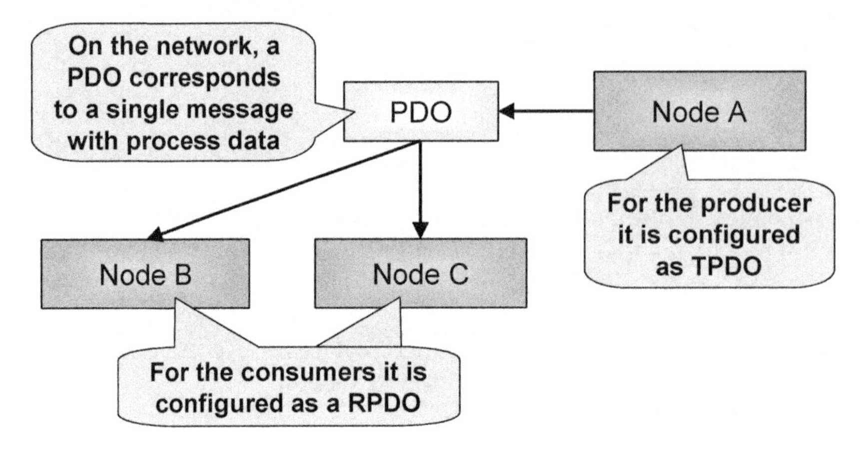

Figure 2.3 PDO: TPDO and RPDO

There are two sets of configuration parameters for a PDO. The communication parameters (indicating which CAN message is used for the PDO and how is it triggered) and the mapping parameters (indicating which Object Dictionary entries are contained in the PDO).

For each TPDO and RPDO that are transmitted and received by a node, the Object Dictionary of that node contains one set of configuration parameters called the PDO communication parameters.

The communication parameters for a TPDO differ slightly from those for an RPDO, as more parameters are required for transmitting a message than for receiving a message. Transmit trigger options, for example, determine when to send the message with the PDO.

2.5.1 TPDO Transmit Trigger Options

One of the advanced features of CANopen is that it supports all generally known transmission and communication methods used in communication networks. CANopen nodes can not only transmit their data individually (either event or time driven) they can also be polled individually or synchronized in groups. In addition, any of these methods can be combined.

Integration Tip:
All these different communication methods contribute to a lengthy test process, especially if combinations of these methods are allowed in a system. When integrating a CANopen network the communication method used should be chosen as early as possible and adopted by all nodes in the system. That allows developers to focus all test procedures on the chosen method, avoiding the additional test procedures required if multiple communication methods are mixed in one network.

Implementation Tip:
If you develop your own CANopen node and it is for a specific system that only supports a chosen set of communication methods, you do not need to implement the unused communication methods. This will reduce your code size, and will also contribute to shorter test cycles. In terms of code and data memory sizes, a strictly time driven implementation typically has the fewest requirements.

There are four major transmit trigger methods supported by CANopen:

- Event driven (COS, Change-Of-State)
- Time driven
- Individual polling
- Synchronized, group polling

Which of these is used by a specific PDO is selected by the PDO communication parameter "transmission type" which is explained further in the individual communication parameter sections below.

2.5.1.1 Event driven

The event driven or change-of-state transmission method simply transmits a TPDO message if the process data in it changes. What exactly is defined as an "event" is typically specified by the Device Profile. It could be *any* change to the data as well as *specific* change to the data (like reaching a certain limit or reaching a minimum difference).

If a TPDO contains a set of digital inputs and the event is "any" change then the TPDO gets transmitted as soon as the data in it changes. If there is no change in the data there will be no transmission until the data actually changes.

There is one worst-case scenario for event driven communication that needs to be handled properly: if one of the inputs changes constantly the TPDO would be transmitted back-to-back (as soon as a TPDO is transmitted the data will have changed again). Such a behavior would occupy 100% of the available bandwidth as illustrated in the top portion of Figure 2.4.

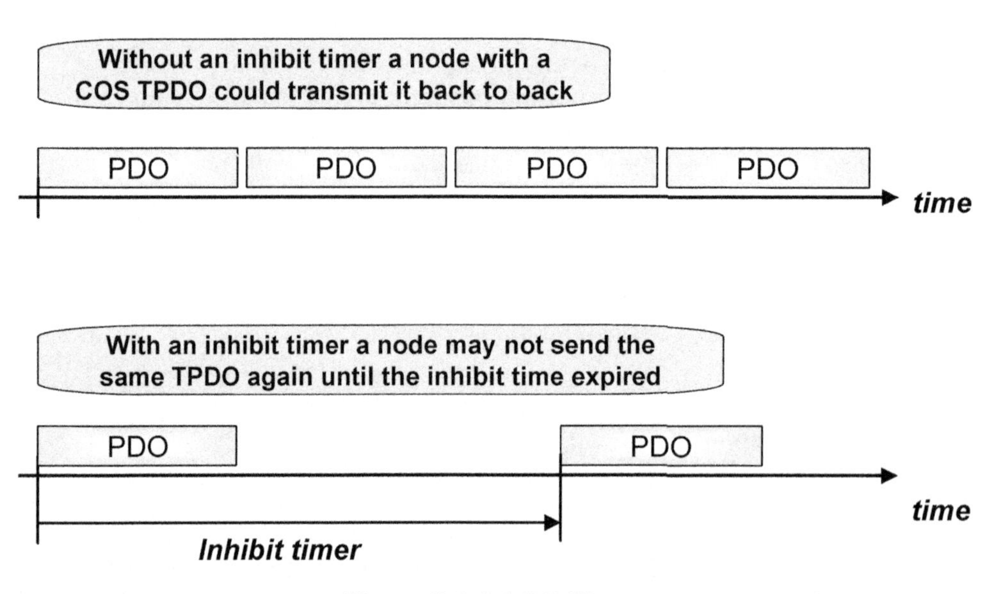

Figure 2.4 Inhibit Timer

CANopen handles this worst-case scenario by introducing the "Inhibit Timer." This is a configurable timeout in multiples of hundreds of microseconds. After starting the transmission of the TPDO the Inhibit Timer must expire before the TPDO may be transmitted again. So the maximum frequency with which a TPDO could occur is directly specified by the Inhibit Time.

> One problem with any event driven communication is the indeterminism: it is very hard to predict the worst-case scenarios of how often messages will get transmitted. By using the Inhibit Timer the worst-case becomes predictable again as the worst-case is directly determined by the Inhibit Time. If it is set to 10 milliseconds the worst-case is that the message will be transmitted every 10 milliseconds.

The Inhibit Timer always affects the entire TPDO and all process variables contained in it. So if the TPDO is transmitted because one process variable in it changed, the Inhibit Time applies even if another process variable in the TPDO changes.

Whenever a TPDO is inhibited from transmission it means that potentially some process data is lost. If the process data actually changes several times while the timer is running not all of these changes will be transmitted.

In some instances what exactly constitutes an event change may vary. The Device Profile for Generic I/O [CiADS401] introduces an extended event detection mechanism for analog values. Some analog values such as a temperature value might only be needed if they either changed "considerably" or reached a certain minima or maxima. CiA specification [CiADS401] supports both configurable delta detection (the system only recognizes an event change if the analog value changes by a user-defined delta) and minima and maxima detection.

In general, a TPDO can contain multiple process variables and potentially also a mixture of digital and analog data. This makes the event change detection a complex process since it can be different for every single process variable contained in a TPDO. Some CANopen nodes try to simplify this by either not allowing the mixing of analog and digital data in one TPDO, or by only implementing simple event change detection (values changed) without the extended detection mechanisms of [CiADS401].

2.5.1.2 Time driven

In the time driven communication method a TPDO is transmitted on a fixed time basis, the Event Timer. The Event Timer for a TPDO is specified in milliseconds. If, for example, the Event Timer is specified to be 50 milliseconds, the TPDO will be transmitted every 50 milliseconds.

The Event Timer is a local timer on each CANopen node. Per default these timers are not synchronized. If multiple nodes use an Event Timer of 20 milliseconds the actual occurrence of these TPDOs on the network may all be within the same millisecond as well as randomly distributed in a 20 millisecond time window.

On one hand the time driven communication method simplifies performance, bandwidth and latency calculations. On the other, it produces more overhead than the event driven communication since data will get transmitted even if it did not change at all.

2.5.1.3 Individual polling

Although it is possible to use individual polling in CANopen, it is recommend that this communication method not be used.

Individual polling is implemented via a CAN feature called "Remote Request." Unfortunately Remote Request has certain disadvantages, including the fact that not all chip manufacturers implement it the same way in their CAN controllers. In other words, it could be that nodes implemented with different CAN controllers are not compatible when using Remote Request.

If an application requires the implementation of a polling mechanism (a message is used to trigger a node to actually transmit its TPDO), the synchronized communication method described below should be used.

2.5.1.4 Synchronized or grouped polling

The main idea behind the synchronized communication mode is to provide motion oriented systems such as robots with "parallelized" inputs and outputs. To avoid jitter effects and ensure smooth movements it is necessary to get all inputs at the same moment in time and to apply all outputs at the same time.

In CANopen a synchronized communication method is implemented using a SYNC signal. The SYNC signal is a specific message without any data that is used only for synchronization purposes. Figure 2.5 below illustrates how sensor data (for example from encoders measuring the positioning of a moving robot arm) is synchronized. Because the SYNC signal is typically produced on a fixed time basis, this triggering mode can also be regarded as using a global timer for triggering instead of the event time local to each node.

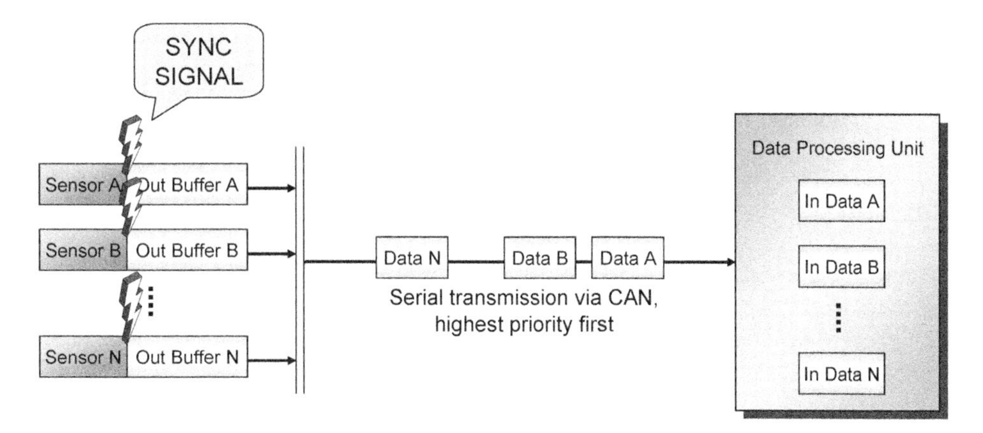

Figure 2.5 SYNC – Synchronized Communication for Sensors

The sensors constantly read their input data and keep a current copy in the message transmit buffer. Upon reception of the SYNC message, all sensors stop updating the buffer and start transmitting the data. Although all messages are transmitted serially via CANopen, once the data arrives in the main processing unit all these inputs will be from the same moment in time, i.e. the time the SYNC signal was transmitted.

The synchronization of outputs works similarly, as illustrated by Figure 2.6.

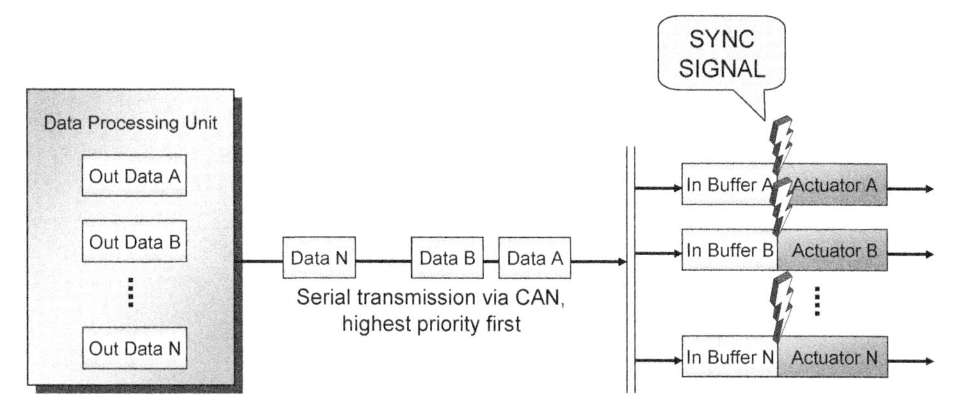

Figure 2.6 SYNC – Synchronized Communication for Actuators

Once the processing unit has new values for the outputs or actuators it transmits the data serially via the network. The actuators receiving the messages keep the received data in their receive buffers *without applying the data to their outputs*. They wait for the next SYNC signal and only upon reception of the SYNC signal will they actually apply their outputs in parallel.

2.5.1.5 How good is the synchronization?

The quality of the synchronization is measured by the maximum time variance that can still occur between the different nodes.

As an example consider a network running at 250kbps. The bit time on such a network is 4 microseconds. Without going into the message details of CAN (see Section 5.2.7 for more details) assume the length of a single message varies between 200 microseconds and 450 microseconds.

If no synchronization method is used and all inputs use their local Event Timer for TPDO triggering, the time variance for the inputs depends pretty much on the Event Timer. If all nodes transmit their data every 50 milliseconds then the worst-case time variance is 50 milliseconds. This means that two outputs that should be applied "in parallel" might actually be applied with a difference of 50 milliseconds.

If synchronization is used, the only delay in each individual node is the time it takes the node to process the receipt of the SYNC signal until either sending the TPDO (for sensors) or applying the data from the RPDO (for actuators). This time is highly dependent on the code quality and microcontroller used in implementing the individual CANopen nodes. However, with proper implementation these times should be less than 100 microseconds. This means that two outputs that should be applied "in parallel" might actually be applied with a difference of 100 microseconds.

Another delay not yet accounted for is that of the SYNC signal itself. In some applications it might not matter if the SYNC signal itself is delayed, because all I/Os are affected the same way and individual delays are relative to the SYNC. However, some applications might require some sort of absolute timing in which case the potential delay of the SYNC itself might be a problem. The typical worst-case delay of the SYNC message itself is the time the longest message can occupy the bus. In the example above this was assumed to be some 450 microseconds. So suddenly the total variance adds up to more than half a millisecond.

Depending on how much more effort is put into the system correcting these variances (they can be measured and corrected), much higher accuracies can be achieved. The paper "High Precision Drive Synchronisation with CANopen" [Rostan02] describes a method that achieves a one microsecond accuracy using internal re-calculations.

Applications that require a very high-resolution synchronization signal typically provide an extra line on the network cable in order to be able to send synchronization pulses directly (and not via the network). With such a mechanism it is possible to bring the variance down to a microsecond or less. However, its implementation is "manufacturer specific" meaning there is currently no CANopen standard covering this type of high-resolution synchronization mechanism.

2.5.2 SYNC Terminology

There are a few terms associated with the synchronization communication method that need to be known in order to configure a system to use the synchronization feature.

2.5.2.1 SYNC COB ID

This is a configurable parameter (OD entry [1005h,00h]) in all nodes that supports synchronized communication. The connection object ID specifies which CAN message identifier is used as the SYNC signal, with the default at 80h. This parameter is individually configurable in each node; thus it is possible to have multiple SYNC signals in a system. This allows developers to group nodes together, with some working with one SYNC signal and others working with another SYNC signal.

2.5.2.2 SYNC Producer

There is only one node that produces the SYNC signal. Although it could be the NMT Master producing the SYNC, it does not need to be. Any node can be a SYNC producer, *but only one node can be the producer of a specific SYNC in a system.*

2.5.2.3 Communication Cycle Period

This is the time period in microseconds with which the SYNC signal occurs. Nodes supporting synchronized communication have this value available in the OD entry [1006h,00h].

2.5.2.4 Synchronous Window Length

This is the time window in microseconds in which all communication triggered by a SYNC signal must occur. Nodes supporting synchronized communication may have this optional value available in the OD entry [1007h,00h].

2.5.3 Combining Transmit Trigger Options

In general, CANopen allows any of the communication methods specified to be combined. As an example, a TPDO could be transmitted synchronized with change-of-state. This results in a TPDO that is only transmitted in response to a SYNC signal if any of the data in it changed since the last transmission.

The possible combinations are not part of the CANopen specification itself and they are usually implemented using the transmission type "manufacturer specific." However, in some cases a device profile might request a specific combination in which case the transmission type "device profile specific" is used.

A frequently used combination is that of event driven and time driven. In an event driven system, there might be long periods of "silence" if data does not change. This might have side effects in cases such as:

- A new node is added to the system and it does not know the "last data transmitted."
- In the rare case of an erroneous message periodic re-transmissions ensure that erroneous data is not valid for an extensive period of time.

Through the combination of event driven and time driven it is possible to specify a time window within which a TPDO is re-transmitted. If the data changes frequently, the TPDO will be re-transmitted within the time period specified by the Inhibit Timer. If it does not change at all, it will still be transmitted at least every Event Time. If the Inhibit Time is 50 (in multiples of 100 microseconds, so 5 milliseconds) and the Event Time is 250 (in multiples of milliseconds, so 250 milliseconds) the TPDO will be at least transmitted every 250 milliseconds but never more frequently than every 5 milliseconds.

2.5.4 PDO Linking and Pre-defined Connection Set

From the network perspective a PDO is nothing more than a message with a message identifier and up to 8 data bytes. In CANopen almost everything is configurable and this includes which message identifier (COB ID) is used for each PDO.

In order to establish a common ground a default usage of the message identifiers is typically implemented. It is called the "pre-defined connection set" and determines which COB IDs should be used by which node by default.

Table 2.12 below shows the identifier ranges assigned for the PDOs.

CAN ID			
From	To	Communication Objects	Comment
0h	--	NMT Service	From NMT Master
80h	--	SYNC Message	From SYNC Producer
81h	FFh	Emergency Message	From nodes 1 to 127
100h	--	Time Stamp Message	From timestamp producer
181h	1FFh	1st Transmit PDO	From nodes 1 to 127
201h	27Fh	1st Receive PDO	For nodes 1 to 127
281h	2FFh	2nd Transmit PDO	From nodes 1 to 127
301h	37Fh	2nd Receive PDO	For nodes 1 to 127
381h	3FFh	3rd Transmit PDO	From nodes 1 to 127
401h	47Fh	3rd Receive PDO	For nodes 1 to 127
481h	4FFh	4th Transmit PDO	From nodes 1 to 127
501h	57Fh	4th Receive PDO	For nodes 1 to 127
581h	5FFh	Transmit SDO	From nodes 1 to 127
601h	67Fh	Receive SDO	For nodes 1 to 127
701h	77Fh	NMT Error Control	From nodes 1 to 127

Table 2.12 The Pre-defined Connection Set

As an example, consider a node (call it node 5) which has the following pre-defined COB IDs for its PDOs:

PDO	**COB ID**
TPDO1	185h
RPDO1	205h
TPDO2	285h
RPDO2	305h
TPDO3	385h

PDO	COB ID
RPDO3	405h
TPDO4	485h
RPDO4	505h

This default connection has no "over-lapping" of any TPDOs and RPDOs specified. This means that by default no RPDO specified uses the same identifier as any TPDO specified, and thus no PDO is directly "linked." (A link is where the COB ID of a TPDO from one node is identical to the COB ID of an RPDO of any other node). Only the NMT Master would be able to listen to all the TPDOs and only a master would be able to generate the RPDOs received by the nodes.

However, because the COB IDs used for the TPDOs and RPDOs are configurable, direct links can be established, for example by changing the COB ID for an RPDO to the same used by another TPDO (also see Section 1.4.5).

What if an application requires that more than four TPDOs and four RPDOs be used for specific nodes?

There are several solutions to this problem. In general the pre-defined connection set is just that – a pre-defined default that can be reconfigured at any time. Another approach can be to simply modify the pre-defined connection set for a specific application. An example for such an approach is described in Section 6.9.4.

2.5.5 RPDO Communication Parameters

In the Object Dictionary the Index area from 1400h to 15FFh is reserved for the RPDO communication parameters. The Index range of 512 (200h) ensures that a maximum of 512 RPDOs can be configured in the Object Dictionary of a single CANopen node. The parameters for the first RPDO (RPDO1) are located at Index 1400h, the parameters for the second at 1401h (RPDO2), for the third at 1402h (RPDO3) and so on.

The parameters for each RPDO are accessible via the Subindex. The table below shows the parameters that are available for every RPDO.

Subindex	Name	Data type
0	Number of entries	UNSIGNED8
1	COB ID	UNSIGNED32
2	Transmission type	UNSIGNED8
3	Inhibit Time	UNSIGNED16
4	Reserved	UNSIGNED8
5	Event Timer	UNSIGNED16

Table 2.13 RPDO Communication Parameters

The "Number of Entries" for a RPDO can be 5 if the Event Timer is supported. The most popular configuration of RPDOs does not use the Event Timer and therefore the Number of Entries is 2.

The "COB ID" is the connection object identifier which is the CAN message identifier used for this RPDO. This parameter determines which CAN message is received and interpreted as the RPDO belonging to this set of parameters.

The "Transmission Type" determines if this RPDO is to be processed immediately upon reception or if a node needs to wait for a synchronization signal (SYNC), before it may process the data received.

The "Inhibit Time" is not used for RPDOs and if implemented should have the value zero.

The "Reserved" parameter is a legacy value from previous CANopen versions and must not be implemented in nodes conforming to the current standard [CiADS302].

The "Event Timer" may be used to generate an emergency if this RPDO is not received before the event timer expires. The event timer is reset upon reception of the RPDO. Implementation of the Event Timer for RPDOs is not very common.

> Although it is possible to have a single node receive up to 512 different RPDOs such a setup is the exception. Many CANopen slave nodes only support a limited number of RPDOs. Just imagine a simple temperature sensor - if it only has a temperature to report it will not need more than 1 TPDO and no RPDOs at all.
>
> A typical number for more generic I/O nodes is up to four RPDOs, as the so-called pre-defined connection set of CANopen (also see section 1.4.5) predefines the COB IDs used for the first four RPDOs in a CANopen slave node.

2.5.6 TPDO Communication Parameters

In the Object Dictionary the Index area from 1800h to 19FFh is reserved for the TPDO communication parameters. As with the RPDOs, the Index range ensures that a maximum of 512 TPDOs can be configured for a single CANopen node. The parameters for the first TPDO (TPDO1) are located at Index 1800h, the parameters for the second at 1801h (TPDO2), for the third at 1802h (TPDO3) and so on.

The parameters for each TPDO are accessible via the Subindex. The table below shows the parameters that are available for every TPDO.

Subindex	Name	Data type
0	Number of entries	UNSIGNED8
1	COB ID	UNSIGNED32
2	Transmission type	UNSIGNED8
3	Inhibit Time	UNSIGNED16
4	Reserved	UNSIGNED8
5	Event Time	UNSIGNED16

Table 2.14 TPDO Communication Parameters

The "Number of entries" for a TPDO is 5, as five parameters are available for the configuration of each TPDO. Only entries zero through two are mandatory, three and five are optional.

As with the RPDO, the COB ID specifies the CAN message identifier used when transmitting this TPDO. The transmission type selects the TPDO trigger behavior. When is the message transmitted? Upon a change-of-state (COS) of any of the process data variables contained in the TPDO? Or is the transmission strictly time driven, occurring every so many milliseconds? An additional listing of all the available values can be found in the Object Dictionary Reference section for the entries [14xxh,02h].

For change-of-state transmission the Inhibit Time specifies a timeout period that must pass before this TPDO can be re-transmitted again. This minimum timeout between two transmissions of a TPDO are specified in multiples of 100 microseconds.

The "Reserved" parameter is a legacy value from previous CANopen versions and must not be implemented in nodes conforming to the current standard [CiADS302].

For event time driven TPDOs the Event Time specifies the time period used for this TPDO. The Event Time is specified in multiples of milliseconds. If it is set to 100 the TPDO is transmitted every 100 milliseconds.

Using a combination of both Inhibit Time and Event Time creates a time window for the transmission of the TPDO. It will be transmitted at least every "Event Time" but not more often then defined by the "Inhibit Time."

2.5.7 PDO Mapping Parameters

As discussed earlier, a PDO can contain data from several Object Dictionary entries in order to be able to exchange multiple process data variables with one message.

The PDO mapping parameters determine which Object Dictionary entries are contained in a PDO. Single Object Dictionary entries are "mapped" or "placed into" a PDO.

The maximum number of data bits available in a PDO is 64. Because the mapping process works on the bit-level a total of 64 Object Dictionary entries can be mapped into a PDO, if each entry is just one bit long. No matter what the length of an individual Object Dictionary entry is, if all the lengths of the mapped entries are added up, the total cannot exceed 64 bits.

A single mapping parameter identifies one specific Object Dictionary entry with its parameters Index, Subindex and length (in bits). These three parameters get coded into one 32-bit value as shown in the table below.

Index	Subindex	Length (bits)
Bits 31 .. 16	Bits 15 .. 8	Bits 7 .. 0

Table 2.15 Content of a 32-bit Mapping Parameter

The PDO mapping consists of an array of such single mapping parameters. The 64 bits available in a PDO are filled entry-by-entry with the data from the Object Dictionary entries specified in the single mapping parameters. The table below shows the format of a PDO mapping record.

Subindex	Name	Data type
0	Number of entries	UNSIGNED8
1	1st OD entry mapped	UNSIGNED32
2	2nd OD entry mapped	UNSIGNED32
3	3rd OD entry mapped	UNSIGNED32
4	4th OD entry mapped	UNSIGNED32
--	--	--
64	64th OD entry mapped	UNSIGNED32

Table 2.16 PDO Mapping Parameters

The "Number of entries" value indicates how many single mapping entries are available in this record (0-64). What follows is an array of single mapping entries that is as long as specified by the "Number of entries" value.

2.5.7.1 PDO Mapping OD Index Ranges

In the Object Dictionary the Index area from 1600h to 17FFh is reserved for the RPDO mapping parameters, and the area from 1A00h to 1BFFh is reserved for the TPDO mapping parameters. The Index range sizes are the same as used by the PDO communication parameters and directly correlate to each other. For example, the RPDO1 communication parameters are at Index 1400h and the mapping parameters at Index

1600h, the TPDO3 communication parameters are at Index 1802h and the mapping parameters at Index 1A02h.

Index	SubIdx	Type	Description
1A02h			3rd Transmit PDO - Mapping
	0	UNSIGNED8	= 4 (Number of used map entries)
	1	**UNSIGNED32**	**= 2010 01 08h (Idx 2010h, SubIdx 1, 8 bit)**
	2	**UNSIGNED32**	**= 2010 02 08h (Idx 2010h, SubIdx 2, 8 bit)**
	3	**UNSIGNED32**	**= 2010 03 10h (Idx 2010h, SubIdx 3, 16 bit)**
	4	**UNSIGNED32**	**= 2010 04 10h (Idx 2010h, SubIdx 4, 16 bit)**
2010h			Manufacturer Specific Inputs
	0	UNSIGNED8	= 4 (Number of sub-index entries)
	1	**UNSIGNED8**	**8-bit variable 'status'**
	2	**UNSIGNED8**	**8-bit variable 'temp'**
	3	**UNSIGNED16**	**16-bit variable 'speed'**
	4	**UNSIGNED16**	**16-bit variable 'rpm'**

TPDO3	status	temp	speed		rpm		Unused	
	Byte 1	Byte 2	Byte 3	Byte 4	Byte 5	Byte 6	Byte 7	Byte 8

Figure 2.7 PDO Mapping Example

Figure 2.7 illustrates an example that maps a total of four manufacturer specific variables into TPDO3. The device in this example has four variables named status, temp, speed and rpm located in the Object Dictionary at [2010h,00h-04h]. The mapping entries for TPDO3 are at location [1A02h] and the four entries at that location map those 4 variables one-by-one into the TPDO3.

2.5.7.2 Dynamic Mapping vs. Static Mapping

The PDO mapping of CANopen is a powerful feature, as it allows the content of single messages to be customized. It makes CANopen very flexible, especially if "Dynamic Mapping" is implemented – meaning the PDO mapping of a node can be re-configured by a configuration tool or master.

However, in many deeply embedded applications this sort of flexibility might not be needed – or it might even be seen as a safety risk factor since in theory "something" could change what is contained in a message. That's why several embedded applications use "Static Mapping." In this case the PDO mapping is hard-coded into the software of a CANopen node and it cannot be re-configured.

Some systems further distinguish between Dynamic Mapping and Variable Mapping. Systems with Variable Mapping can only be configured while they are in the so-called "pre-operational" state which means the network is not currently transmitting PDOs with process data. In contrast, Dynamic Mapping allows a re-configuration while a node is "operational," meaning it is actively transmitting and receiving PDOs with process data.

Another benefit of Static Mapping is that fewer resources (code and data memory, CPU process time) are required for its implementation. Readers with embedded programming experience will recognize that the resources required to implement Dynamic Mapping on an 8-bit microcontroller are substantial.

2.5.7.3 PDO Mapping Practice

The PDO mapping is designed to work on the single bit level. Object Dictionary (OD) entries with single bits could be mapped individually. Theoretically one could create a PDO containing 2 single bit OD entries followed by an 8-bit OD entry, followed by a single bit OD entry followed by a 16-bit OD entry.

Obviously this kind of "bit-juggling" is not very microcontroller oriented where the straight-forward approach would be to use data widths of 8, 16 or 32 bits. In order to keep things simple and manageable many CANopen implementations limit themselves to an 8-bit-oriented PDO mapping, meaning that the length of any Object Dictionary entry that can be mapped to a PDO must be a multiple of 8 bits.

2.5.7.4 TPDO vs. RPDO Mapping and Dummy Entries

It should be pointed out that the mapping for a PDO is typically different on the TPDO and the RPDO side. In general, the transmitting node maps Object Dictionary entries into the TPDO which are "inputs" – process data coming into the node that needs to be communicated to others. The receiver looks at this message as an RPDO and will have its own usage for the data in it. Some of the data might directly be used as "output" to an actuator or it might get stored locally for further processing, resulting in a completely different set of PDO mappings on the receiver side.

The receiver might not even have a use for all the data contained in an RPDO. It could very well be that there are multiple process data variables available in an RPDO, but the local receiver only needs one of those process data variables. In that case the "unwanted" process data variables must still be mapped. CANopen supports multiple "dummy" entries that can be used for mapping such "unwanted" data in RPDOs. The Object Dictionary entries used for such dummy mapping are those from Index 1 to 7 which are also used to specify some basic data types (see Section 2.2.3).

2.6 Network Management (NMT)

Objective

This section explains the Network Management services available in CANopen. These include the NMT Master message to start/stop nodes, node guarding, heartbeat and emergencies.

We will also show which NMT services must be implemented on every CANopen slave. This includes the NMT slave state machine.

2.6.1 NMT Slave State Diagram

Every CANopen slave node must implement an NMT state machine that allows the slave to be in different operating states. The diagram in Figure 2.8 illustrates the major states a slave node can be in. It should be noted that some of these state transitions can be made automatically (by the slave themselves) where others can only be made upon receiving the corresponding NMT Master message. The NMT Master message can be directed at either individual nodes or at all nodes simultaneously. It contains the new state that the addressed node(s) should switch to.

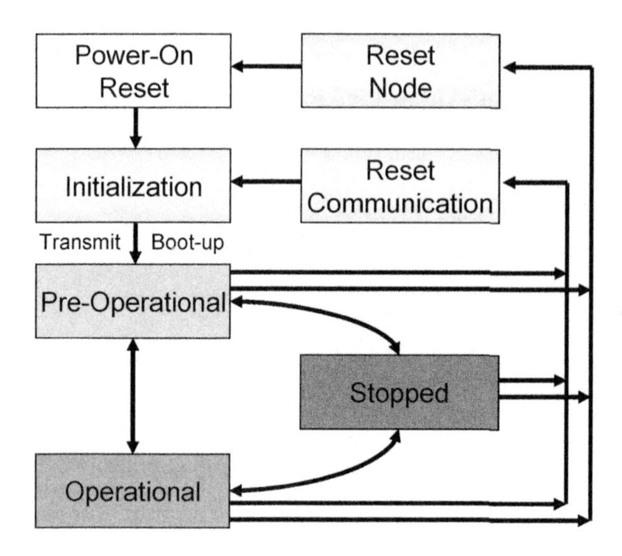

Figure 2.8 The Network Management States of a Slave

Upon power-up a CANopen slave node comes out of "Power-On Reset" and goes into Initialization. It initializes the entire application and the CAN/CANopen interfaces and communication. At the end of the initialization the node tries to transmit its boot-up message. As soon as it is transmitted successfully, the node switches to the Pre-operational state.

Using the NMT Master message, an NMT Master can switch individual nodes or all nodes back and forth between the three major states: Pre-operational, Operational and Stopped. In addition, the NMT Master has the option to request two different reset actions. Upon receiving the "Reset Communication" command, a CANopen slave node will reset the CAN/CANopen communication interfaces. A "Reset Node" command, however, results in a reset of the entire node with all peripherals and all software. Both reset states result in a new boot-up message being transmitted by the node and the node reverting back to the Pre-operational state, where it will wait for further NMT message commands.

Additional States

The NMT state diagram introduced by the DS301 CANopen standard only covers the basic requirements that apply to all CANopen nodes. This standard state machine is sufficient to achieve CANopen conformance. However, some of the frameworks or device profiles published today require additional states to be implemented to allow services like a Node ID claiming procedure (which is executed before a node can "boot-up").

2.6.1.1 CANopen Messages Produced and Consumed

The main difference between the various NMT states is that not all types of CANopen communication are actively used in each state. Table 2.17 shows which communication a node may perform when it is in a particular NMT state.

	Initializing	Pre-operational	Operational	Stopped
Boot-Up	◆			
SDO		◆	◆	
Emergency		◆	◆	
SYNC/TIME		◆	◆	
Heartbeat/ Nodeguard		◆	◆	◆
PDO			◆	

Table 2.17 NMT State Dependent Communication

In the Initializing state a node may only produce the boot-up message and it does not consume any messages.

In the Pre-operational state a node actively participates in all communication related to SDOs, Emergencies (if used by the node), Timestamps (if used by the node) and Heartbeat/Node Guarding.

There is only one difference between the Pre-operational state and the Operational state. The operational state adds PDO communication, allowing the node to exchange and work with process data. Only in Operational mode does a CANopen node truly run, meaning it executes all the input and output functions that it was designed to do.

In the Stopped state a node literally stops all communication, except for the minimal NMT services.

The DS301 CANopen specification does not provide any means for a node to autostart, which means there is no way for a node to switch into Operational without waiting for a message from the NMT Master.

However, many deeply embedded CANopen networks do not have an NMT Master and two work-a-rounds were common in the past. One was to implement a "minimal" NMT Master – one of the slave nodes would simply transmit the NMT "go to operational" message for everybody. Alternatively, some applications ignored the standard and just allowed the nodes to autostart – after switching to Pre-operational they would go straight into Operational by themselves.

Because this was a common problem to NMT Master-less systems, a solution was standardized with DSP302: CiA Draft Standard Proposal – the Framework for CANopen Managers and Programmable CANopen Devices. Now Object Dictionary entry 1F80 NMT Start-up offers a bit to allow autostart of nodes.

In other words, autostart is now in accordance with the standard but only if the Object Dictionary entry 1F80 is implemented to report that the node is autostarting.

2.6.2 Heartbeat or Node Guarding

In order to be CANopen compliant, every CANopen slave node must implement either the Heartbeat or the Node Guarding services. Today, the recommendation is to use heartbeat instead of node guarding as heartbeat consumes less bandwidth, is more flexible and is safer.

With node guarding it is the NMT Master's responsibility to poll ("guard") all slaves for their current NMT state information. If a node does not respond within a specified time, the NMT Master may assume that this node was lost and can take appropriate action (for example re-initialize or shut-down the system). In addition, individual nodes can also monitor the guarding messages from the NMT Master and take the absence of the poll message (exceeding a configurable time limit) as an indication that the connection to the master was lost.

The monitoring options become more flexible when using heartbeat. With the heartbeat method, each slave node by itself transmits a heartbeat, consisting of a 1-byte CAN message containing the current NMT state a node is in. The heartbeat producer time is configurable (entry [1017h,00h], UNSIGNED16, in milliseconds). Figure 2.9 illustrates two heartbeat messages repeatedly produced by two individual nodes at individual heartbeat times.

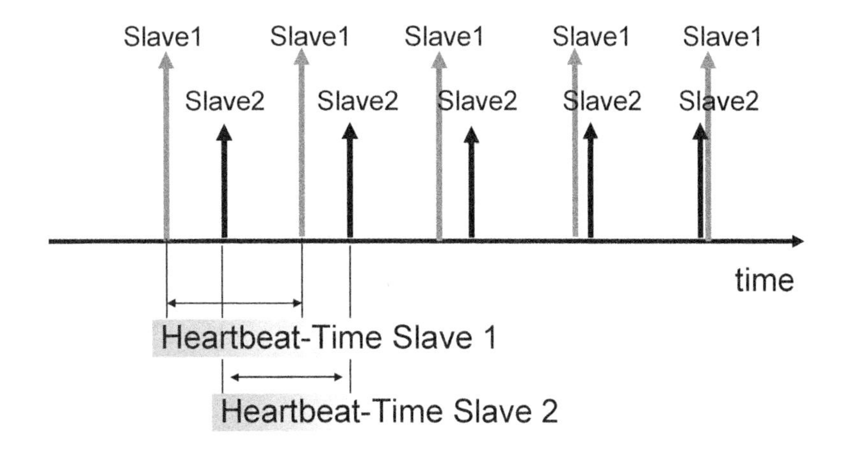

Figure 2.9 Heartbeat

One obvious benefit of the heartbeat method versus the node guarding is that the bandwidth used for the monitoring is cut by half (no polling required). In addition, each node can decide by itself which heartbeats it would like to monitor. A common practice is to monitor all the heartbeats from the direct communication partners. This would allow a node transmitting a PDO to listen to the heartbeat of all the consumers of that PDO to ensure that they are still "alive" and operational.

Another aspect is that of safety (see the following chapter for more details on "safety-related" systems). One first step towards a "safer" system is that no single node should be essential to the system. Using node guarding the NMT Master becomes essential; if it fails, all nodes will be affected. However, with the heartbeat method, no single node is essential for the heartbeat mechanism. Failure of a single node does not necessarily result in the failure of all nodes.

2.6.3 Emergencies (EMCY)

Each CANopen slave node is assigned one emergency message, sometimes simply referred to as EMCY. The CAN identifier used for these is 80h plus the CANopen

Node ID. So node number 5 uses the CAN identifier 85h for transmitting emergency messages.

An emergency message always contains 8 data bytes out of which the first 2 bytes are used for a CANopen error code (see the Reference Section for the list of all CANopen error codes defined). The third byte contains a copy of the error register (same value as at OD entry [1001h,00h]). The remaining 5 bytes are available for manufacturer specific error codes.

In general, emergencies are only reported once. The reported emergency is considered to "still be there" until the node uses another emergency message to clear/reset that specific emergency.

If, for example, a node reports a temperature emergency (measured temperature exceeds the limits) it will only report it once. Only when the temperature has returned within limits will the node transmit another emergency message, this time clearing/resetting the temperature emergency.

If the high-byte of the CANopen error code is 00h the message is not an emergency but a reset of an emergency.

For a listing of the defined error codes, see the Appendices in the Reference Section.

2.7 CANopen Example Configurations and Exercises

Objective

In this section we would like to give the reader a few configuration examples and exercises. Although the focus is on the TPDO and RPDO configuration, we will also cover configuration of a heartbeat producer and consumer.

2.7.1 Heartbeat Producer and Consumer Configuration Example

Summary of Object Dictionary entries controlling the heartbeat mechanism:

[1016h,00h] Consumer Heartbeat Time, number of elements in array
[1016h,xxh] Single 32-bit entries for each heartbeat monitored
Bit 0-15: Heartbeat time

Bit 16-23: Node ID monitored
Bit 24-31: Reserved

[1017h,00h] Producer Heartbeat Time, in milliseconds

2.7.1.1 Exercise

What entries need to be made to the Object Dictionary of node number 5 if that node needs to:

1. produce a heartbeat of 250ms and
2. monitor the heartbeat of node 7 (produced every 500ms) and
3. monitor the heartbeat of node 9 (produced every 1,000ms)?

2.7.1.2 Solution

1. Write the value 250d into OD entry [1017h,00h] of node 5.
2. To monitor the heartbeat of a node, the consumer's time (which is a time-out, meaning a heartbeat is considered 'lost' if it does not appear within that time) must be set to higher value than the producer's time. A reasonable value is some 1.5 to 2 times the producer's time.

   Write the value 750d (500 times 1.5) into OD entry [1016h,01h] of node 5 (first heartbeat consumer entry).

   Write the value 1 into OD entry [1016h,00h] to indicate that one heartbeat is monitored.
3. Write the value 1,500d (1,000 times 1.5) into OD entry [1016h,02h] of node 5 (second heartbeat consumer entry). Write the value 2 into OD entry [1016h,00h] to indicate that two heartbeats are monitored.

2.7.2 PDO Linking Example

Summary of Object Dictionary entries controlling the PDO linking (indicating which CAN identifier is used for each RPDO and TPDO):

[1400h,01h] COB ID of RPDO1 (default is 200h + Node ID)
[1401h,01h] COB ID of RPDO2 (default is 300h + Node ID)
[1402h,01h] COB ID of RPDO3 (default is 400h + Node ID)
[1403h,01h] COB ID of RPDO4 (default is 500h + Node ID)

[1800h,01h] COB ID of TPDO1 (default is 180h + Node ID)
[1801h,01h] COB ID of TPDO2 (default is 280h + Node ID)
[1802h,01h] COB ID of TPDO3 (default is 380h + Node ID)
[1803h,01h] COB ID of TPDO4 (default is 480h + Node ID)

2.7.2.1 Exercise

1. Node 5 needs to be configured to directly listen for the default TPDO1 transmitted by node number 6. RPDO1 of node 5 should be used to receive TPDO1 of node 6 (for illustration see Figure 2.10).

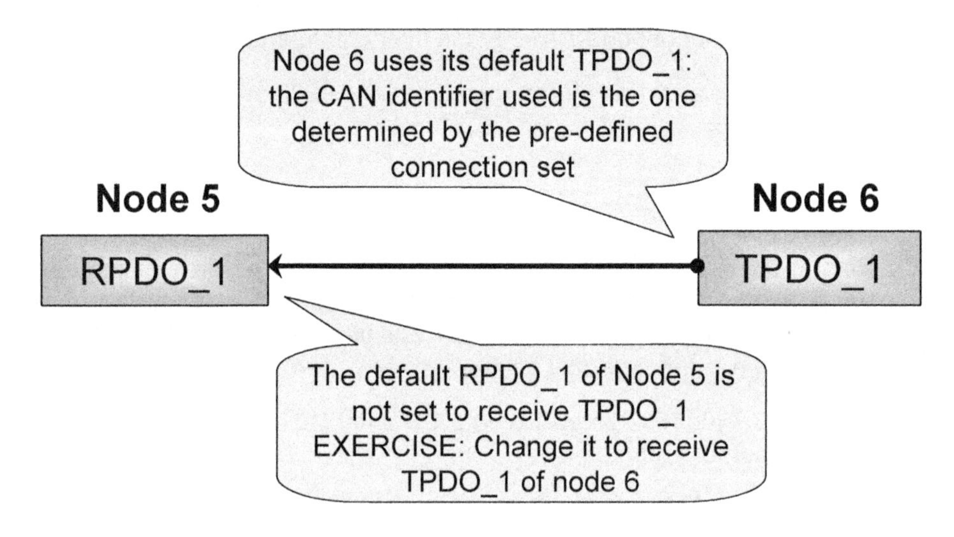

Figure 2.10 PDO Linking Exercise

2.7.2.2 Solution

1. The default CAN identifier used by node 6 for TPDO1 is 186h (180h base address plus 6 for Node ID 6).

   Write the value 186h (390d) into OD entry [1400h,01h] of node 5.

2.7.3 PDO Linking and Mapping Example

Summary of Object Dictionary entries controlling the PDO mapping (indicating which OD entries are used for each PDO):

[1600h,00h] RPDO1 Mapping, number of entries mapped
[1600h,xxh] Index, Subindex and length (in bits) of a single entry mapped
[1601h,00h] RPDO2 Mapping, number of entries mapped
[1601h,xxh] Index, Subindex and length (in bits) of a single entry mapped

[1A00h,00h] TPDO1 Mapping, number of entries mapped
[1A00h,xxh] Index, Subindex and length (in bits) of a single entry mapped
[1A01h,00h] TPDO2 Mapping, number of entries mapped
[1A01h,xxh] Index, Subindex and length (in bits) of a single entry mapped

2.7.3.1 Exercise

Node 2Ah transmits two 16-bit analog values in its TPDO2 (two UNSIGNED16 values mapped into TPDO2 of node 2Ah) using the default CAN identifier.

Node 2Dh transmits two 16-bit analog values in its TPDO3 (two UNSIGNED16 values mapped into TPDO3 of node 2Dh) using the default CAN identifier.

1. Node 1Fh should be configured to receive in its RPDO2 the TPDO2 from node 2Ah and in RPDO3 the TPDO3 of node 2Dh.
2. Node 1Fh has an array of 4 UNSIGNED16 values at [6411h,01h-04h]. Configure the mapping of RPDO2 and RPDO3 so that the values from RPDO2 go into [6411h,01h-02h] and the values from RPDO3 into [6411h,03h-04h].

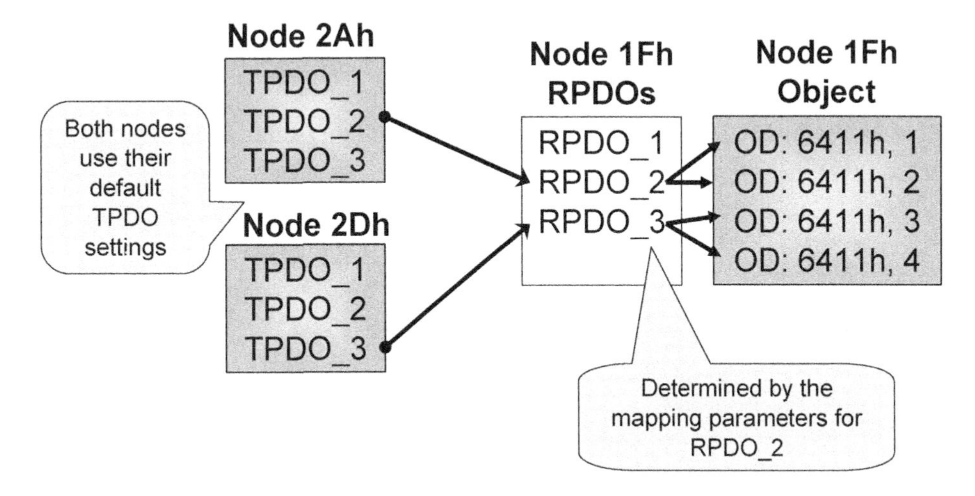

Figure 2.11 PDO Linking and Mapping Exercise

2.7.3.2 Solution

Note: In order to change PDO parameters, a PDO typically needs to be disabled. This can be achieved by setting bit 31 in the COB ID.

1. The default CAN identifier used for TPDO2 of node 2Ah is 2AAh. The default CAN identifier used for TPDO3 of node 2Dh is 3ADh.

   To configure node 1Fh to receive these: Write the value 2AAh into OD entry [1401h,01h] of node 1Fh. Write the value 3ADh into OD entry [1402h,01h] of node 1Fh.

2. To configure the mapping:

   Write the value '0' into OD entry [1601h,00h] of node 1Fh (informs node that mapping will be changed).

   Write the value 64110110h into OD entry [1601h,01h] of node 1Fh (first mapping entry for RPDO2, Index 6411h, Subindex 01h, length 10h).

   Write the value 64110210h into OD entry [1601h,02h] of node 1Fh (second mapping, Index 6411h, Subindex 02h, length 10h).

   Write the value '2' into OD entry [1601h,00h] of node 1Fh (total number of entries mapped is 2).

   Write the value '0' into OD entry [1602h,00h] of node 1Fh (informs node that mapping will be changed).

   Write the value 64110310h into OD entry [1602h,01h] of node 1Fh (first mapping entry for RPDO3, Index 6411h, Subindex 03h, length 10h).

   Write the value 64110410h into OD entry [1602h,02h] of node 1Fh (second mapping, Index 6411h, Subindex 04h, length 10h).

   Write the value '2' into OD entry [1602h,00h] of node 1Fh (total number of entries mapped is 2).

2.8 Contents of CANopen Messages

> Objective
>
> This section is only for those readers who need to have an understanding of CANopen on the individual message basis. Do you need to be able to interpret individual CAN messages for their CANopen content? Well, this section is for you!
>
> If you do not need to know this level of detail, feel free to skip this section and proceed to the next chapter.

2.8.1 Endianess

All numerical data types consisting of multiple bytes are transferred in CANopen (whether in SDO or PDO) in the "Little Endian" format. Bytes are ordered by significance and the lower significant bytes come first.

For example, a 2-byte word would be transmitted low-byte first, followed by the high-byte. A 4-byte word is transmitted with the least significant byte first, followed by the bytes of next higher significance and the most significant byte transmitted last.

> When implementing CANopen software on a specific microcontroller, developers must pay attention to the byte ordering.
>
> With 8-bit architectures the byte ordering is determined by the compiler alone and not the architecture. Some compilers for 8-bit architectures are able to support both Little and Big Endian formats, so in these cases a simple compiler switch might select the correct implementation.
>
> If a 16-bit architecture based on Big Endian is used, however, an appropriate byte swapping must be implemented. Most commercial CANopen stacks can automatically activate byte swapping via a #define statement that enables an appropriate byte swapping macro.

2.8.2 SDO Communication

When using SDO communication, one needs to differentiate between two major communication modes, typically referred to as "expedited transfer" and "segmented transfer." A third, optional mode is the "block transfer", an optimized method to transfer large data amounts. Section 2.8.5 explains the messages used for block transfer.

An expedited transfer consists of one SDO request and one SDO response.

A segmented transfer consists of an SDO initiation request and response and then one pair of request and response for each 7-byte segment.

With expedited transfer up to four bytes of data can be directly embedded in an SDO request or response, suitable for accesses to Object Dictionary entries that are up to 4-bytes long. The segmented transfer allows for transmission of data bigger than 4-bytes and is required to access Object Dictionary entries that are longer than 4-bytes.

> When implementing CANopen on microcontrollers with "limited resources" it is desirable to only implement expedited transfer and to omit the segmented transfer. Some of the latest CiA drafts actually take that into account; for example, the device profile (CiADSP418) for batteries (such as those used in electrical vehicles). The only Object Dictionary entries that would exceed the 4-byte limit would be extended identification strings with up to 20 characters.
>
> In order to utilize "expedited transfer only" for battery implementations these entries have to be divided into several Subentries of 4 bytes each. So at Subindex 1, one would find the first 4 characters of the string, at Subindex 2 the next 4 characters and so on.

2.8.2.1 The Initiate SDO Download – Request

The client (typically the node trying to configure a CANopen slave) sends this request to a SDO server (implemented within a CANopen slave) by using the CAN identifier 600h plus the Node ID of the CANopen slave addressed. The download request is a request to write to a specific Object Dictionary entry.

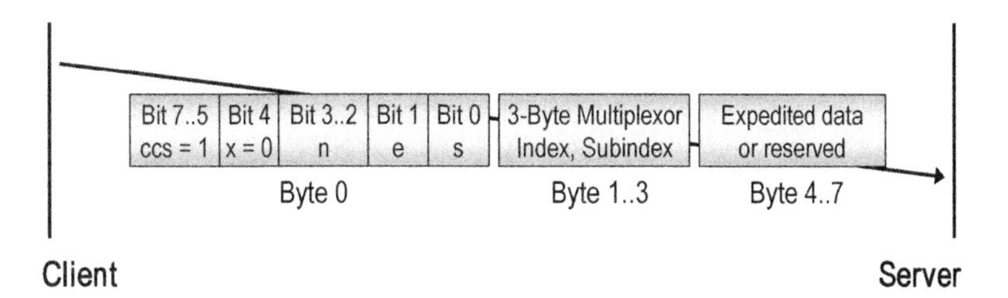

Figure 2.12 Initiate SDO Download – Request

Message contents:

- ccs: Client Command Specifier = 1
- e: set to 1 for expedited transfer (data is in bytes 4-7)
- s: set to 1 if data size is indicated
- n: if e=s=1, number of data bytes in Byte 4..7 that do *not* contain data
- x: reserved
- The Multiplexor contains the Index and Subindex of the OD entry this write access should go to

2.8.2.2 *The Initiate SDO Download – Response*

This is the response sent back from the SDO server to the client indicating that the previously received download (write) request was processed successfully. The default CAN identifier used for this message is 580h plus the Node ID of the node implementing the SDO server.

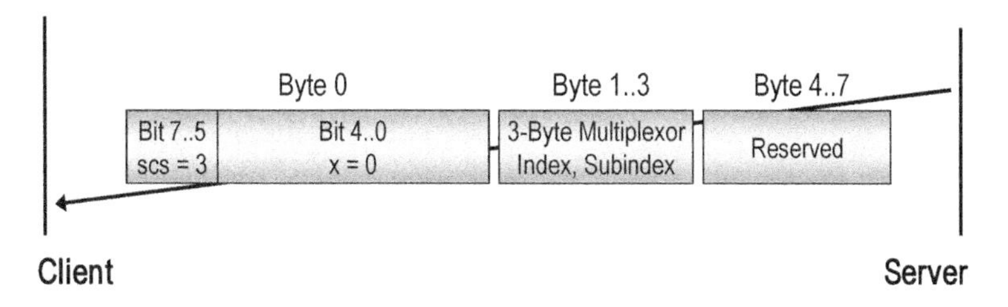

Figure 2.13 Initiate SDO Download - Response

Message contents:

- scs: Server Command Specifier = 3
- x: reserved
- The Multiplexor contains the Index and Subindex of the OD entry that received the write access

2.8.2.3 The Download SDO Segment – Request

If in the initiation sequence a segmented transfer was negotiated, this message is used to transmit the next segment (of up to 7 bytes) from client to SDO server.

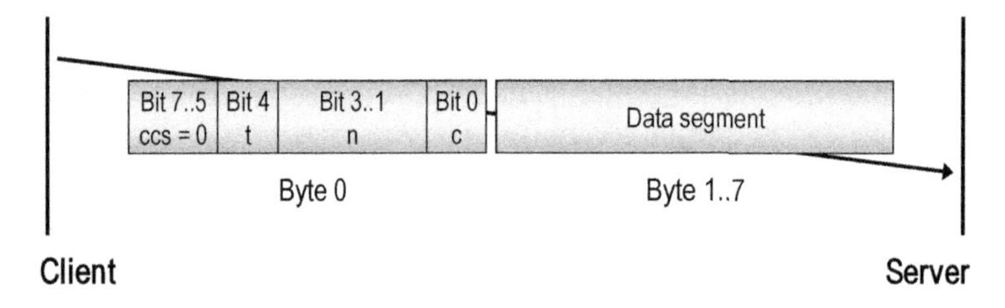

Figure 2.14 Download SDO Segment - Request

Message contents:

- ccs: Client Command Specifier = 0
- c: set to 1 if this is the last segment/fragment
- n: number of data bytes in Byte 1..7 that do *not* contain data
- t: toggle bit – set to 0 in first segment, toggled with each subsequent request

2.8.2.4 The Download SDO Segment – Response

This is the response sent back from the SDO server to the client indicating that the previously received download (write) segment request was processed successfully.

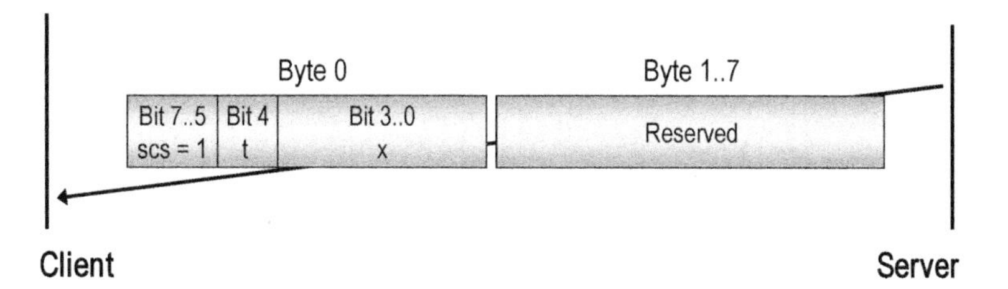

Figure 2.15 Download SDO Segment - Response

Message contents:

- scs: Server Command Specifier = 1
- x: reserved
- t: toggle bit – set to 0 in first segment response, toggled with each subsequent response

2.8.2.5 The Initiate SDO Upload – Request

The client (typically the node trying to configure a CANopen slave) sends this request to an SDO server (implemented within a CANopen slave) by using the CAN identifier 600h plus the Node ID of the CANopen slave addressed. The upload request is a request to read from a specific Object Dictionary entry.

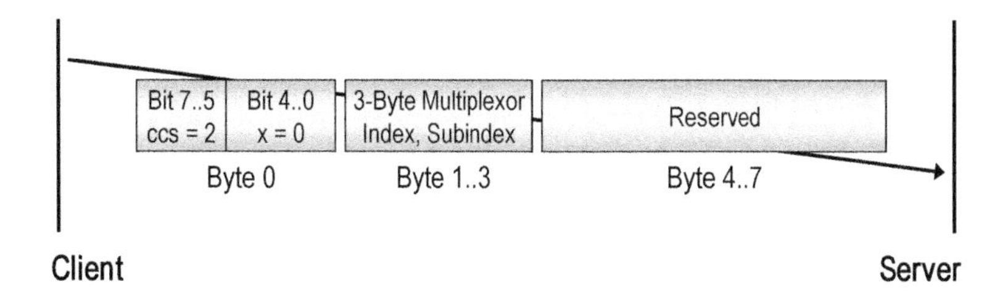

Figure 2.16 Initiate SDO Upload – Request

Message contents:

- ccs: Client Command Specifier = 2
- x: reserved

- The Multiplexor contains the Index and Subindex of the OD entry that the client wants to read

2.8.2.6 The Initiate SDO Upload – Response

This is the response sent back from the SDO server to the client indicating that the previously received upload (read) request can be processed. If expedited transfer is used, the data read from the Object Dictionary is part of the response, otherwise additional segmented transfers are used. The default CAN identifier used for this message is 580h plus the Node ID of the node implementing the SDO server.

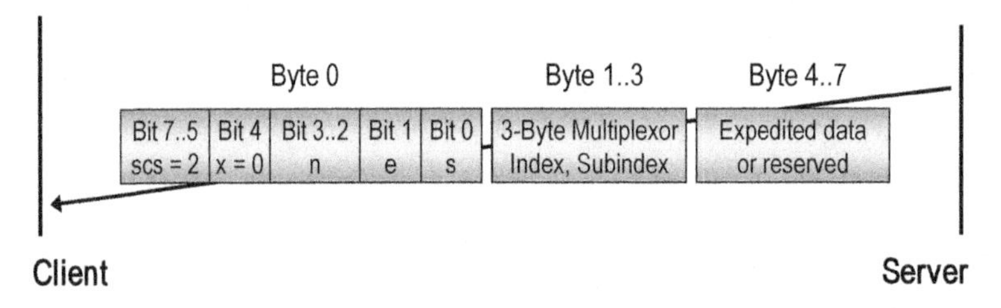

Figure 2.17 Initiate SDO Upload – Response

Message contents:

- scs: Server Command Specifier = 2
- e: set to 1 for expedited transfer (data is in bytes 4-7)
- s: set to 1 if data size is indicated
- n: if e=s=1, number of data bytes in Byte 4..7 that do *not* contain data
- x: reserved
- The Multiplexor contains the Index and Subindex of the OD entry this write access should go to

2.8.2.7 The Upload SDO Segment – Request

If in the initiation sequence a segmented transfer was negotiated, this message is used to request that the next segment (of up to 7 bytes) be transmitted from SDO server to client.

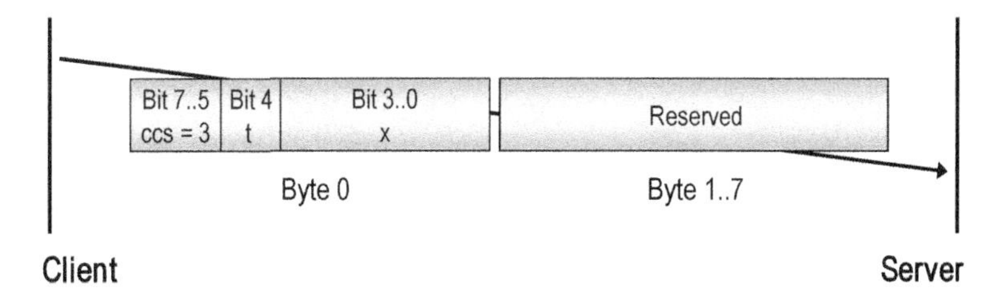

Figure 2.18 Upload SDO Segment – Request

Message contents:

- ccs: Client Command Specifier = 3
- x: reserved
- t: toggle bit – set to 0 in first segment request, toggled with each subsequent request

2.8.2.8 The Upload SDO Segment – Response

This is the response sent back from the SDO server to the client indicating that the previously received upload (read) segment request was processed successfully. The data segment is part of this message.

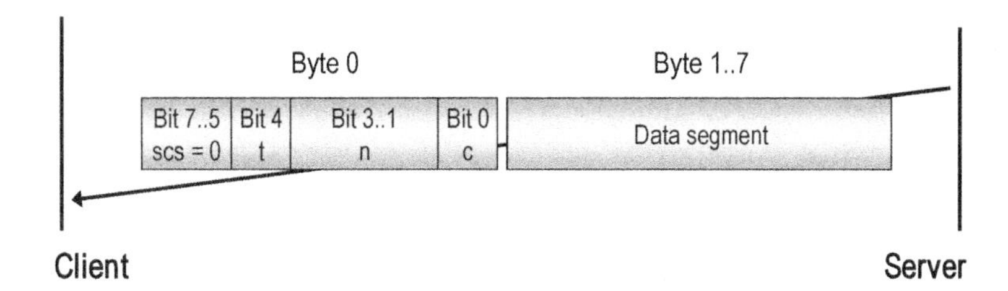

Figure 2.19 Upload SDO Segment - Response

Message contents:

- scs: Server Command Specifier = 0
- c: set to 1 if this is the last segment/fragment
- n: number of data bytes in Byte 1..7 that do *not* contain data

- t: toggle bit – set to 0 in first segment, toggled with each subsequent response

2.8.2.9 The Abort SDO Transfer

At any time the client or server may abort an SDO transmission. The error code gives an indication as to why the transfer was aborted. Typical errors are that a desired Object Dictionary entry is not implemented by the SDO server or that the entry is of a different length (for example writing a 2-byte value to a 4-byte entry). For a listing of the possible error codes see the Reference Section.

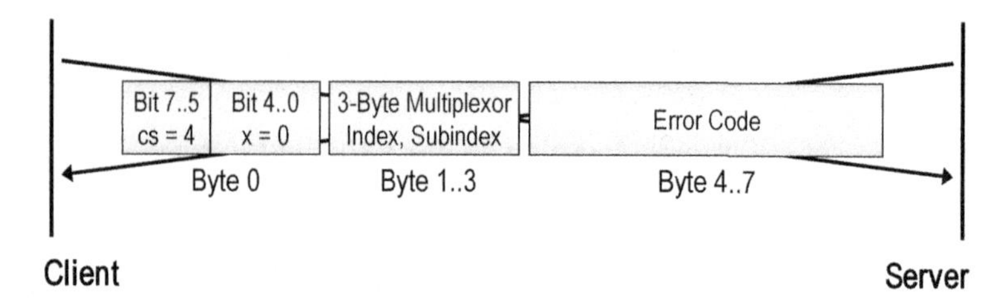

Figure 2.20 Abort SDO Transfer

Message contents:

- cs: Command Specifier = 4
- x: reserved
- The Multiplexor contains the Index and Subindex of the OD entry that was affected
- The Error Code gives an indication of what went wrong

2.8.3 Network Management (NMT) Communication

2.8.3.1 The NMT Master Message

The NMT Master message has the CAN message identifier 0 (zero) and contains 2 bytes. All CANopen slave nodes must be able to receive this message and act upon its content.

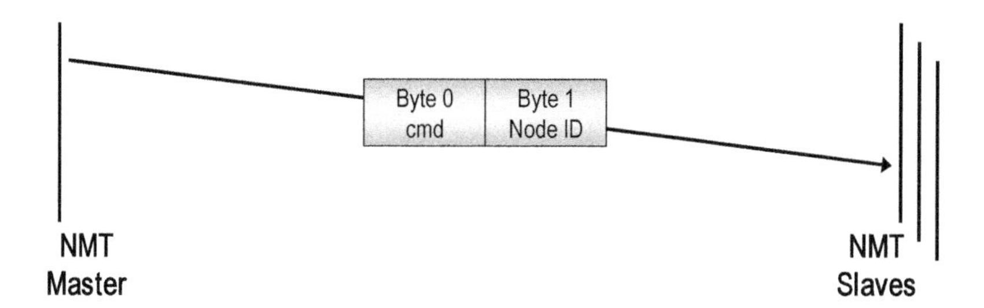

Figure 2.21 NMT Master Message

Message contents:

- cmd: One of the following commands to switch into the specified NMT state: 1 = Operational, 2 = Stopped, 128 = Pre-operational, 129 = Reset Node, 130 = Reset Communication
- Node ID: zero if addressed at all nodes, or the specific Node ID of the single node addressed with this message

2.8.3.2 The Heartbeat

The heartbeat message sent by an individual node has the CAN message identifier 700h plus the Node ID. It only contains one byte showing the current NMT state of that node.

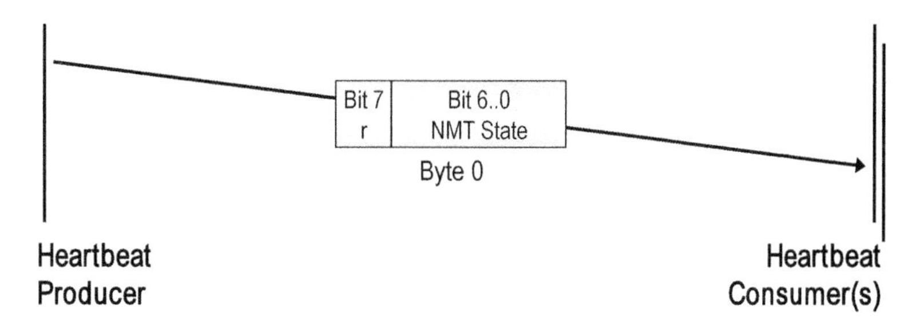

Figure 2.22 Heartbeat

Message contents:

- NMT State: Reports the current NMT state the node is in: 0 = Boot-up, 4 = Stopped, 5 = Operational, 127 = Pre-operational

- r: reserved

2.8.4 Emergency Communication

The emergency message sent by an individual node has a CAN identifier of 80h plus the Node ID.

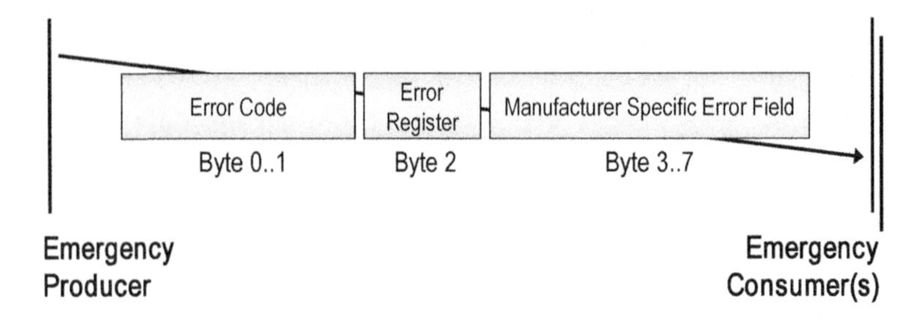

Figure 2.23 Emergency

Message contents:

- Error Code: 2-byte error code – see Table 2.18
- Error Register: copy of the 1-byte error register at [1001h,00h]
- Manufacturer Specific Error Field: Up to 5 bytes for manufacturer specific error codes

Error Code	Decsription
00xx	Error Reset or No Error
10xx	Generic Error
20xx	Current
21xx	Current, device input side
22xx	Current inside the device
23xx	Current, device output side
30xx	Voltage

Table 2.18 Emergency Error Codes

Error Code	Decsription
31xx	Mains Voltage
32xx	Voltage inside the device
33xx	Output Voltage
40xx	Temperature
41xx	Ambient Temperature
42xx	Device Temperature
50xx	Device Hardware
60xx	Device Software
61xx	Internal Software
62xx	User Software
63xx	Data Set
70xx	Additional Modules
80xx	Monitoring
81xx	Communication
8110	CAN Overrun (Objects Lost)
8120	CAN in Passive Error Mode
8130	Life Guard Error or Heartbeat Error
8140	Recovered from Bus Off
8150	Transmit COB ID Collision
82xx	Protocol Error
8210	PDO not processed due to length of error
8220	PDO length exceeded
90xx	External Error
F0xx	Additional Functions
FFxx	Device Specific

Table 2.18 (Continued) Emergency Error Codes

2.8.5 SDO Block Transfer

The block transfer mode is an optimized transfer mode for Object Dictionary entries that contain large amounts of data. In this transfer mode, up to 889 bytes (segmented into 127 messages with each 7 bytes) are combined into one data block and are transmitted using back-to-back messages.

The block transfer mode is optional and can only be used if both client and server support this communication mode. If one of the nodes does not support block transfer, the segmented or expedited transfer has to be used.

A download is divided into the following communication stages:

- Initiate Block Download - Client requests from Server to use block transfer mode for a download.
- Download Blocks - Client sends data blocks to Server and expects one response per block. Each block contains up to 127 segments.
- End of Download Block - Client and Server confirm that the transmission is now complete.

At any stage, any of the two communication partners may abort the transfer by sending an abort SDO transfer message.

2.8.5.1 Initiate Block Download

To initiate a block download the client sends the request shown in Figure 2.24 to the server to which the server sends a response, also shown in Figure 2.24.

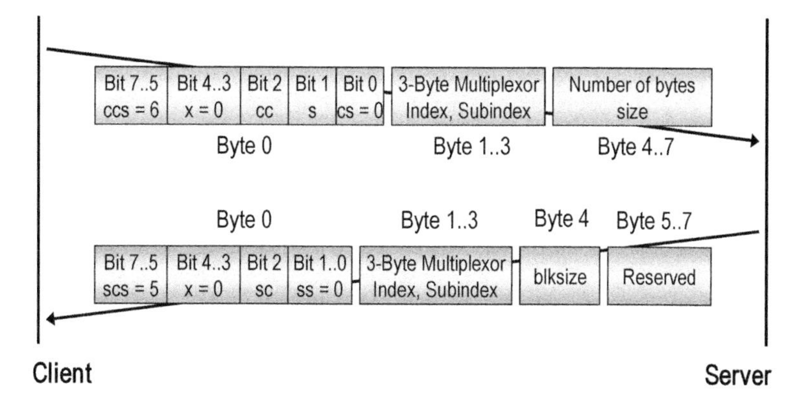

Figure 2.24 Initiate Block Download

Message contents of the request:

- ccs: Client Command Specifier = 6
- x: Reserved
- cc: Client CRC support, set to 1 if client supports CRC
- s: Size indicator, set if size of data to transmit is indicated
- cs: Client subcommand = 0
- The Multiplexor contains the Index and Subindex of the OD entry that the client wants to write to
- size: Contains the size of the data block in bytes, if s is set

Message contents of the response:

- scs: Server Command Specifier = 5
- x: Reserved
- sc: Server CRC support, set to 1 if server supports CRC
- ss: Server Subcommand = 0
- The Multiplexor contains the Index and Subindex of the OD entry that the client wants to write to
- blksize: The number of segments per block (1-127)

2.8.5.2 Download Blocks

After successful initiation, the client starts transmitting the blocks. Each block consists of as many segments as specified by "blksize" during initiation. At the end of a block, the client expects the server to send a response.

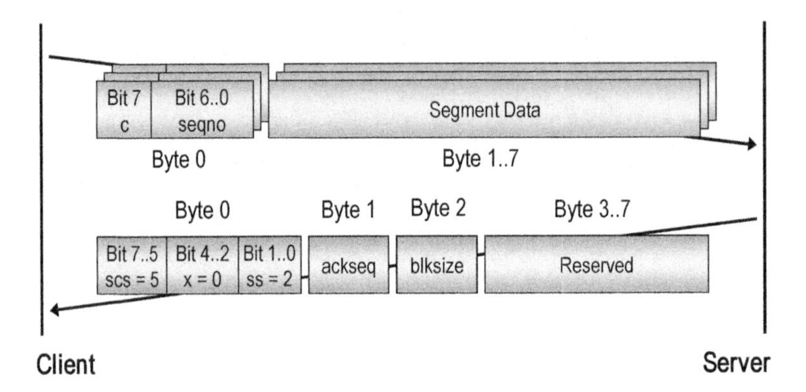

Figure 2.25 Download Blocks

Message contents of the request:

- c: Set to 1 if this is the last segment of the block
- seqno: Sequence counter from 1 to blksize (see initiation)

Message contents of the response:

- scs: Server Command Specifier = 5
- x: Reserved
- ss: Server Subcommand = 2
- ackseq: Number of segments acknowledged (received correctly) - the Client must re-transmit those that are not acknowledged
- blksize: The number of segments per block (1-127) that the Client must use for the next block

2.8.5.3 End of Download Block

After the client transmitted all blocks and the server acknowledged all blocks, the client and server confirm to each other if the transmission was successful.

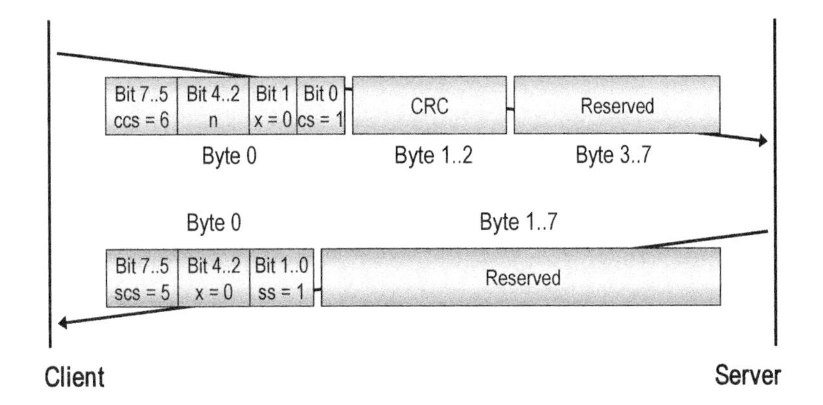

Figure 2.26 End of Download Block

Message contents of the request:

- ccs: Client Command Specifier = 6
- n: Number of bytes in last segment that do not contain data
- x: Reserved
- cs: Client subcommand = 1
- crc: Cyclic Redundancy Checksum of the transferred data, leave at zero if CRC is not used (details about CRC generation are at the end of this chapter)

Message contents of the response:

- scs: Server Command Specifier = 5
- x: Reserved
- ss: Server Subcommand = 1

An upload is divided into the following communication stages:

- Initiate Block Upload - Client requests from Server to use block transfer mode for an upload.
- Upload Blocks - Client receives data blocks from Server and returns one response per block. Each block contains up to 127 segments.

- End of Download Block - Client and Server confirm that transmission is now complete.

At any stage, any of the two communication partners may abort the transfer by sending an abort SDO transfer message.

2.8.5.4 Initiate Block Upload

To initiate a block upload a total of three messages are exchanged as shown in Figure 2.27.

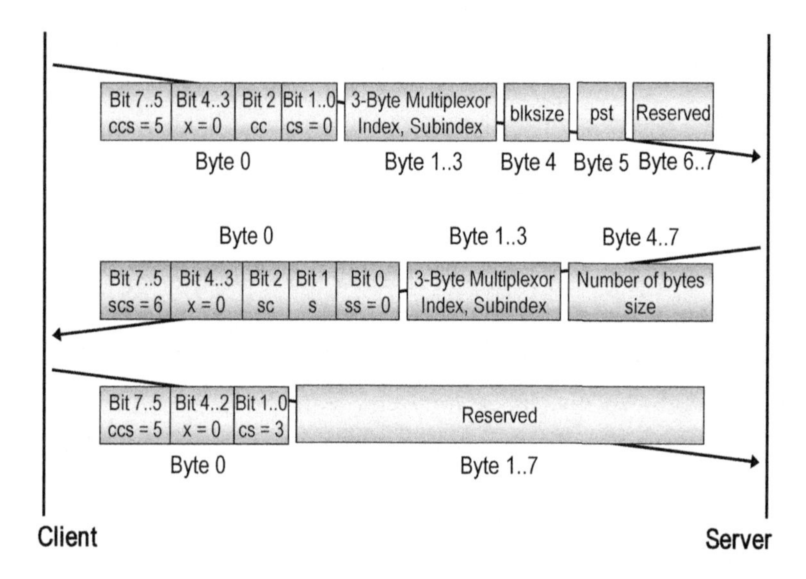

Figure 2.27 Initiate Block Upload

Message contents of the first request:

- ccs: Client Command Specifier = 5
- x: Reserved
- cc: Client CRC support, set to 1 if client supports CRC
- cs: Client subcommand = 0
- The Multiplexor contains the Index and Subindex of the OD entry that the client wants to read from
- blksize: The number of segments per block (1-127)

- pst: If set to a non-zero value, the server may switch back to the regular segmented SDO transfer if the total data to be transmitted is less than or equal to the number of bytes defined by pst

Message contents of the response:

- scs: Server Command Specifier = 6
- x: Reserved
- sc: Server CRC support, set to 1 if server supports CRC
- s: Size indicator, set if total size of data transfer is indicated
- ss: Server Subcommand = 0
- The Multiplexor contains the Index and Subindex of the OD entry that the client wants to read from
- size: The total number of bytes that need to be transmitted (possibly using multiple blocks)

Message contents of the second request:

- ccs: Client Command Specifier = 5
- x: Reserved
- cs: Client subcommand = 3

2.8.5.5 Upload Blocks

After successful initiation, the server starts transmitting the blocks. Each block consists of as many segments as specified by "blksize" during initiation. At the end of each block, the server expects the client to send a response.

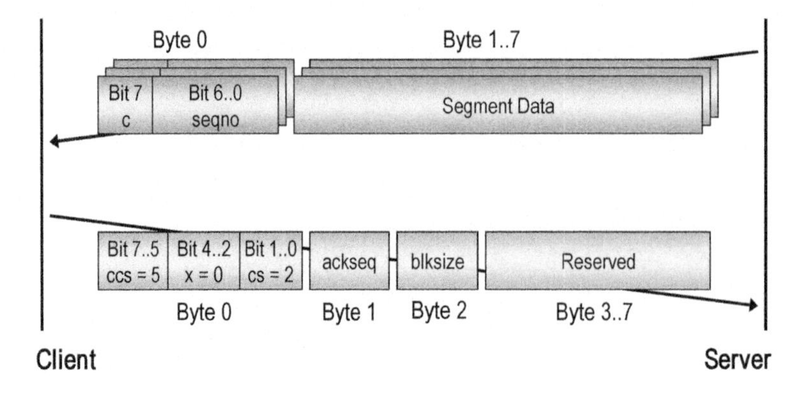

Figure 2.28 Upload Blocks

Message contents of the block segments:

- c: Set to 1 if this is the last segment of the block
- seqno: Sequence counter from 1 to blksize (see initiation)

Message contents of the acknowledge:

- ccs: Client Command Specifier = 5
- x: Reserved
- cs: Client Subcommand = 2
- ackseq: Number of segments acknowledged (received correctly) - the Server must re-transmit those that are not acknowledged
- blksize: The number of segments per block (1-127) that the Server must use for the next block

2.8.5.6 End of Upload Block

After the server transmits all blocks and the client acknowledges all blocks, the client and server confirm to each other if the transmission was successful.

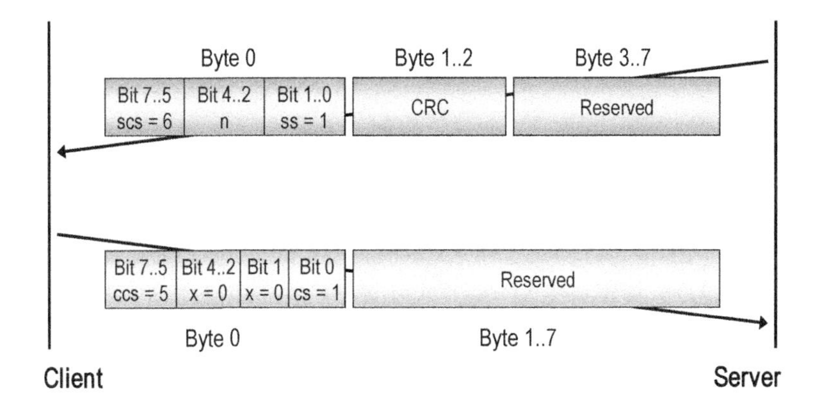

Figure 2.29 End of Upload Block

Message contents of the server's confirmation:

- scs: Server Command Specifier = 6
- n: Number of bytes in last segment that do not contain data
- ss: Server subcommand = 1
- crc: Cyclic Redundancy Checksum of the transferred data, leave at zero if CRC is not used

Message contents of the client's confirmation:

- ccs: Client Command Specifier = 5
- x: Reserved
- cs: Client Subcommand = 1

2.8.5.7 CRC Calculation

The Cyclic Redundancy Checksum used for the block transfer has 16 bits and is calculated over the entire data range of each block. The polynomial used for the calculation is:

$$x^{16} + x^{12} + x^{5} + 1$$

3 CANopen Beyond DS301

"Get your facts first and then you can distort them as much as you wish."
Mark Twain

The part of the CANopen standard that was covered in the previous chapter [CiADS301] lays down the foundation that any CANopen application builds upon. It describes the basic communication, data structuring, and network management methods used in the network.

However, one of the advantages of CANopen is its "openness" which enables it to incorporate additional specifications and standards which cover application or device-specific aspects of a CANopen implementation. To avoid the need to further modify or enhance existing standards, the approach is to include all enhancements in additional documents, especially if they are device or application-specific. These can be maintained by the CiA, but there are also proprietary profiles that specify how CANopen is used in one specific product.

Although the creation of proprietary profiles is acceptable for many deeply embedded networks, one of the main reasons for using a standard is to avoid re-inventing the wheel. So it is highly recommended that developers get all the facts about existing profiles and frameworks first (to see which elements can be adapted) before inventing a new proprietary profile from scratch.

3.1 Frameworks and Profiles Overview

> Objective
>
> In this section we want to make the reader familiar with the various types of documents that together make up the CANopen standard. We'll also explain what the documents are for and how they are numbered.
>
> Because many new profiles are in the process of being developed, this can only be a "snap shot" of current developments. With time, documents that are currently considered "proposals" will become "standards" and new proposals will be available.

One of the huge advantages of a higher-layer protocol is the guarantee of interchangeability between the same type of off-the-shelf devices from different manufacturers. This ensures interoperability between all devices that comply with this networking standard, thus simplifying the task of system integration. The documents that constitute the CANopen Device Profiles describe, in detail, how to use CANopen for a particular type of device, what communication parameters are available, and how the Object Dictionary is set up.

Device Profile	Title
DS401	CANopen device profile for generic I/O modules
DSP402	CANopen device profile for drives and motion control
DS404	CANopen device profile for measuring devices and closed loop controllers
DS405	CANopen interface and device profile for IEC 61131-3 programmable devices
DS406	CANopen device profile for encoders
DSP408	CANopen device profile for fluid power technology proportional valves and hydrostatic transmissions
DSP410	CANopen device profile for inclinometers
DSP413	CANopen device profiles for truck gateways

Table 3.1 List of Selected Device Profiles

Device Profile	Title
DSP414	CANopen device profiles for weaving machines
DSP418	CANopen device profile for battery modules
DSP419	CANopen device profile for battery chargers
DSP420	CANopen profiles for extruder downstream devices

Table 3.1 (Continued) List of Selected Device Profiles

Taking this thought one step further, a networking standard can also describe the communication aspect of a complete application, including not only the individual devices that are part of this application, but all interfaces between them as well. In CANopen, the documents describing this are called Application Profiles.

Application Profile	Title
DSP407	Application Profile for Passenger Information
WD416	Application Profile for Building Door Control
DSP417	Application Profile for Lift Control Systems

Table 3.2 List of Application Profiles

There are many instances where CANopen devices or applications require more mechanisms for configuration, data access or transport than what is covered by the Communication Profile DS301. Some of these will actually change what is seen transmitted on the CAN bus. These "extensions" to CANopen are called Frameworks. Some Device Profiles, but not all, build on top of them, whereas all Application Profiles so far refer to at least one Framework.

The Frameworks, although targeted at the communication requirements of particular applications, are nevertheless "open" to use in other applications. For instance, an application that requires safety-relevant communication could utilize either of the very different methods that the Framework for Safety-relevant Communication DSP304 or the Framework for Maritime Electronics DSP307 describes. More details will be covered in Section 3.5.

Framework	Title
DSP302	CANopen Framework for CANopen Managers and Programmable CANopen Devices
DSP304	CANopen Framework for Safety-relevant Communication
DSP305	CANopen Layer Setting Services and Protocols (LSS)
DSP307	CANopen Framework for Maritime Electronics

Table 3.3 List of CANopen Frameworks

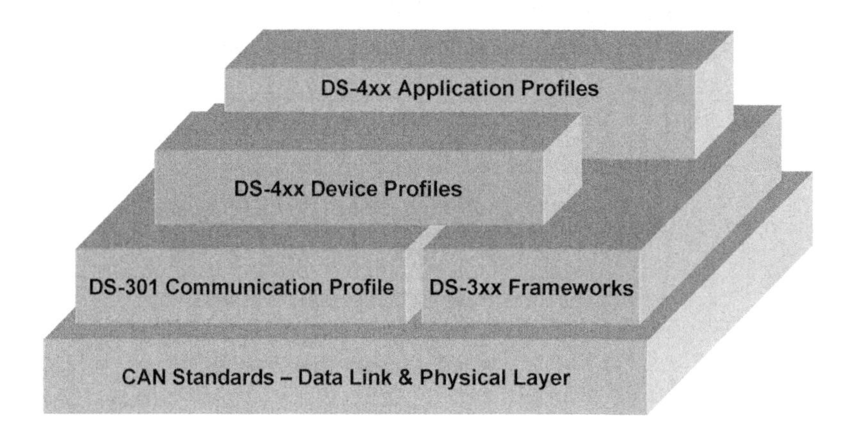

Figure 3.1 The CANopen Standard Documents

Figure 3.1 illustrates how the different standard documents build on or complement each other. While the Device and Application Profiles carry 4xx numbers, the Frameworks use 3xx to indicate that they are on about the same level as the Communication Profile. The CAN bus as the underlying technology is described in CiA standard documents DS-1xx and 2xx as well as several EN and ISO standards.

What does the "DS" in "DS301" stand for?

When browsing the CANopen standard documents you will notice that some of the documents start with "DS", but there are other initials used as well. The following are all the acronyms that have been used so far:

- DS - Draft Specification
- DSP - Draft Standard Proposal
- WD - Working Draft or Work Draft Proposal
- TR - Technical Report
- DR - Draft Recommendation

A *Draft Specification* describes an essentially fixed standard that presumably will not undergo major changes in the future.

Draft Standard Proposals are released standard documents as well, but they may be simplified or expanded to accommodate changed requirements in the future. Officially still considered draft documents, you will nevertheless find many implementations in the CANopen world that comply with these "best possible standards."

The *Working Draft* or *Work Draft Proposal* documents describe parts of the CANopen standard that are still very much "in the works." They often carry 0.xx version numbers, may change a great deal before being released, and are not recommended at all for actual implementations. They are for informational purposes only and in most cases document the work of *SIGs*, the Special-Interest Groups within the CiA.

A *Technical Report* will give definitions and guidelines, for example on the implementation and testing aspect of the network.

Draft Recommendations describe a "best recommended practice" for hardware aspects of a CANopen implementation such as connectors, cabling, and indicator LEDs.

3.2 About Masters and Managers (DS302)

Objective

This section explains the different master and manager services available in CANopen. Terms covered include the NMT Master, the SDO Manager, the Configuration Manager and the CANopen Manager as described in [CiADS302].

Flying Masters are also introduced. For details on Flying Master operation, see [CiADS302].

Unlike other fieldbus systems, CANopen does not require a single master that combines all "intelligence" in the network. Instead, there are several different functionalities that provide application-supporting services.

The following sections summarize and clarify some of the terms described in DSP302, the Framework for CANopen Managers and Programmable CANopen Devices.

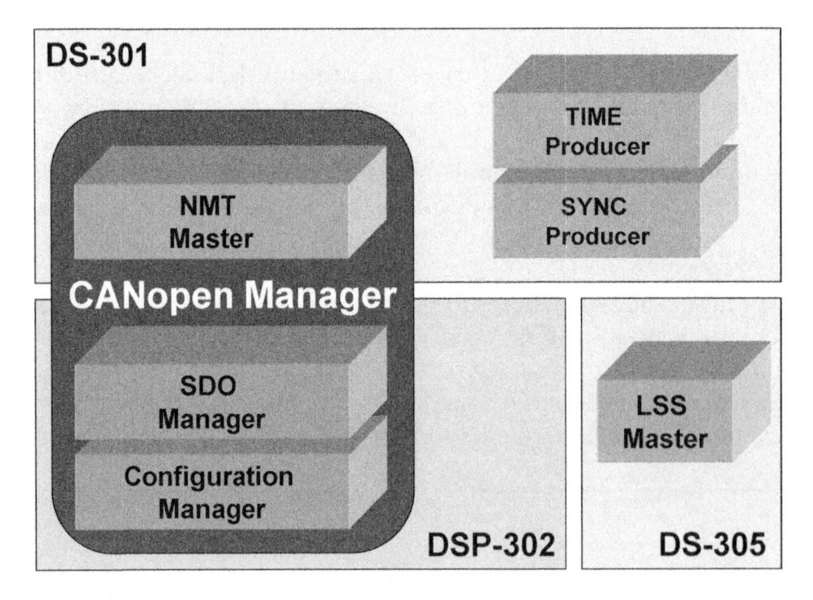

Figure 3.2 Application-supporting Functionalities in CANopen

As illustrated by Figure 3.2, the manager functionality specified by DS301 is the NMT Master, which is primarily responsible for starting and stopping the network. There are additional message producing functions that are often executed by a master, but can also be executed by another node. These functions are the time stamp production and the SYNC signal production. The only requirement for each of these functions is that they must be executed exclusively on one node for one SYNC or time stamp. A particular time stamp or SYNC message may only be produced by one node in an entire network.

In addition to the NMT Master, DS302 defines the functionality provided by an SDO Manager and a Configuration Manager. A CANopen Manager is simply the term specified for a device that provides the NMT Master function and at least one of the management functions for configuration or SDO management.

The LSS (Layer Setting Services) Master is only listed for completeness. The layer setting services allow the assignment of Node IDs and selection of the bit rate used.

Another related term is the "Flying Master." In any CANopen network, there may be only one active NMT Master or CANopen Manager at any time. If a backup is required in the event of a failing NMT Master or CANopen Manager, Flying Masters can be used. Flying Masters are NMT Masters or CANopen Managers that monitor each other and ensure that only one of them is active at any time. Upon failure or disconnection of the currently active Flying Master, the dormant Flying Master automatically wakes up and takes over.

3.2.1 The NMT Master

DS301 defines the NMT Master as a service to provide mechanisms that control and monitor the state of nodes and their behavior in the network. The primary command message used is the "NMT Master Message" that can either address an individual node or address all nodes at once. The commands that can be issued with the NMT Master messages are requests to change the NMT state of a node as explained in Section 2.6.1 and shown in Figure 2.9.

There is only one active NMT Master allowed in a CANopen network, but since there may be more than one device capable of performing the NMT service, DSP302 defines Object Dictionary entries to make it configurable. In addition to enabling/disabling the NMT service, these configuration entries contain the Network List that tells the NMT Master what types of nodes are in the network, how they are to be treated during boot-up and when there is an Error Control Event for a particular node.

If another node wants to generate an NMT command on the network, it can ask the NMT Master to generate it by writing to its Object Dictionary, see Figure 3.3.

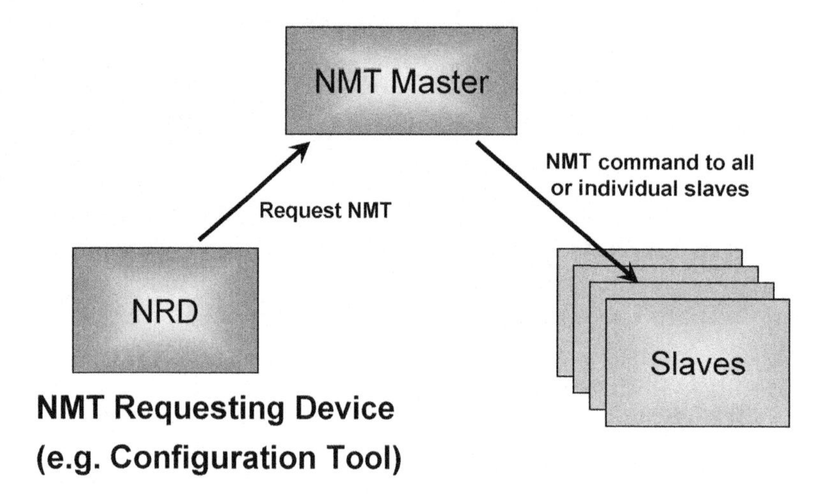

Figure 3.3 Request NMT Function

The NMT Master recognizes the request (for example issued by a configuration tool) and then sends the requested NMT command.

Although this is the recommended procedure, not all NMT Masters support this functionality. Because of this, most CANopen configuration tools directly generate the desired NMT message themselves. It is up to the user of these tools to ensure that an existing NMT Master does not interfere with the messages generated by the configuration tool.

The following is a description of the primary Object Dictionary entries used to configure the NMT Master.

3.2.1.1 [1F80h]: NMT Startup

Contains an UNSIGNED32 value to control the NMT behavior of a slave node or the NMT Master.

The bits specified for this entry are:

Bit	Description
0	If 0 the device is not the NMT Master. If 1 the device is the NMT Master.
1	If 0 then start only explicitly assigned nodes. If 1 then start all nodes. If bit 3 is 1 then this bit is ignored.
2	If 0 then automatically enter the Operational state on boot-up. If 1 then do not automatically enter the Operational state on boot-up. NMT Slave: may be read-write or read-only with fixed value.
3	If 0 then the NMT Master may automatically start nodes. The behavior is configured using bit 1. If 1 then the NMT Master may not automatically start nodes. Bit 1 is ignored.
4	If 0 then when a node fails to respond to node guarding or heartbeat, then reset only that node. If 1 then when a node fails to respond to node guarding or heartbeat, reset all nodes. If bit 6 is 1 then this bit is ignored.
5	If 0 then the NMT Master will not participate in the Flying Master process. If 1 then the NMT Master will participate in the Flying Master process.
6	If 0 then use the configuration specified by bit 4. If 1 then ignore bit 4 and if a node fails to response to node guarding or heartbeat, stop all nodes.
7 – 31	Reserved. Always zero.

Table 3.4 Control Bits for NMT Startup

For NMT slaves, only bit 2 of 1F80h is of interest. All other bits should be implemented read-only and set to zero. NMT slaves that always autostart will have bit 2 set and implemented as read-only. NMT slaves that can be configured to either autostart or not autostart need to have bit 2 implemented as read-write.

Index	**1F80h**
Name	NMT Startup
Mandatory	No, recommended for NMT Masters
Subindex	**00h**
Name	NMT Startup
Type	UNSIGNED32
Default Value	Not defined
Access	Read/Write
Mandatory	No
Map to PDO	No

Table 3.5 Object Dictionary Entry NMT Startup

3.2.1.2 [1F81h,xxh]: Slave Assignment

For each node of the network, these entries specify which management or master functions have to be executed on them. Are they mandatory for the network operation? Do they have to be of a specific type (vendor ID and/or product ID match)? What are their configuration and startup options?

The Subindex range representing the individual nodes on the network is 01h to 7fh. The Subindex directly corresponds to the Node ID.

Each slave assignment entry in [1F81h,xxh] is of type UNSIGNED32. The individual bits give the following information about the node the entry refers to.

Bit	Description
0	0 if the node is not a slave for this NMT Master. 1 if the node is a slave for this NMT Master.
1	Reserved.

Table 3.6 Control Bits for Slave Assignment

Bit	Description
2	0 if the node should not be automatically configured and started when a boot-up message is detected being transmitted from the node. 1 if the node should be automatically configured and started when a boot-up message is detected being transmitted from the node.
3	0 if the node is an optional slave. The network may be started if this node can not be contacted. 1 if the node is a mandatory slave. Do not start the network if this node can not be contacted.
4	0 if the node may be reset regardless of the current state of the node. 1 if the node may only be reset if the node is currently not operational.
5	0 if application software version verification is not required for the node. 1 if application software version verification is required for the node.
6	0 if automatic software update of the node is not allowed. 1 if automatic software update of the node is allowed.
7	Reserved.
8 – 15	Retry Factor, if node guarding is used.
16 – 31	Guard Time in milliseconds, if node guarding is used.

Table 3.6 (Continued) Control Bits for Slave Assignment

When node guarding is used and a node guarding request is not answered by a node, then the master will re-send the guarding request (Retry Factor -1) times. The time interval between the re-tries is Guard Time.

Index	**1F81h**
Name	Slave Assignment
Mandatory	No
Subindex	**00h**
Name	Number of Entries
Type	UNSIGNED8
Default Value	7Fh

Table 3.7 Object Dictionary Entry Slave Assignment

Access	Read Only
Mandatory	No
Map to PDO	No
Subindex	**01h – 7Fh**
Name	Slave Assigment Node 1 – 127
Type	UNSIGNED32
Default Value	Not defined
Access	Read/Write
Mandatory	No
Map to PDO	No

Table 3.7 (Continued) Object Dictionary Entry Slave Assignment

3.2.1.3 [1F82h,xxh]: Request NMT

Writing an NMT command to this entry is a request to the NMT Master to send its NMT Master message to the select node(s). The Subindex directly corresponds to the Node ID number to be addressed with the NMT Master message. Subindex 80h represents "all nodes" meaning the NMT message will be sent to all nodes.

The values written to these Object Dictionary entries are the same command values as used by the NMT Master message:

- 04h: Enter Stopped state
- 05h: Enter Operational state
- 06h: Reset Application
- 07h: Reset Communication
- 7Fh: Enter Pre-operational state

Reading one of the entries returns the last reported state of the selected node (reported by the heartbeat or node guarding message). Because that information is available on the network anyway it is redundant to have a copy in the NMT Master. However, it gives an indication about what the NMT Master thinks the current state of a node is. The possible values are:

- 00h: Unknown state
- 01h: Node missing
- 04h: Stopped
- 05h: Operational
- 7fh: Pre-operational

Index	**1F82h**
Name	Request NMT
Mandatory	No
Subindex	**00h**
Name	Number of Entries
Type	UNSIGNED8
Default Value	80h
Access	Read Only
Mandatory	Yes if 1F82h entry is implemented
Map to PDO	No
Subindex	**01h – 7Fh**
Name	Request NMT for Node 1 - 127
Type	UNSIGNED8
Default Value	Not defined
Access	Read/Write
Mandatory	Yes if 1F82h is implemented
Map to PDO	No
Subindex	**80h**
Name	Request NMT for All Nodes
Type	UNSIGNED8
Default Value	Node defined
Access	Write Only

Table 3.8 Object Dictionary Entry Request NMT

Mandatory	Yes if 1F82h is implemented
Map to PDO	No

Table 3.8 (Continued) Object Dictionary Entry Request NMT

3.2.1.4 [1F84h,xxh] to [1F88h,xxh]: Network List

The network list allows the NMT Master to keep a local copy of the Device Type [1000h] and the Identity Object [1018h,xxh] values of each node. The Subindex used in the Network List directly corresponds to the Node ID number of the node for which ID information has been stored in the Network List.

These entries are useful if the NMT Master needs to check if all nodes are in place with the correct Node ID and to verify that no nodes have been exchanged. If a value in the network list is set to zero, then a "don't care" is assumed for that node and it does not matter what value the node reports.

Depending on the level of detail required, an NMT Master could only implement parts of the Network List. If only the device type information needs to be confirmed, the implementation of [1F84h,xxh] is sufficient. If the level of detail required goes all the way down to the serial number, then the entire network list must be implemented.

Index	**1F84h**
Name	Device Type Identification
Mandatory	No
Subindex	**00h**
Name	Number of Entries
Type	UNSIGNED8
Default Value	7Fh
Access	Read Only
Mandatory	Yes if this entry is implemented
Map to PDO	No
Subindex	**01h – 7Fh**

Table 3.9 Object Dictionary Entry Network List: Device Type

Name	Device Type Identification for Node 1 - 127
Type	UNSIGNED32
Default Value	Not defined
Access	Read/Write
Mandatory	Yes if this entry is implemented
Map to PDO	No

Table 3.9 (Continued) Object Dictionary Entry Network List: Device Type

Index	**1F85h**
Name	Vendor Identification
Mandatory	No
Subindex	**00h**
Name	Number of Entries
Type	UNSIGNED8
Default Value	7Fh
Access	Read Only
Mandatory	Yes if this entry is implemented
Map to PDO	No
Subindex	**01h – 7Fh**
Name	Vendor Identification for Node 1 - 127
Type	UNSIGNED32
Default Value	Not defined
Access	Read/Write
Mandatory	Yes if this entry is implemented
Map to PDO	No

Table 3.10 Object Dictionary Entry Network List: Vendor ID

Index	**1F86h**
Name	Product Code
Mandatory	No
Subindex	**00h**
Name	Number of Entries
Type	UNSIGNED8
Default Value	7Fh
Access	Read Only
Mandatory	Yes if this entry is implemented
Map to PDO	No
Subindex	**01h – 7Fh**
Name	Product Code for Node 1 – 127
Type	UNSIGNED32
Default Value	Not defined
Access	Read/Write
Mandatory	Yes if this entry is implemented
Map to PDO	No

Table 3.11 Object Dictionary Entry Network List: Product Code

Index	**1F87h**
Name	Revision Number
Mandatory	No
Subindex	**00h**
Name	Number of Entries
Type	UNSIGNED8
Default Value	7Fh

Table 3.12 Object Dictionary Entry Network List: Revision Number

Access	Read Only
Mandatory	Yes if this entry is implemented
Map to PDO	No
Subindex	**01h – 7Fh**
Name	Revision Number for Node 1 – 127
Type	UNSIGNED32
Default Value	Not defined
Access	Read/Write
Mandatory	Yes if this entry is implemented
Map to PDO	No

Table 3.12 (Continued) Object Dictionary Entry Network List: Revision

Index	**1F88h**
Name	Serial Number
Mandatory	No
Subindex	**00h**
Name	Number of Entries
Type	UNSIGNED8
Default Value	7Fh
Access	Read Only
Mandatory	Yes if this entry is implemented
Map to PDO	No
Subindex	**01h – 7Fh**
Name	Serial Number for Node 1 – 127
Type	UNSIGNED32
Default Value	Not defined
Access	Read/Write

Table 3.13 Object Dictionary Entry Network List: Serial Number

Mandatory	Yes if this entry is implemented
Map to PDO	No

Table 3.13 (Continued) Object Dictionary Entry Network List: Serial Number

3.2.1.5 [1F89h]: BootTime

The time in milliseconds a NMT Master waits after sending a Reset Command for all mandatory nodes to start-up. If this time expires and one of the mandatory nodes was not found, the NMT Master will go into an error state.

Index	**1F89h**
Name	Boot Time
Mandatory	No but recommended for NMT Masters
Subindex	**00h**
Name	Boot Time
Type	UNSIGNED32
Default Value	0h
Access	Read/Write
Mandatory	No
Map to PDO	No

Table 3.14 Object Dictionary Entry Boot Time

3.2.2 The SDO Manager and Dynamic SDO Connections

The Pre-defined Connection Set in CANopen specifies only one SDO channel for every Node ID. This means that by default every CANopen slave node implements exactly one SDO server, and only one node (typically a master or configuration tool) will act as an SDO client to access the Object Dictionaries of the slaves. No other node can use the same SDO channels to talk to any of the slaves without risking collisions with an SDO request message with the same identifier from another SDO client. Even if the CAN messages do not collide, the additional SDO requests from a different source can easily interfere with the other SDO communication that is going on. Clearly this has to be avoided.

Therefore, an SDO Manager is specified that is in charge of all SDO channels and that has exclusive access to them. If implemented, the SDO Manager and the NMT Master are on one and the same node.

A node, for instance a configuration or diagnosis tool that needs to talk to any of the slaves, has to request a channel from the SDO Manager first. The channel can only be used after that request has been granted or it has been confirmed that there is no SDO Manager present in the network.

> Multiple SDO Clients and Servers
>
> Even though by default all nodes implement only one SDO client, DS301 specifies that each node in the network may support up to 128 SDO servers, and just as many clients (objects 1200h-12FFh, SDO Server and SDO Client Parameters). An SDO Manager makes use of the additional SDO clients and can *dynamically* configure them when other devices request them. A second option is to *statically* configure additional SDO channels during network configuration.

The procedure for the dynamic assignment of an SDO channel can be summarized as follows:

1. The SDO Requesting Device (SRD) sends the "Dynamic SDO Request" message. Because this message has the fixed CAN message identifier 6E0h and data length zero, it can be sent by any node at any time.
2. The SDO Manager recognizes the "Dynamic SDO Request" and starts scanning the network. It reads the 1F10h "Dynamic SDO Connection State" entry of each node until it finds the one that issued the request.
3. Once the SRD is found the SDO Manager enables the SDO client functionality in the SRD to allow the node client access to the Object Dictionary of the SDO Manager.
4. The SRD can now request or release channels by writing to the 1F00h and 1F01h entries.
5. The SDO Manager will act on the requests and try to establish the SDO channel by writing to the Object Dictionary entries for SDO channel configuration of both the SRD and the target node that the SRD wants to connect with.

To allow for a more efficient method for configuration tools to get access to an entire network, there is also a mechanism to simply request all default SDO channels. If the SDO Manager grants this request, it will stay away from all default SDO channels.

Figure 3.4 illustrates the messages involved when a SRD requests all SDO channels from the SDO Manager.

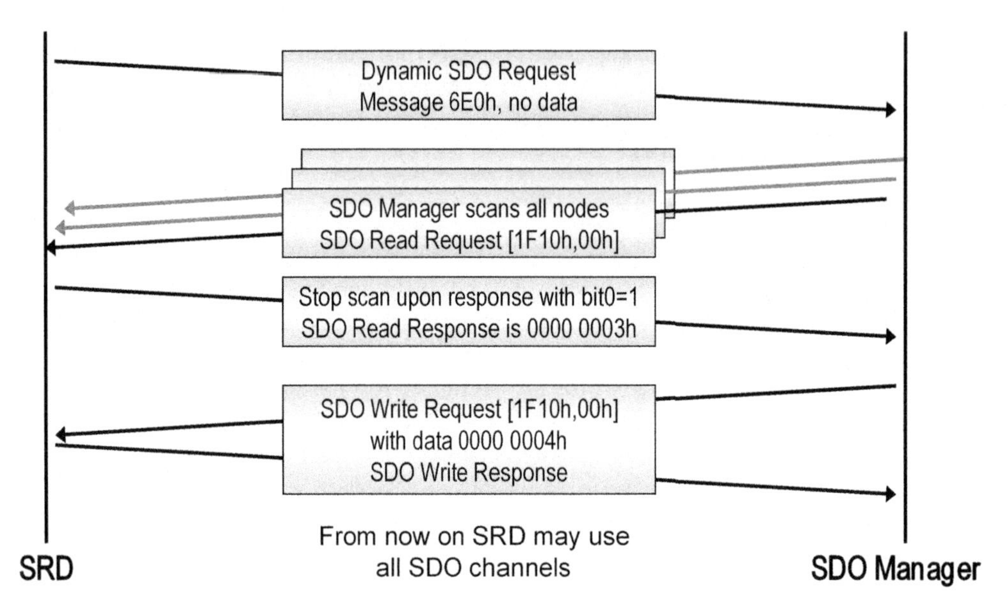

Figure 3.4 Dynamic Request for all SDO Channels

When the SDO Manager reads [1F10h,00h] from the SRD, the SRD replies with 00000003h to indicate that it desires to use all SDO channels. The SDO Manager will overwrite [1F10h,00h] with 00000004h to indicate that the request was granted.

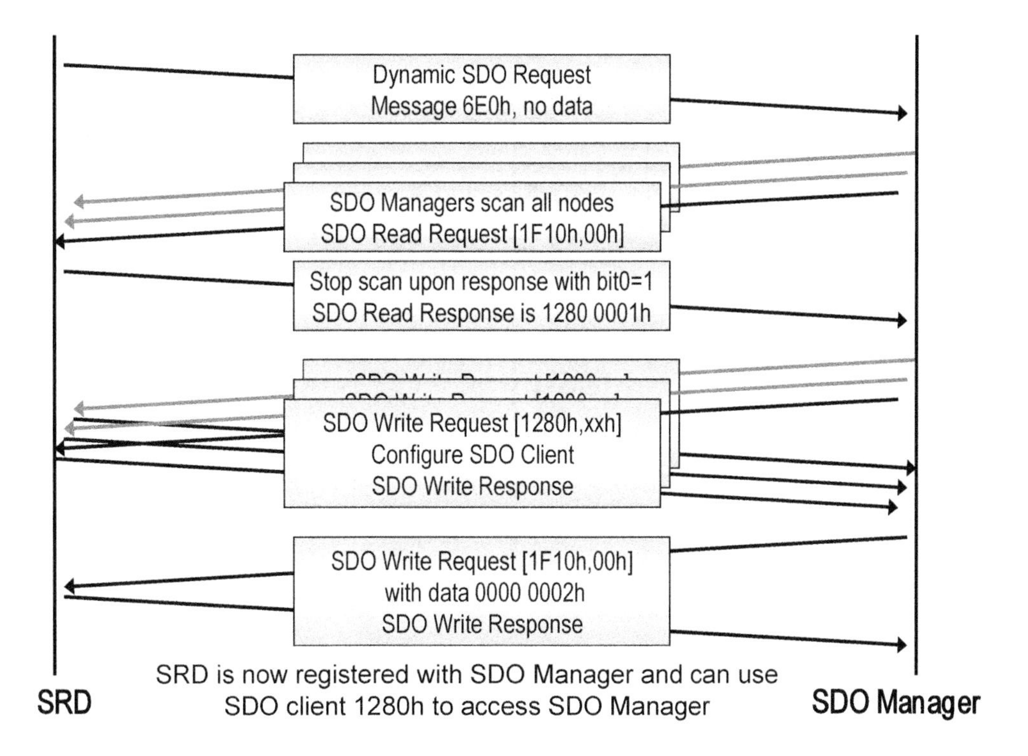

Figure 3.5 Register as SRD

When the SRD only wants to request single SDO channels, it needs to register as a SRD with the SDO Manager first. The sequence for such an SRD registration process is illustrated in Figure 3.5. This time the SRD replies to the SDO Manager's read request of [1F10h,00h] with the value 12800001h indicating that it wants to register with the SDO Manager and requires an SDO channel to the SDO Manager. In this case the SRD also informs the SDO Manager that the next available SDO Client within the SRD is [1280h] (this is only an example; it could also have been one of the other SDO clients in the range from 1280h to 12FFh).

The SDO Manager now executes several SDO Write Requests to 1280h to configure the SDO Client to link to an SDO Server within the SDO Manager. Once the SDO Client is configured, the SRD has an SDO channel to the SDO Manager to execute read or write accesses to the Object Dictionary in the SDO Manager. The process is completed by the SDO Manager writing 00000002h to [1F10h,00h] of the SRD.

Once the SRD is registered it can write to the 1F00h and 1F01h entries of the SDO Manager to request or release single SDO channels.

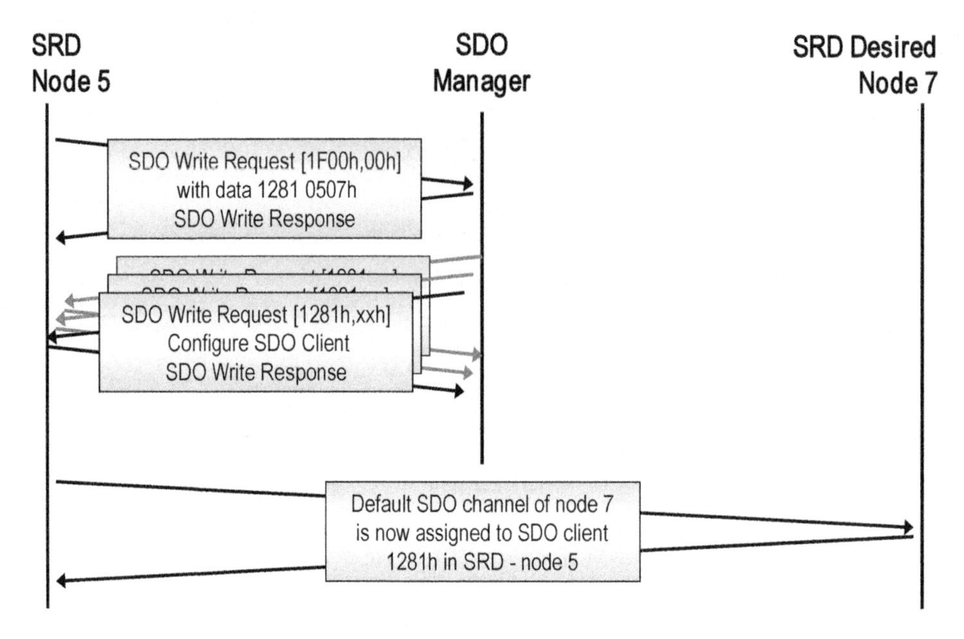

Figure 3.6 SRD Requesting a SDO Channel

Figure 3.6 shows the messages involved in the request of a single SDO channel. In this example, the SRD has the node ID 5 and desires an SDO channel to node 7 with node 7 being the server and node 5 the client. The SRD uses its SDO channel to the SDO Manager to write to [1F00h,00h] of the SDO Manager. The value written contains the next available SDO Client in the SRD (here 1281h), the Node ID of the SRD and the Node ID of the target node.

Assuming that node 7 only has one SDO server implemented (the default), the SDO Manager would set up the SDO Client 1281h in the SRD to use the default SDO channel and would itself refrain from using that channel further.

However, if node 7 has multiple SDO servers implemented, the SDO Manager would not assign the default SDO channel to node 5 but would configure both nodes to use a new channel. So in addition to configuring the SDO Client in the SRD it would also configure the additional SDO Server in node 7 as illustrated in Figure 3.7.

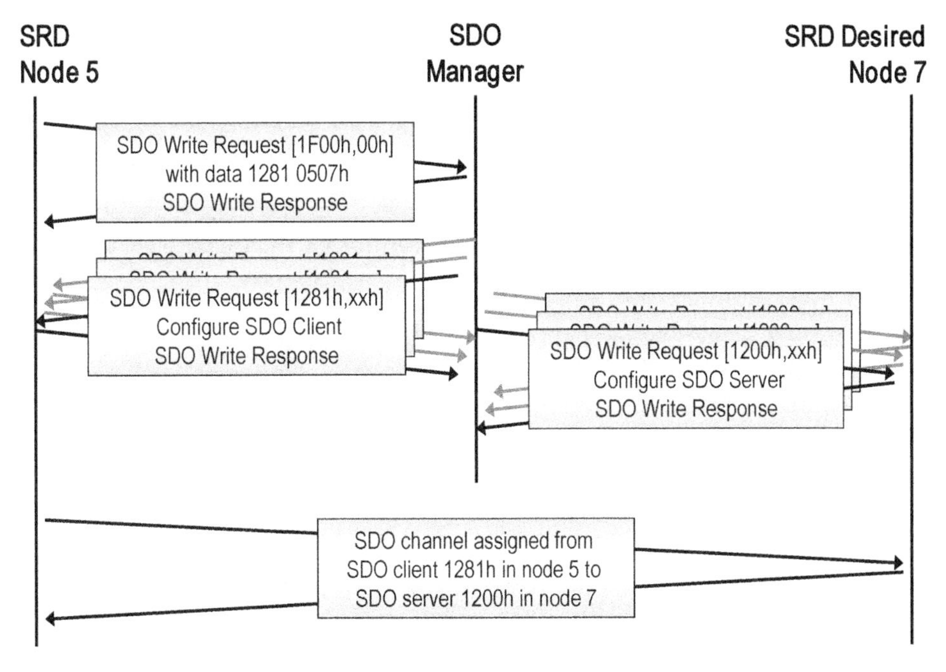

Figure 3.7 Fully Dynamic SDO Channel

The following is a description of the primary Object Dictionary entries used to configure the SDO Manager.

3.2.2.1 [1F00h] and [1F01h]: Request and Release SDO Channel

After it has been assigned an SDO channel by the SDO Manager, a device that wants to request or release SDO channels (the SRD) can do so by writing to these Object Dictionary entries in the SDO Manager. The entries contain a value of type UNSIGNED32 that contains the following bits:

Bit	Description
0 – 7	Node ID of the node the SRD wants an SDO channel to; when writing to 1F01h and this is zero, then request to release all connections and cease to be a SRD.

Table 3.15 Control Bits for Request and Release SDO Channel

Bit	Description
8 – 15	Node ID of the SRD.
16 – 31	Index of a free Client SDO Entry in the SRD's Object Dictionary (1280h – 12FFh); when writing to 1F01h and this is zero, then this is a request to release all connections.

Table 3.15 (Continued) Control Bits for Request and Release SDO Channel

The value contains the Node ID of the device that the SRD wants to establish an SDO channel to, as well as the Node ID of the SRD itself. The SDO channel requested would make the SRD the SDO client, and the node would be the SDO server. The third value available defines the Index of a free SDO client configuration Index within the SRD. The SDO Manager will use that entry to configure the SDO client within the SRD.

The Object Dictionary entries are specified as follows:

Index	**1F00h**
Name	Request SDO
Mandatory	Yes for SDO Managers, not used on other nodes
Subindex	**00h**
Name	Request SDO
Type	UNSIGNED32
Default Value	Not defined
Access	Write Only
Mandatory	Yes for SDO Managers
Map to PDO	No

Table 3.16 Object Dictionary Entry Request SDO

Index	**1F01h**
Name	Release SDO

Table 3.17 Object Dictionary Entry Release SDO

Mandatory	Yes for SDO Managers, not used on other nodes
Subindex	**00h**
Name	Release SDO
Type	UNSIGNED32
Default Value	Not defined
Access	Write Only
Mandatory	Yes for SDO Managers
Map to PDO	No

Table 3.17 (Continued) Object Dictionary Entry Release SDO

3.2.2.2 [1F10h]: Dynamic SDO Connection State

This entry is not implemented by the SDO Manager, but by the devices that want to request dynamic SDO channels. After an SDO Manager receives the request indicating that there are nodes on the network which are requesting an SDO channel, the SDO Manager scans all 1F10h entries of all nodes to find out which node(s) requested an SDO channel.

Bit	Description
0	Request Indication.
1 - 2	Connection State.
3	Request Error Control.
4 - 7	Reserved. Always zero.
8 - 15	Error code.
16 - 31	Index of a free Client SDO Entry in the SRD's Object Dictionary (1280h – 12FFh).

Table 3.18 Control Bits Dynamic SDO Connection State

Read requests from this entry have to return zero if the node does not request an SDO channel. If the node sent a "Dynamic SDO Request" message (COB ID 6E0h, no data field), it must report the following values upon a read access:

- Set Request Indication bit to signal that an SDO channel is requested.

- Set Connection State to one if access to all SDO channels is desired, otherwise leave zero. This is useful for configuration tools that temporarily want to have access to all nodes.
- Set Request Error Control to one, if the SDO Manager should continue to provide error control services (heartbeat monitoring or node guarding) for the nodes that the SRD gets connected to.
- Set the Index (range 1280h to 12FFh) of the SRD's SDO client entry to be used for the communication with the SDO Manager. This value can remain zero if access to all SDO channels is requested.

The SDO Manager will write to this entry to confirm or deny the request:

- Request Indication bit is cleared to signal successful registration as SRD.
- Connection State is set to one if registration as an SRD was successful or to two if all SDO channels were assigned to the SRD. A value of three indicates that the dynamic SDO channel assignment is completed (SDO client and server configured on both ends).This value will be set to zero if an error occurred in which case an error code will be reported (see Table 3.19).
- Request Error Control will be set to one if the SDO Manager continues to perform the error control for the node that the SRD established an SDO channel with.

Error Code	Description
00h	Unspecified error.
01h	There was no free SDO channel to create a connection between the SDO Manager and SRD.
02h	There were no more free SDO channels in the CANopen network.
03h	The Slave does not have any free Server SDOs.
04h	The Slave node is not available.
05h – FFh	Reserved.

Table 3.19 Error Codes Used with Dynamic SDO Channel Assignment

Index	**1F10h**
Name	Dynamic SDO Connection State
Mandatory	Yes for Nodes using Dynamic SDO Channels
Subindex	**00h**
Name	Dynamic SDO Connection State
Type	UNSIGNED32
Default Value	Not defined
Access	Read/Write
Mandatory	Yes for Nodes using Dynamic SDO Channels
Map to PDO	No

Table 3.20 Object Dictionary Entry Dynamic SDO Connection State

3.2.2.3 [1F02h,xxh]: SDO Manager COB IDs

In order to assign CAN message IDs for additional SDO channels, the SDO Manager needs to know which IDs are still available in the system. This is the configurable list of IDs that the SDO Manager can use for dynamic channel assignments. Each list entry is of type UNSIGNED32 and has the following bits defined for 11-bit COB IDs:

Bit	Description
0 – 10	COB ID
11 - 29	Set to 0
30	0 if the COB ID is free to be used for an SDO channel. 1 if the COB ID is currently in use for an SDO channel.
31	0 if the COB ID is valid, this Subentry is used. 1 if the COB ID is not valid, this Subentry is not used.

Table 3.21 Control Bits of SDO Manager COB ID Entries

The important bits are 30 and 31. If 31 is set, the entry is used by the SDO Manager. The COB ID is "owned" by the SDO Manager and only it is allowed to use or assign it.

This also applies if bit 30 is cleared. If bit 30 is cleared, the COB ID is currently not used for any SDO channel, however, the SDO Manager can assign it anytime.

Index	**1F02h**
Name	SDO Manager COB IDs
Mandatory	Yes for SDO Managers
Subindex	**00h**
Name	Number of Entries
Type	UNSIGNED8
Default Value	Not defined
Access	Read/Write
Mandatory	Yes for SDO Managers
Map to PDO	No
Subindex	**01h – FEh**
Name	COB ID 1 – 127
Type	UNSIGNED32
Default Value	Not defined
Access	Read/Write
Mandatory	No
Map to PDO	No

Table 3.22 Object Dictionary Entry SDO Manager COB ID

3.2.2.4 [1F03h,xxh]: SDO Connections

These entries contain the table of the current dynamic SDO channels assigned by the SDO Manager. The data type is UNSIGNED32 and the following bits are defined for these entries:

Bit	Description
0 – 7	SDO Server Node ID
8 – 15	SDO Server Communication Parameter Offset
16 – 23	SDO Client Node ID
24 – 31	SDO Client Communication Parameter Offset

Table 3.23 Control Bits for SDO Connections Entry

The offsets are added to the base Index address for the SDO communication parameters. The Index with the SDO Server Communication Parameters in the SDO Server is 1200h plus the offset. The Index with the SDO Client Communication Parameters in the SDO Client is 1280h plus the offset.

Index	**1F03h**
Name	SDO Connections
Mandatory	Yes for SDO Managers
Subindex	**00h**
Name	Number of Entries
Type	UNSIGNED8
Default Value	Not defined
Access	Read Only
Mandatory	Yes for SDO Managers
Map to PDO	No
Subindex	**01h - FEh**
Name	SDO Connection 1-254

Table 3.24 Object Dictionary Entry SDO Connections

Type	UNSIGNED32
Default Value	Not defined
Access	Read Only
Mandatory	No
Map to PDO	No

Table 3.24 (Continued) Object Dictionary Entry SDO Connections

3.2.3 The Configuration Manager

In a network where the individual nodes are not pre-configured and must be configured after every power-up, a Configuration Manager is required. The Configuration Manager's task is to locally store the configuration of each node and transfer that information to the nodes upon each power-up of the system. If implemented, the Configuration Manager and the NMT Master are on the same node.

If the Configuration Manager runs on a PC-style computer it stores the configuration information by having local copies of the DCF files with the configuration for each node in the system. It stores these files in an array of Object Dictionary entries located at 1F20h (type of the entry is DOMAIN, which has an unspecified length). The Subindex directly represents the Node ID. So the entry at [1F20h,07h] holds the DCF for node number 7.

To optimize memory usage in Configuration Managers that run without a file system, a concise version of the DCF is typically stored in 1F22h. The concise DCF format is compressed in two ways. First, it only contains the Object Dictionary entries that need to be configured (if some default values are actually used in a node, these do not get configured). Secondly, a binary format is used instead of an ASCII format.

The concise DCF format is straight-forward:

- The first entry is a variable of type UNSIGNED32 specifying how many entries are in this DCF: "Number of entries."
- For each entry, the following record is stored:

    o Index (UNSIGNED16)

    o Subindex (UNSIGNED8)

    o Length (UNSIGNED32), length of the data field to follow, in bytes o Data (DOMAIN), the data field

Although the regular ASCII DCF format and the concise DCF format are used when accessing the entries at 1F20h or 1F22h, this does not tell us anything about the true storage format within the Configuration Manager.

There might be some managers with enough intelligence that they do some compression of their own. For example, they may accept regular DCF formats, but internally store the information in the concise format. Others might further compress the concise format internally by not allowing any values greater than UNSIGNED16 for the number of entries or the length of an entry.

Because the configuration process for an entire system can take multiple seconds to execute for each node, several functions have been provided to shorten the configuration cycle. If the individual nodes support the storage of their last configuration in non-volatile memory, a Configuration Manager would only need to double check to see if each node still has the valid last configuration stored.

The following is an overview of the primary Object Dictionary entries used to configure the Configuration Manager.

3.2.3.1 [1F20h-1F22h,xxh]: DCF Storage

These Object Dictionary entries store the DCF configuration files for the nodes that need to be handled by the Configuration Manager. The Subindex directly relates to the CANopen Node ID number of the node to which a DCF belongs. If 1F20h and 1F21h are implemented, 1F22h does not need to be implemented. The concise format used in 1F22h is intended as an alternative for Configuration Managers that do not have enough physical resources (memory storage capacity, CPU performance to interpret ASCII DCF) to implement 1F20h and 1F22h.

Index	**1F20h**
Name	Store DCF
Mandatory	No
Subindex	**00h**
Name	Number of Entries

Table 3.25 Object Dictionary Entry Store DCF

Type	UNSIGNED8
Default Value	7Fh (highest Node ID available)
Access	Read Only
Mandatory	Yes if 1F20h is implemented
Map to PDO	No
Subindex	**01h – 7Fh**
Name	Store DCF Node 1 – 127
Type	DOMAIN
Default Value	Not defined
Access	Read/Write
Mandatory	Yes if 1F20h is implemented
Map to PDO	No

Table 3.25 (Continued) Object Dictionary Entry Store DCF

Read attempts from a 1F20h Subindex with no DCF stored result in a SDO Abort with the error code 08000024h "Data Set Empty."

Index	**1F21h**
Name	Storage Format
Mandatory	No
Subindex	**00h**
Name	Number of Entries
Type	UNSIGNED8
Default Value	7Fh (highest Node ID available)
Access	Read Only
Mandatory	Yes if 1F20h is implemented
Map to PDO	No
Subindex	**01h – 7Fh**

Table 3.26 Object Dictionary Entry Storage Format

Name	Storage Format Node 1 – 127
Type	UNSIGNED8
Default Value	00h: uncompressed ASCII
Access	Read/Write
Mandatory	Yes if 1F20h is implemented
Map to PDO	No

Table 3.26 (Continued) Object Dictionary Entry Storage Format

Index	**1F22h**
Name	Concise DCF
Mandatory	No
Subindex	**00h**
Name	Number of Entries
Type	UNSIGNED8
Default Value	7Fh (highest Node ID available)
Access	Read Only
Mandatory	Yes if this entry is implemented
Map to PDO	No
Subindex	**01h – 7Fh**
Name	Concise DCF Node 1 – 127
Type	DOMAIN
Default Value	Not defined
Access	Read/Write
Mandatory	Yes if this entry is implemented
Map to PDO	No

Table 3.27 Object Dictionary Entry Concise DCF

Read attempts from a 1F22h Subindex with no DCF stored result in the return of an "empty" concise stream where the first 32-bit entry (Number of Entries) is zero. So the response would be 00000000h.

3.2.3.2 [1F26h-1F27h,xxh]: Expected Configuration Date and Time Stamp

If the individual nodes on the network support the "Store Parameter Functionality" at 1010h (storing a configuration in non-volatile memory locally), the entry 1020h "Verify Configuration" of these nodes will be set to the date and time of the last configuration.

The entries at 1F26h and 1F27h of the Configuration Manager contain a copy of these entries so that the Configuration Manager can quickly confirm if the last configuration saved is still the one to be used.

Index	**1F26h**
Name	Expected Configuration Date
Mandatory	No, required for handling nodes that use 1010h
Subindex	**00h**
Name	Number of Entries
Type	UNSIGNED8
Default Value	7Fh (highest available Node ID)
Access	Read Only
Mandatory	Yes if 1F26 is implemented
Map to PDO	No
Subindex	**01h – 7Fh**
Name	Expected Configuration Date Node 1 – 127
Type	UNSIGNED32
Default Value	Not defined
Access	Read/Write
Mandatory	Yes if 1F26 is implemented
Map to PDO	No

Table 3.28 Object Dictionary Entry Expected Configuration Date

Index	**1F27h**
Name	Expected Configuration Time
Mandatory	No, required for handling nodes that use 1010h
Subindex	**00h**
Name	Number of Entries
Type	UNSIGNED8
Default Value	7Fh (highest available Node ID)
Access	Read Only
Mandatory	Yes if 1F27 is implemented
Map to PDO	No
Subindex	**01h – 7Fh**
Name	Expected Configuration Time Node 1 – 127
Type	UNSIGNED32
Default Value	Not defined
Access	Read/Write
Mandatory	Yes if 1F27 is implemented
Map to PDO	No

Table 3.29 Object Dictionary Entry Expected Configuration Time

3.2.4 The CANopen Manager

The term "CANopen Manager" was created in order to have a single term for the combination of master and manager functionalities. A node is called a CANopen Manager if it provides the NMT Master functionality and at least one of the functions of an SDO Manager or Configuration Manager.

For details about the functionality provided by these masters and managers, see the previous sections.

3.2.5 The Boot-up Process

Due to the many configuration options available, the boot-up process can vary greatly in CANopen. The complexity of the boot-up process can vary from simple pre-configured, master-less systems that start-up themselves, to complex dynamic systems with Flying Masters and an elaborate configuration process depending on the components hooked up to the network. [CiADS302] uses a number of flow diagrams to illustrate the possible boot-up options. For simplicity this book stays with two common examples.

3.2.5.1 Minimal NMT Master Boot-up

In a system with only a minimal NMT Master, the individual nodes transmit their boot-up message and then stay in pre-operational mode. The NMT Master continuously scans the network, either by passively waiting for the boot-up messages or actively by trying to read Object Dictionary entries, such as 1000h device type information. Alternatively it could ensure a synchronous start-up by issuing a "reset all nodes" command to make sure that it did not miss a boot-up message.

Once the NMT Master determines that all nodes are available that are required for smooth network operation, it will send the NMT Master message "start all nodes" to start the network and the devices.

Optionally, it would continue monitoring the network and then react to network failures like nodes disappearing or changing their operating state.

3.2.5.2 CANopen Manager Boot-up

If a system contains nodes that need to be configured before they can start operation, a CANopen Manager with Configuration Management is required.

The NMT Master related boot-up procedure of the CANopen Manager is basically identical to the NMT Master boot-up described previously. However, before starting the network, it needs to be confirmed that each node is configured correctly. The CANopen Manager verifies the configuration of each node by first checking to see if the node supports the "save parameters functionality" (saving a configuration in non-volatile memory). If it does not support this function, it needs to be configured.

If it does support the save parameters function, it just needs to be verified that the configuration is still valid. That can be done by comparing the date and time stamp of the last configuration stored in the node with that stored in the CANopen Manager. If they are identical, no further configuration is required. If they do not match, the node needs to be configured.

Once all nodes are processed, the NMT Master message "start all nodes" can be sent by the CANopen Manager.

3.3 Device Profile for Encoder (DS406)

Objective

There is hardly any CANopen device that could be simpler than a single channel encoder reporting exactly one position value to the network. That's why this example was chosen as a practical example of what is specified in a device profile.

The content of this section is based on [CiADS406]. The definitions and requirements shown are those for a "C1 class" encoder, a basic encoder without scaling or other extended functionality, which simply reports one position value.

3.3.1 Introduction

In order to distinguish between encoders with "basic functionality" and "extended functionality" the device profile introduces two classes: "C1" and "C2." The "C1 class" encoders are basic encoders reporting one position value. "C2 class" encoders can not only have advanced functions like scaling, they can also consist of multiple encoders and report the values from multiple encoders.

Although not pointed out directly in the specification, using only absolute position values when transmitting positions via a network based on CAN is recommended. One of the known problems of CAN communication is that in some rare cases the error detection and re-transmission scheme can cause the duplication of messages. In other words, a node can conceivably receive a message twice.

If the data in that message is an absolute position value nothing happens. However, if the data is incremental and it is received twice the receiver would now assume an incorrect position for the encoder.

3.3.2 Object Dictionary Entries

The following Object Dictionary entries are mandatory for "C1 class" encoders. They are in addition to the regular mandatory entries like error register and Identity Object.

3.3.2.1 Index [1000h]: Device Type

One of the first things a device profile defines is the details about how Object Dictionary entry [1000h] has to be implemented. This is the device type entry, typically the first entry read by CANopen Masters or configuration tools that scan the network for connected nodes.

The device profile for encoders specifies that the low word of the 32-bit device type field contains 0196h (= 406d, the device profile number).

The high word contains the encoder type, which can be one of the values in Table 3.30.

Encoder Type	Description
1	Absolute single-turn rotary encoder
2	Absolute multi-turn rotary encoder
3	Absolute single-turn rotary encoder with counter
4	Incremental rotary encoder
5	Incremental rotary encoder with counter
6	Incremental linear encoder
7	Incremental linear encoder with counter
8	Absolute linear encoder
9	Absolute linear encoder with cyclic coding
10	Multi-sensor encoder interface

Table 3.30 Encoder Types

3.3.2.2 Index [1800h,xxh] and [1A00h,xxh]: 1st TPDO Parameters

The first default transmit PDO contains exactly one variable: the 4-byte encoder position value stored in [6004h]. The mapping entry in [1A00h,01h] is 60040020h. The transmission type [1800h,02h] is set to 254: device profile specific. It is transmitted asynchronously using the event timer. Older implementations use the device profile specific event timer at [6200h], newer implementations will adapt the device profile independent event timer [1800h,05h]. In case both are implemented, they must always be identical (writing to one also changes the other).

3.3.2.3 Index [1801h,xxh] and [1A01h,xxh]: 2nd TPDO Parameters

The contents of the second default transmit PDO are identical to the first. The only difference is that it has a different default for the transmission type [1801h,02h], which is 1. It is set to synchronous transmission with every SYNC signal received.

3.3.2.4 Index [6000h]: Operating Parameters

This is an UNSIGNED16 read-write value where individual bits report some of the operating parameters like measuring direction or scaling capabilities available in this encoder. For "C1 class" rotary encoders only bit 0 "Code Sequence" is mandatory. It has to be set to 1 for clockwise operation, meaning turning the encoder clockwise increments the position value. It has to be set to 0 if turning it counterclockwise increments the position value.

3.3.2.5 Index [6004h]: Position Value

The 32-bit read-only position value stored as UNSIGNED32.

3.3.2.6 Index [6500h]: Operating Status

The operating status is a read-only version of entry [6000h].

3.3.2.7 Index [6501h]: Resolution

This UNSIGNED32 read-only value is used slightly differently on rotary and linear encoders. For rotary encoders it shows the number of measuring steps reported by a single 360 degree turn of the encoder. For linear encoders it shows the length of a single measuring step in nanometers.

3.3.2.8 Index [6502h]: Revolutions

This UNSIGNED16 read-only entry is used for rotary encoders. It contains the number of full 360 degree turns the encoder can count. For single turn rotary encoders this value is 1. The total measuring range reported in the position value is Revolutions [6502h] multiplied by Resolution [6501h].

3.3.3 Encoder Object Dictionary Example

The following DS406 related Object Dictionary entries would be implemented for a basic C1 class absolute multi-turn rotary encoder.

Index	Sub index	Description	Data Type	Default Value
1000h	00h	Device Type	UNSIGNED32	00020196h
1800h		1st TPDO Communication Parameters		
1800h	00h	Number of Entries	UNSIGNED8	5
1800h	01h	COB ID	UNSIGNED32	180h + Node ID
1800h	02h	Transmission Type	UNSIGNED8	FEh
1800h	03h	Inhibit Time	UNSIGNED16	0
1800h	05h	Event Time	UNSIGNED16	0
1A00h		1st TPDO Mapping Parameters		
1A00h	00h	Number of Entries	UNSIGNED8	1
1A00h	01h	1st Mapping Entry: Position Value	UNSIGNED32	60040020h
1801h		2nd TPDO Communication Parameters		
1801h	00h	Number of Entries	UNSIGNED8	5
1801h	01h	COB ID	UNSIGNED32	280h + Node ID
1801h	02h	Transmission Type	UNSIGNED8	1
1801h	03h	Inhibit Time	UNSIGNED16	0
1801h	05h	Event Time	UNSIGNED16	0
1A01h		2nd TPDO Mapping Parameters		

Table 3.31 An Object Dictionary Example for Encoders

Index	Sub index	Description	Data Type	Default Value
1A01h	00h	Number of Entries	UNSIGNED8	1
1A01h	01h	2nd Mapping Entry: Position Value	UNSIGNED32	60040020h
6000h	00h	Operating Parameters	UNSIGNED16	(no default)
6004h	00h	Position Value	UNSIGNED32	(no default)
6500h	00h	Operating Status	UNSIGNED16	(no default)
6501h	00h	Resolution	UNSIGNED32	(no default)
6502h	00h	Revolution	UNSIGNED16	(no default)

Table 3.31 (Continued) An Object Dictionary Example for Encoders

3.4 Device Profile for Generic I/O (DS401)

Objective

The device profile for generic I/O is one of the most often implemented CANopen device profiles. By default it supports a total of up to 64 digital input channels and up to 64 digital output channels. The analog channels provided by default are a total of up to 24 (12 channels for input and 12 for output), each 16-bit resolution.

The content of this section is based on [CiADS401]. The definitions and requirements shown are those that are mandatory and must be supported by a generic I/O device in order to be able to claim DS401 compliance.

3.4.1 Introduction to Generic I/O

As with all device profiles, DS401 defines a small set of mandatory functionality that a device must have in order to be able to claim DS401 compliance. In addition, it specifies a much larger set of optional functionality that may be implemented if it is required. In the case of DS401 the optional functions can add up to the point where it is hard to implement all of them if the target is an 8-bit microcontroller.

For example, there can be about ten configurable parameters for each analog input channel. These can include an offset and a scaling and several parameters for the change-of-state detection. The change-of-state detection could include detection involving an upper and lower limit/threshold as well as a negative or positive value difference.

Another indicator of the potential for complexity is the number of device profile specific Object Dictionary entries specified. The mandatory entries are about one for every 8 bits of digital input or output data, one for every 16 bits of output data and two for each 16 bits of analog input data.

In addition, the device profile specifies hundreds (or thousands if all the entries specified for single-bit access are counted) of optional Object Dictionary entries that either contain configuration parameters or alternate access to the process data (for example 16-bit access instead of 8-bit access to the digital data).

For the scope of this book we focus on the mandatory function set and some selected optional functions that are commonly implemented in many devices. For a complete listing of the optional functions see [CiADS401].

3.4.2 Object Dictionary Entries

The following Object Dictionary entries are mandatory for "Generic I/O" devices. They are in addition to the regular mandatory entries like error register and Identity Object.

3.4.2.1 Index [1000h]: Device Type

The device profile for generic I/O specifies that the low word of the 32-bit device type field contains 0191h (= 401d, the device profile number).

Bits 16 through 19 provide information about the type of I/O provided. There is one bit each that can be set to signal the support of a specific I/O type as listed in Table 3.32.

Bit	Description
16	Digital Input
17	Digital Output

Table 3.32 I/O Types

Bit	Description
18	Analog Input
19	Analog Output

Table 3.32 I/O Types

In addition, bits 24-31 are used to report special functionality. So far only one value has been specified. If bits 24-31 contain a 1 the device is a joystick.

3.4.2.2 Index [140xh,xxh] and [160xh,xxh]: RPDO Parameters

By default a total of up to four RPDOs are configured. The transmission type [140xh,02h] is set to 255: manufacturer specific. The default behavior is that upon receiving a RPDO the data contained in the RPDO is immediately applied to the outputs.

The default RPDO mapping is illustrated in Figure 3.8. RPDO1 contains 8 digital output bytes that, upon receiving, will be copied to the Object Dictionary entries [6200h,01h-08h]. RPDO2, 3 and 4 each contain four 16-bit analog values that are mapped to the Object Dictionary entries [6411h,00h-0Ch].

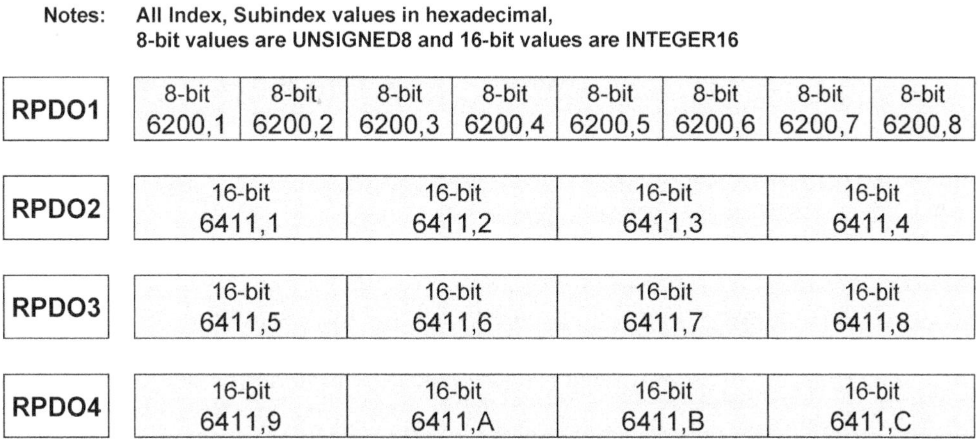

Notes: **All Index, Subindex values in hexadecimal,**
8-bit values are UNSIGNED8 and 16-bit values are INTEGER16

RPDO1	8-bit 6200,1	8-bit 6200,2	8-bit 6200,3	8-bit 6200,4	8-bit 6200,5	8-bit 6200,6	8-bit 6200,7	8-bit 6200,8

RPDO2	16-bit 6411,1	16-bit 6411,2	16-bit 6411,3	16-bit 6411,4
RPDO3	16-bit 6411,5	16-bit 6411,6	16-bit 6411,7	16-bit 6411,8
RPDO4	16-bit 6411,9	16-bit 6411,A	16-bit 6411,B	16-bit 6411,C

Figure 3.8 Default RPDO Mapping of DS401

3.4.2.3 Index [180xh,xxh] and [1A0xh,xxh]: TPDO Parameters

By default a total of up to four TPDOs are configured. The transmission type [180xh, 02h] is set to 255: manufacturer specific. The default behavior is a change-of-state transmission – input data gets transmitted whenever the inputs change. Both inhibit and event times have a default of 0.

The default TPDO mapping is illustrated in Figure 3.9. TPDO1 contains 8 digital input bytes that are taken from the Object Dictionary entries [6000h,01h-08h]. TPDO2, 3 and 4 each contain four 16-bit analog values that are taken from the Object Dictionary entries [6401h,00h-0Ch].

Notes: All Index, Subindex values in hexadecimal,
8-bit values are UNSIGNED8 and 16-bit values are INTEGER16

TPDO1	8-bit 6000,1	8-bit 6000,2	8-bit 6000,3	8-bit 6000,4	8-bit 6000,5	8-bit 6000,6	8-bit 6000,7	8-bit 6000,8

TPDO2	16-bit 6401,1	16-bit 6401,2	16-bit 6401,3	16-bit 6401,4
TPDO3	16-bit 6401,5	16-bit 6401,6	16-bit 6401,7	16-bit 6401,8
TPDO4	16-bit 6401,9	16-bit 6401,A	16-bit 6401,B	16-bit 6401,C

Figure 3.9 Default TPDO Mapping of DS401

3.4.2.4 Index [6000h,xxh]: Read Digital Inputs

This array is mandatory for devices that support digital inputs. It is an array of UNSIGNED8 read-only values that contain the digital inputs. Subindex 0 specifies how many Subentries are implemented. The default is 8 providing a total of 8x8 = 64 digital input bits. The maximum value allowed is FEh allowing for a total of 254x8 = 2032 digital inputs.

3.4.2.5 Index [6002h,xxh]: Polarity of Inputs

Although not mandatory, this is an Object Dictionary entry supported by many generic I/O devices. If implemented, it is an array of UNSIGNED8 read-write values that is exactly as long as the array in [6000h,xxh].

If implemented, each bit in this array defines the polarity inversion of the bits in the [6000h,xxh] array. If a bit in this array is set, the corresponding bit in [6000h,xxh] is inverted. If a bit is cleared, the corresponding bit is not changed.

3.4.2.6 Index [6200h,xxh]: Write Digital Outputs

This array is mandatory for devices that support digital outputs. It is an array of UNSIGNED8 read-write values that contain the digital outputs. The entry is specified as "read-write" in order to be able to read-back the last value written to the output. However, these entries can only be mapped to RPDOs, not to TPDOs.

Subindex 0 specifies how many Subindexes are implemented. The default is 8 providing a total of 8x8 = 64 digital output bits. The maximum value allowed is FEh allowing for a total of 254x8 = 2032 digital outputs.

3.4.2.7 Index [6202h,xxh]: Polarity of Outputs

Although not mandatory, this is an Object Dictionary entry supported by many generic I/O devices. If implemented this is an array of UNSIGNED8 read-write values that is exactly as long as the array in [6200h,xxh].

If implemented, each bit in this array defines the polarity inversion of the bits in the [6200h,xxh] array. If a bit in this array is set, the corresponding bit in [6200h,xxh] is inverted. If a bit is cleared, the corresponding bit is not changed.

3.4.2.8 Index [6206h,xxh] and [6207h,xxh]: Error Mode and Error Value for Outputs

Although not mandatory, these are Object Dictionary entries supported by many generic I/O devices. If implemented, these are arrays of UNSIGNED8 read-write values exactly as long as the array in [6200h,xxh].

If implemented, each bit in [6206h,xxh] determines if a default value should be applied to the corresponding output upon detecting an error condition or if the node is stopped. The entries in [6207h,xxh] are the default values that should be applied if this function is enabled and an error or stop condition is detected.

3.4.2.9 Index [6401h,xxh]: Read Analog Inputs

This array is mandatory for devices that support analog inputs. It is an array of INTEGER16 read-only values that contain the analog inputs. Subindex 0 specifies how many Subentries are implemented. The default is 12 analog inputs. The maximum value allowed is FEh allowing for a total of 254 analog inputs.

If the resolution of the inputs is less then the 16 bits provided (for example, some just have a 10-bit resolution), then the value must be shifted to the most significant bits and the least significant, unused bits must be filled with zeros.

3.4.2.10 Index [6411h,xxh]: Write Analog Outputs

This array is mandatory for devices that support analog outputs. It is an array of INTEGER16 read-write values that contain the analog outputs. The entry is specified as "read-write" in order to be able to read-back the last value written to the output. However, these entries can only be mapped to RPDOs, not to TPDOs.

Subindex 0 specifies how many Subentries are implemented. The default is 12 analog outputs. The maximum value allowed is FEh allowing for a total of 254 analog outputs.

If the resolution of the outputs is less then the 16 bits provided (for example just have a 10-bit resolution), then the value received must be shifted in order to use the most significant bits and ignore the least significant, unused bits.

3.4.2.11 Index [6443,xxh] and [6444,xxh]: Error Mode and Error Value for Outputs

Although not mandatory, these are Object Dictionary entries supported by many generic I/O devices. If implemented these are an array of UNSIGNED8 read-write values for [6443h,xxh] and an array of INTEGER32 read-write values for [6444h,xxh]. The length of both arrays is the same as the length of array [6411h,xxh].

If implemented, each entry in [6443h,xxh] determines if a default value should be applied to the corresponding output upon detecting an error condition or if the node is stopped. If the entry is "1" the corresponding default error value is used. The entries in [6444h,xxh] are the default values that should be applied if this function is enabled and an error or stop condition is detected.

The data type of the default error values is INTEGER32 in order to be usable for any integer based output (including INTEGER8 and INTEGER16). When used for another data type output the lower bytes are ignored. In the case of an INTEGER16 output, only the two most significant bytes of the INTEGER32 default error value are used.

3.4.3 Illustrations

Figure 3.10 shows the mandatory Object Dictionary entries involved for handling a digital or analog output. The Object Dictionary entry with the process output data can be modified from the network side of a device by a SDO request or by a RPDO that has this process output data mapped into it. With each change of the process output data, the data gets immediately applied to the application side of the device. The process data is not modified in any way and the last applied output value will be continuously applied until it gets changed via the network. It will not change in the event of an error.

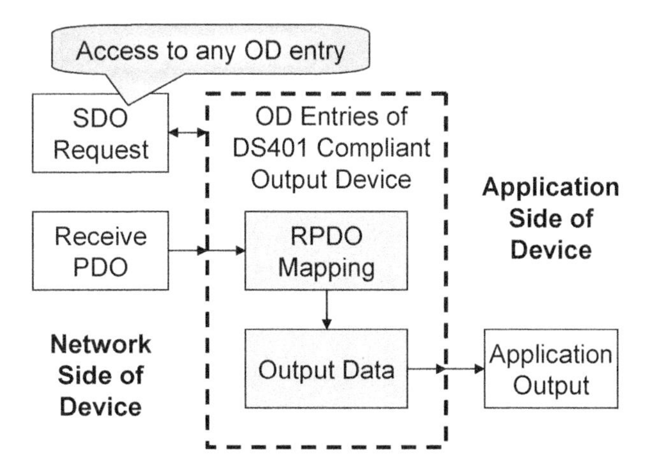

Figure 3.10 Basic Setup of Output Processing

A more advanced setup of an output device is shown in Figure 3.11. Here the Object Dictionary entries for error mode and values and some data manipulation entries such as polarity change or scaling are implemented.

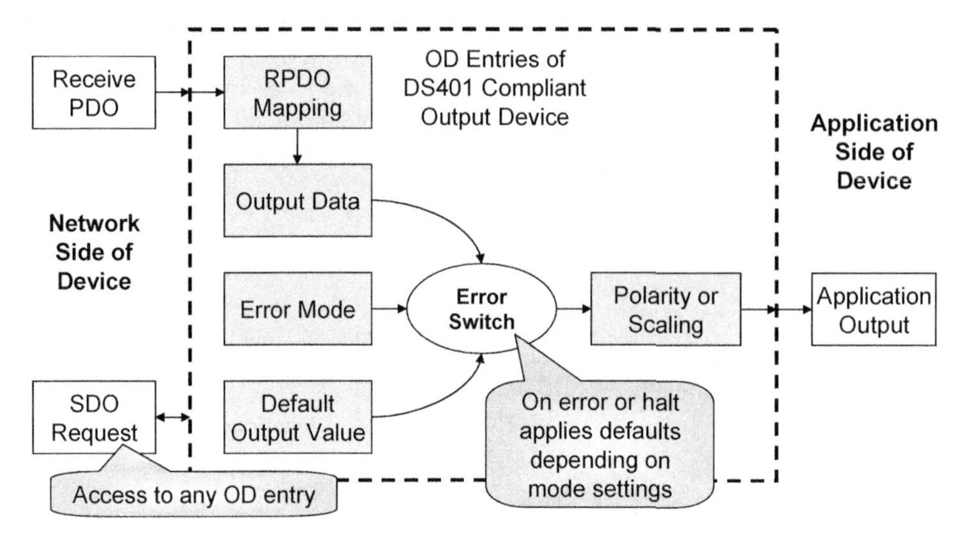

Figure 3.11 Output with Optional Error Mode and Data Manipulation

On the network side, the process data is still received via RPDOs or by a SDO request to the Object Dictionary entry with the process data. However, if (and how) it gets applied to the application side of the device depends on several settings. During regular operation, the data will be manipulated as specified before it gets applied. For digital data the manipulation option is a potential change in polarity. For analog data the manipulation is an optional scaling with a multiplication factor and an offset.

If an error occurs or the device is halted, the output applied depends on the settings of the Error Mode and the Error value. If the Error Mode is enabled for a particular output, then the specified default output value will be applied.

The diagram in Figure 3.12 shows the basic elements of input processing. The TPDO mapping parameters determine into which TPDO an input signal is copied.

For each TPDO, the transmit trigger mechanism in an input device has to check the conditions for the actual transmission of a TPDO. That can be an expiration of a timer or the detection of a COS (change-of-state).

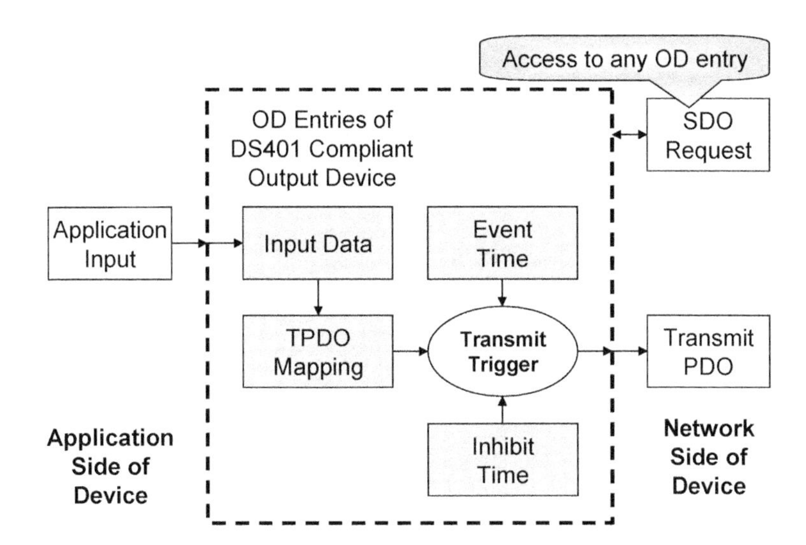

Figure 3.12 Basic Setup of Input Processing

A more advanced setup is shown in Figure 3.13. Here the data gets manipulated before it is stored in the Object Dictionary. For digital data the manipulation option is a potential change in polarity. For analog data the manipulation is an optional scaling with a multiplication factor and an offset.

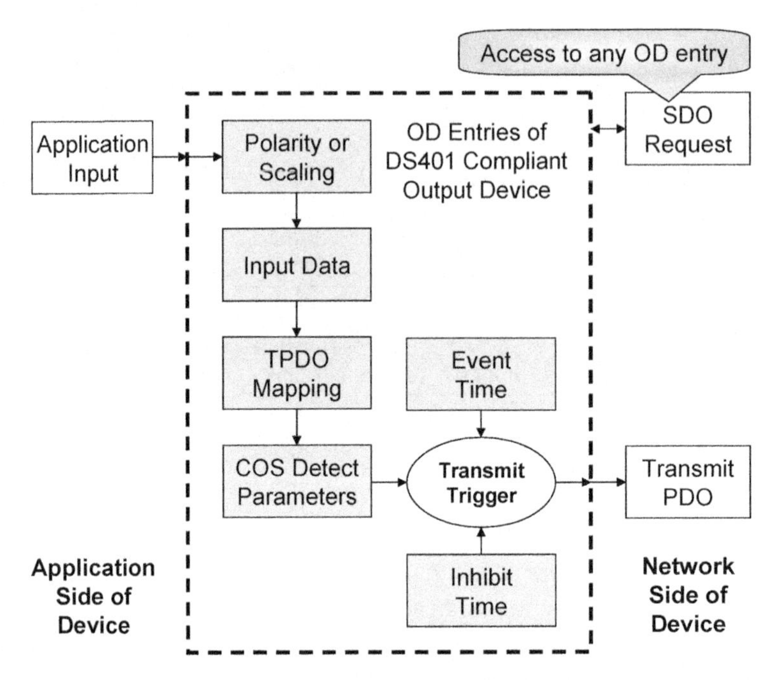

Figure 3.13 Input Processing with Data Manipulation and Advanced COS Detection

The advanced COS detection for digital inputs is an edge detection. A COS can either be recognized on any data change, only on a rising edge (zero to one transition) or only on a falling edge (one to zero transition).

With analog signals, the advanced COS detection can either be a function of reaching a pre-defined limited or a value difference. When using the pre-defined limit, a COS is detected if a certain threshold (upper limit or lower limit) is reached. When using value differences, a configurable value difference relative to the last data transmitted must be reached in order to recognize a COS.

3.4.4 Generic I/O Object Dictionary Example

The following DS401-related Object Dictionary entries would be implemented for a node with 2 bytes (each digital) and 2 words (each analog).

Index	Sub index	Description	Data Type	Default Value
1000h	00h	Device Type	UNSIGNED32	000F0191h
1400h		1st RPDO Communication Parameters		
1400h	00h	Number of Entries	UNSIGNED8	2
1400h	01h	COB ID	UNSIGNED32	200h + Node ID
1400h	02h	Transmission Type	UNSIGNED8	FFh
1401h		2nd RPDO Communication Parameters		
1401h	00h	Number of Entries	UNSIGNED8	2
1401h	01h	COB ID	UNSIGNED32	200h + Node ID
1400h	02h	Transmission Type	UNSIGNED8	FFh
1600h		1st RPDO Mapping Parameters		
1600h	00h	Number of Entries	UNSIGNED8	2
1600h	01h	1st Mapping Entry: 1st Write 8-bit Output	UNSIGNED32	62000108h
1600h	02h	2nd Mapping Entry: 2nd Write 8-bit Output	UNSIGNED32	62000208h
1601h		2nd RPDO Mapping Parameters		
1601h	00h	Number of Entries	UNSIGNED8	2

Table 3.33 DS401 Object Dictionary Example

Index	Sub index	Description	Data Type	Default Value
1601h	01h	1st Mapping Entry: 1st Write 16-bit Output	UNSIGNED32	64110110h
1601h	02h	2nd Mapping Entry: 2nd Write 16-bit Output	UNSIGNED32	64110210h
1800h		1st TPDO Communication Parameters		
1800h	00h	Number of Entries	UNSIGNED8	5
1800h	01h	COB ID	UNSIGNED32	180h + Node ID
1800h	02h	Transmission Type	UNSIGNED8	FFh
1800h	03h	Inhibit Time	UNSIGNED16	0
1800h	05h	Event Time	UNSIGNED16	0
1801h		2nd TPDO Communication Parameters		
1801h	00h	Number of Entries	UNSIGNED8	5
1801h	01h	COB ID	UNSIGNED32	280h + Node ID
1801h	02h	Transmission Type	UNSIGNED8	FFh
1801h	03h	Inhibit Time	UNSIGNED16	0
1801h	05h	Event Time	UNSIGNED16	0
1A00h		1st TPDO Mapping Parameters		
1A00h	00h	Number of Entries	UNSIGNED8	2
1A00h	01h	1st Mapping Entry: 1st Read 8-bit Input	UNSIGNED32	60000108h
1A00h	02h	2nd Mapping Entry: 2nd Read 8-bit Input	UNSIGNED32	60000208h
1A01h		2nd TPDO Mapping Parameters		

Table 3.33 (Continued) DS401 Object Dictionary Example

Index	Sub index	Description	Data Type	Default Value
1A01h	00h	Number of Entries	UNSIGNED8	2
1A01h	01h	1st Mapping Entry: 1st Read 16-bit Input	UNSIGNED32	64010110h
1A01h	02h	2nd Mapping Entry: 2nd Read 16-bit Input	UNSIGNED32	64010210h
6000h		Read Digital Inputs 8-bit		
6000h	00h	Number of Entries	UNSIGNED8	2
6000h	01h	Read Digital Input 1	UNSIGNED8	--
6000h	02h	Read Digital Input 2	UNSIGNED8	--
6002h		Polarity Digital Input 8-bit		
6002h	00h	Number of Entries	UNSIGNED8	2
6002h	01h	Polarity Digital Input 1	UNSIGNED8	0
6002h	02h	Polarity Digital Input 2	UNSIGNED8	0
6200h		Write Digital Outputs 8-bit		
6200h	00h	Number of Entries	UNSIGNED8	2
6200h	01h	Write Digital Output 1	UNSIGNED8	--
6200h	02h	Write Digital Output 2	UNSIGNED8	--
6202h		Polarity Digital Output 8-bit		
6202h	00h	Number of Entries	UNSIGNED8	2
6202h	01h	Polarity Digital Output 1	UNSIGNED8	0
6202h	02h	Polarity Digital Output 2	UNSIGNED8	0
6410h		Read Analog 16-bit Inputs		
6410h	00h	Number of Entries	UNSIGNED8	2
6410h	01h	Read Analog Input 1	INTEGER16	--

Table 3.33 (Continued) DS401 Object Dictionary Example

Index	Sub index	Description	Data Type	Default Value
6410h	02h	Read Analog Input 2	INTEGER16	--
6411h		Write Analog 16-bit Outputs		
6411h	00h	Number of Entries	UNSIGNED8	2
6411h	01h	Write Analog Output 1	INTEGER16	--
6411h	02h	Write Analog Output 2	INTEGER16	--

Table 3.33 (Continued) DS401 Object Dictionary Example

3.5 Safety-Relevant Communication (DSP304, DSP307)

Objective

In this section we outline some of the functions and methods used to implement safety-relevant communication on CANopen.

Because safety aspects also depend on the specific application, implementation methods may vary. For example, there is a CANopen framework for safety-relevant communication [CiADSP304]. However, the maritime industry also had safety requirements specific to their application, resulting in the CANopen framework for maritime electronics [CiADSP307].

3.5.1 Introduction and Terminology

In general, any application that has the potential to "significantly" harm the environment, injure or even kill one or multiple persons is considered "safety-related." All major standardization bodies publish standards defining safety-related systems and specifying safety levels. One of the standards available is IEC 61508 which defines a total of 4 safety-integrity levels (SIL) with 1 being the lowest safety level and 4 the highest.

The safety-integrity levels are a measurement of the worst that can happen if something goes wrong. For example, the controls for a table saw have the potential to contribute to a severe injury of a person. Such an application would be considered a SIL1 application. However, the controls of a chemical plant are in a different safety level as their failure could contribute to multiple fatalities. This is considered at least a SIL3, perhaps even a SIL4 application.

A safety-related system can typically be divided into multiple safety functions, each responsible for a single crucial aspect of the entire system. A system is considered to be functionally safe if all of its safety functions are carried out without failure.

The distinction between a safety-relevant versus a safety-critical application is the existence of a "safe state." If a system has a safe state it can turn to it is considered safety-relevant. This includes all applications with an emergency shut-off switch – if the system is switched off it can do no harm. Safety-critical systems, on the other hand, require continuous control; once airborne a plane needs continuous control and once started a chemical process might need continuous control.

For more information on safety related systems, IEC62508 and related subjects see [Smith01].

3.5.2 Defects Happen

One of the basic rules when designing a safety-related system is that defects happen. However, a single defect may not be allowed to result in the failure of a safety function. As far as electronics are concerned this means that redundancies need to be added. Circuits and wiring can be duplicated on the hardware side. On the software side, activities can also be duplicated; for example, sending a message twice.

There are multiple locations where redundancies can be added to a CANopen node: microcontrollers and software can be duplicated, the CAN controller can be duplicated, the transceiver and wiring can be duplicated and the messages on the bus can be duplicated. Which duplications really make sense depends on both the application and how the duplications are made. Duplicating a microcontroller and using the same software on both does not do much to increase safety. If there is a conceptual defect in either the microcontroller or the software it will be present in the first and the second microcontroller. And typically in a complex system a software bug is far more likely than an electronic component failing.

Another implication of "defects happen" is that no single component in the network should be essential to the operation of the system. Failure of a single node should not prohibit other nodes from continuing their communication. This consideration favors a truly distributed system without a master, which is fully supported by CANopen where the nodes can exchange messages directly, whether the other nodes are alive or not. Furthermore, this is another example of the CANopen heartbeat being preferable to node guarding. Node guarding requires a single node to poll all the slaves, making the node that does the polling essential to the system. With the heartbeat method, however, nodes produce their heartbeat independently and the communication partners can directly monitor that heartbeat.

If for some reason a master is essential to the system then it would need to be duplicated. CANopen provides methods to use one or multiple "Flying Masters" as a backup to the main master. As the term suggests, they can take over the main master's responsibilities "on-the-fly" during operation of the network.

3.5.3 Adding Safety to CANopen

Like many other computerized networks CANopen can be used in safety-relevant applications. The major criteria for safety-relevant communication is met if a system can reliably detect the loss of communication (maybe because the network cable broke or a node went off-line) and fall back to the safe state as a result. This also illustrates the fact that primary responsibility for implementing safety lies with the consumer of the messages and not the producer. Thus the essential device is not the emergency shut-off switch but all the devices consuming that message - they need to be able to both receive messages from the switch, as well as be able to detect the loss of the switch and then act upon that information.

The other criterion is to ensure reliable, secure communication. If an emergency shut-off switch continuously sends the message "We are still all on GO," it must be ensured that there is no way that this node (or any other) produces this message by accident.

When designing software for safety-relevant applications one should carefully consider where it makes sense to design tasks that run independently and where they need to be interlocked.

The worst (but still functional) design for an emergency shut-off switch would have two independent tasks: one copies the current state of the button to the message transmit buffer (setting up the "We are still all on GO" message) the second task would simply trigger the transmit message from a timer interrupt service routine.

If the first task fails for any reason, the second would still continue to transmit "We are still all on GO."

More safety can be added in CANopen by adding redundancies to different levels. Figure 3.14 illustrates the basic communication path of a CANopen node. A microcontroller communicates with a CAN controller that uses a transceiver to exchange signals with the physical media. Software redundancies can be added without changing this structure. Messages can be duplicated, sent twice over the network, and only when both instances are transmitted is the transmission regarded as successful. To further increase the safety, the second message has all data bits inverted and at least two bits in the message identifier field inverted, too.

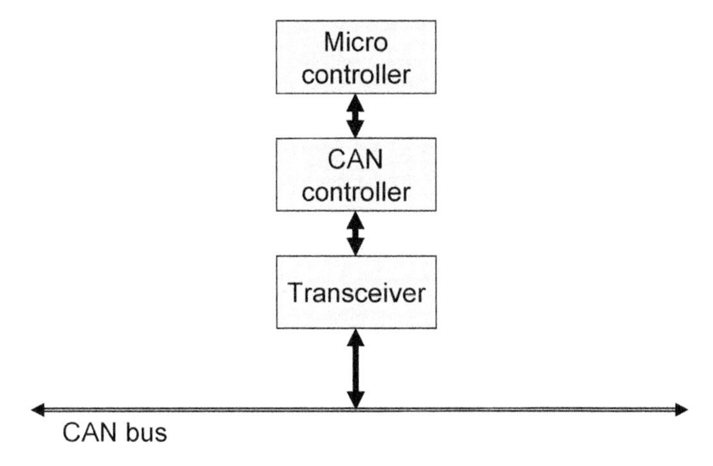

Figure 3.14 Adding Software Redundancy

In addition to adding software redundancies, some applications may require adding hardware redundancies, too. One of the scenarios suggested by the CANopen framework for safety-relevant communication is illustrated in Figure 3.15. Here the micro-controller and the CAN controller are duplicated, but they still share the same transceiver and physical media. This would be primarily used where electronic circuits responsible for going into a safe state need to do a "controlled" shut-down, like switching off components in a certain order or applying additional brakes to a motor.

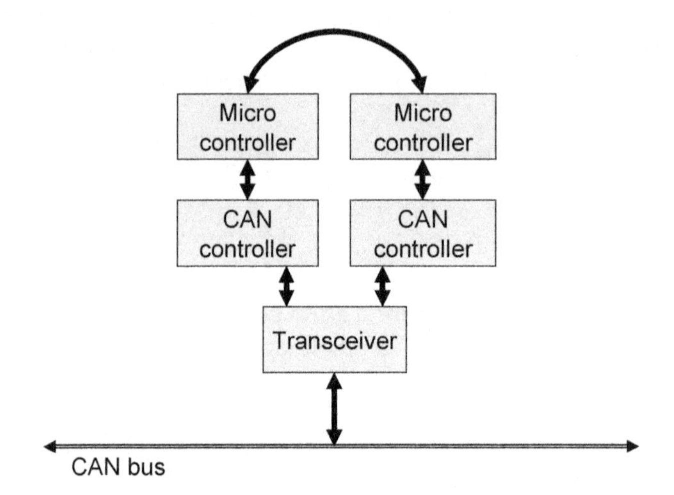

Figure 3.15 Adding Hardware Redundancy

The CANopen framework for maritime electronics uses another approach. The examination of the safety requirements in a ship placed the probability of "something happening to the wiring or an entire segment of the network" above the probability of "something happening to individual electronics." As a result, the framework recommends duplicating the wiring as shown in Figure 3.16.

The idea is to not only duplicate the wiring but also to ensure that the wires take different paths. So the main trunks of the bus would be separated, for instance one going along the starboard side of a ship and one along the port side.

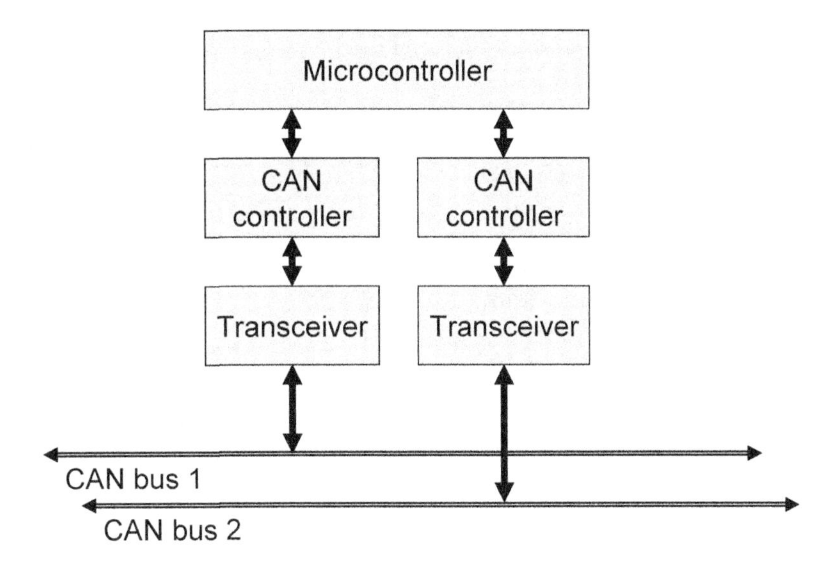

Figure 3.16 Adding Redundant Wiring

3.5.4 CANopen SRDO – Safety-Relevant Data Object

The CANopen framework for safety-relevant communication specifies SRDOs to be used for transmitting safety-relevant data. They key elements of an SRDO are:

- An SRDO consists of two messages.
- The first message's format is the same as used for regular PDOs, meaning one or multiple Object Dictionary entries can be mapped into the SRDO.
- The second message is a duplication of the first with all data bits inverted and at least 2 bits inverted in the message identifier field.

All SRDO transmissions have two essential timings. If any of these are exceeded when receiving a SRDO, the receiver has to take that as an indication that "something happened" and should switch to the safe state immediately.

- **SCT – Safeguard Cycle Time**
  This is the time between multiple SRDOs or, in other words, the "event time" of the SRDO. The SCT is measured between the occurrences of the second message of an SRDO.

- **SRVT – Safety-Relevant Validation Time**
  This is the maximum time allowed between the first and the second message of an SRDO.

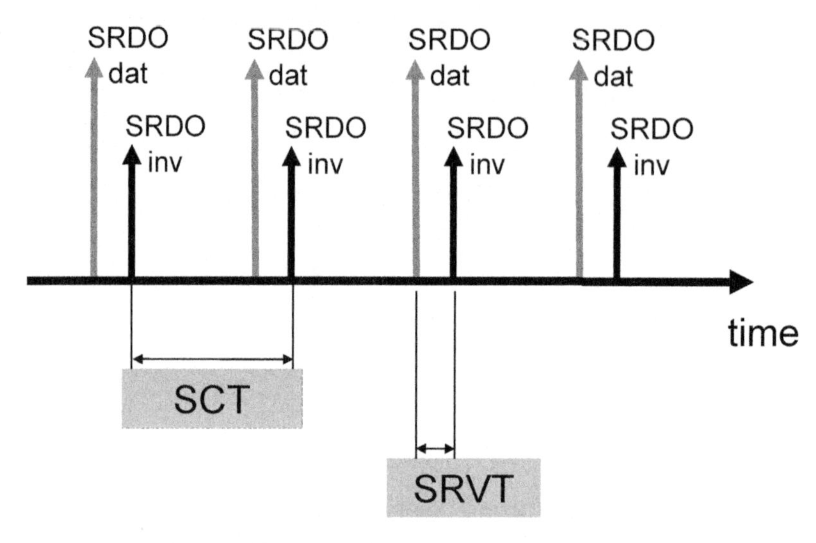

Figure 3.17 SRDO Timing

Figure 3.17 illustrates the SCT and the SRVT which can both be specified in milliseconds. Typically the SRVT is expected to be much shorter than the SCT, as the second message should follow almost back-to-back to the first message.

When configuring the SCT and SRVT great care should be taken as to how to configure the producer and the consumers. Due to potential message delays after error conditions on the bus or timer variations in the different microcontrollers, one should not set the consumer and the producer to the same SCT and SRVT timings. The recommended approach is to set the timings on the consumers first, since that is the timing used to make a decision if the safe state should be entered or not. The corresponding times on the producer side need to be shorter, so that the SRDO is produced more often than expected. This ensures that even with variations in the timing on the producer side the SRDO can still be received by the consumers. Once again, the exact difference depends on the specific application requirements. In general, setting the producer timing to some 80%-90% of the consumer timing is a good start.

3.5.5 Object Dictionary Entries for SRDOs

The following is a summary of selected parameters that are configurable for the up to 63 SRDOs.

3.5.5.1 Index [1301h,xxh] to [1340h,xxh]: Communication Parameter

Similar to PDO configuration, each SRDO has one record of communication parameters. The record has a total of six entries resulting in seven Subentries:

Subindex	Name	Description	Type
0	Number of Entries	Set to '6'	UNSIGNED8
1	Direction	1 for Tx, 2 for Rx	UNSIGNED8
2	SCT	Safeguard Cycle Time in milliseconds	UNSIGNED16
3	SRVT	Safety-Relevant Validation Time in milliseconds	UNSIGNED8
4	Transmission Type	Set to '254' (FEh)	UNSIGNED8
5	COB ID 1	CAN message ID of first message	UNSIGNED32
6	COB ID 2	CAN message ID of second message	UNSIGNED32

Table 3.34 SRDO Communication Parameters

By default the CAN message identifiers 101h to 180h are used for SRDOs.

3.5.5.2 Index [1381h,xxh] to [13C0h,xxh]: Mapping Parameter

The structure and usage of the mapping parameters is identical to that of PDOs. See Section 2.5.6 for details on how PDO mapping works.

3.5.5.3 Index [13FEh]: Configuration Valid

The 8-bit value at [13FEh] is used by a safety device to signal that it has a valid configuration. Only if all SRDO configurations are valid will this value be set to A5h. All other values signal that the current configuration is not valid.

3.5.5.4 Index [13FFh,xxh]: Safety Configuration Checksum

The array at location [13FFh,xxh] contains one 16-bit checksum for each SRDO configured in the local node. The CRC for each SRDO is generated over the communication parameters (excluding Subindex 0 and 4) and mapping parameters.

To configure an SRDO, a CANopen Master or configuration tool first sets the SRDO communication and mapping parameters and then has to write the matching CRC checksum for this configuration into the corresponding checksum field. Reading back the entire configuration before truly declaring a configuration as valid is recommended.

4 CANopen Configuration Example

"Few things are harder to put up with than the annoyance of a good example."
Mark Twain

This chapter uses a fictitious industrial control or automation system to give an example of a CANopen network integration cycle. Although it is a fictitious example it is representative of typical industrial manufacturing machinery that uses a main Programmable Logic Controller (PLC) with some I/O nodes connected to it.

4.1 Evaluating the System Requirements

> Objective
>
> At the beginning of each system design the overall requirements must be evaluated. For our example we assume a central controller (PLC style) with distributed I/O points. On the communication side elements to be evaluated include communication response times, bandwidth, distance, and number and type of communication participants (nodes in the network).

4.1.1 Defining the System

In order to keep the example simple, a total of three nodes (in addition to the PLC) are assumed for this machine.

- The "Left Node" and the "Right Node" have identical communication requirements. They should be DS401 compliant (generic I/O) and have a fixed number of digital inputs and outputs (2 bits) and analog inputs and outputs (2 times 11 bits).
- The "Middle Node" is a specialty module (manufacturer specific) providing some inputs (provided in 2 bytes) from a user panel.

By default the entire communication is controlled by the PLC. All inputs are sent to the PLC, all outputs come from the PLC. As an optional extension, a future version of the control system will have the middle node send its data directly to the Left and Right nodes without the PLC relaying the data.

Because the focus of this section is to establish the required communication between the nodes, the control algorithm used on the PLC is assumed to be already developed and in place.

In order to get an accurate timing, the SYNC signal is used. The SYNC for this example has a 33ms communication cycle period. The control cycle time would be 66ms, as two messages are involved for an input to output transfer (first message from input to PLC, second from PLC to output).

The bus speed is chosen to be 125kbps, although the maximum network cable length for that speed (about 1500 feet) is probably not needed. In general it is a good idea to not use a higher bus speed than required. Keeping the bus speed low decreases EMI and increases overall system stability and tolerance.

4.1.2 Estimated Bandwidth Usage

A rough estimate of bandwidth usage can be calculated as follows:

1. Calculate the number of data bytes transmitted in each SYNC cycle:
   - From Left Node to PLC: 5 (1 byte digital, 2 words analog)
   - From Middle Node to PLC: 2 (2 bytes digital)
   - From Right Node to PLC: 5 (1 byte digital, 2 words analog)
   - From PLC to Left Node: 5
   - From PLC to Right Node: 5

This results in a total of 22 data bytes.

2. Calculate the data bandwidth required:

The SYNC cycle time is 33ms, so 22 bytes transmitted every 30ms is about 100 bytes for every 100ms, or 1,000 bytes per second. Multiplying by 8 (to achieve bits per second not bytes per second) results in about 8kbps.

3. Estimate the total bandwidth:

Besides the data bytes, CAN messages contain message ID information, control bits, a checksum and other overhead information. Unfortunately there is no easy rule of thumb for the relationship between data bytes and overhead bits. The overhead factor can be anywhere from less than 2 to as much as 6 or more if many short messages are used.

Using the overhead factor range of 2 to 6 would result in a bandwidth range of

$$2 * 8\text{kbps} = 16\text{kbps} \quad \text{to} \quad 6 * 8\text{kbps} = 48\text{kbps}$$

Because the chosen bit rate of the network is 125kbps the estimation above would (even for the worst case) result in a bandwidth usage of

$$48\text{kbps} / 125\text{kbps} = 38.4\%$$

This is an acceptable margin for a rough estimate. However, if the system would be switched down to a 50kbps network speed a more detailed calculation would be required.

Advanced development tools such as Vector's proCANopen or CANoe automatically perform these calculations and provide timing reports. The bandwidth statement given by CANoe also calculates bandwidth usage for event driven communication.

4.2 Choosing the Devices and Tools

Objective Once the requirements are set one needs to select the devices and tools used to configure and test the devices and the network. In this section we choose tools and devices for our example.

4.2.1 Choices to Make

Before one can make a selection one needs to know what the available options are. A fairly compete listing of off-the-shelf CANopen products is published by the CiA. There is both a product database and a CANopen product guide available online at www.can-cia.org.

It should be noted that depending on the availability of the "best-match" products, it might be necessary to take a step back and re-evaluate the system requirements. Perhaps a similar product that is not a best-fit for the application has some other advantages, and thus can be considered for the application if the requirements are adjusted.

For this example the following devices were chosen:

- PLC:
  Schneider TSX Compact with CAN communication adapter
- Left and Right Node:
  Schneider Advantys STB modular I/O system
- Middle Node:
  Manufacturer specific solution based on Philips CANopenIA-XA

- Configuration Tool:
  Vector proCANopen
- Simulation and Analyzing Tool:
  Vector CANoe

This is a fictitious example, and many devices and tools from various manufacturers could have been chosen. However, some of the devices and tools listed above have some specific features in conjunction with CANopen.

4.2.2 Modular, Generic I/O

In regards to CANopen, the system shown in Figure 4.1 has two features that stand out.

Figure 4.1 Schneider Advantys Modular I/O System

First, it is a modular I/O system. It allows building CANopen generic I/O devices with exactly the number of digital and analog channels required by the application. As a result, the number of data channels available per device are never fixed and could even be expanded in the future without the need to exchange entire devices. In addition, simple control functions (like Boolean functions, comparators or counters) can be performed within the module, creating a simple automation island.

Second, it uses CANopen on its backplane. The communication used between the individual components of the Advantys STB system is a dedicated CANopen network. The module on the left in Figure 4.1 is called a CANopen NIM (Network Interface Module). It has two CANopen interfaces and acts as a gateway between the upstream network (where it is a CANopen generic I/O module) and the downstream network (where it is a CANopen Master communicating with the individual modules of the local system).

4.2.3 Tools

When it comes to configuration, simulation and test of a CANopen network, the Vector tools proCANopen and CANoe provide powerful features as the following examples will show. These tools can be used to configure, simulate and also test large CANopen networks before they are built physically. Furthermore the simulation engine of CANoe can be used to create the communication of an entire network. This way a new node for a network can be tested without requiring the "live" network for the test phase.

4.3 Configuring Single Devices

Objective
Many CANopen devices offer several setup or configuration options through jumpers, switches or proprietary setup tools that are not necessarily available through regular CANopen configuration tools. This section shows some of the proprietary setup tools available with some devices.

4.3.1 Advantys STB Configuration

As the previous section has shown, some CANopen nodes can be fairly sophisticated, such as the modular Schneider Advantys STB system. In order to manage the modularity of the system, Schneider provides its own setup software that configures a modular system and, depending on the configuration, generates an appropriate Electronic Data Sheet – the electronic specification for the functionality provided by a CANopen node.

Such individual, device-specific setups and configurations need to be performed before the device can be integrated into a CANopen network.

The screen shot in Figure 4.2 shows an example of an Advantys STB system consisting of several Advantys STB modules and additional external third party CANopen modules that can be integrated into the "local" CANopen network. The configuration of a system can be done manually (drag and drop individual modules from the catalog) or read in from a physical system. The connection between the PC and the system is made via a serial link to a special connector on the NIM.

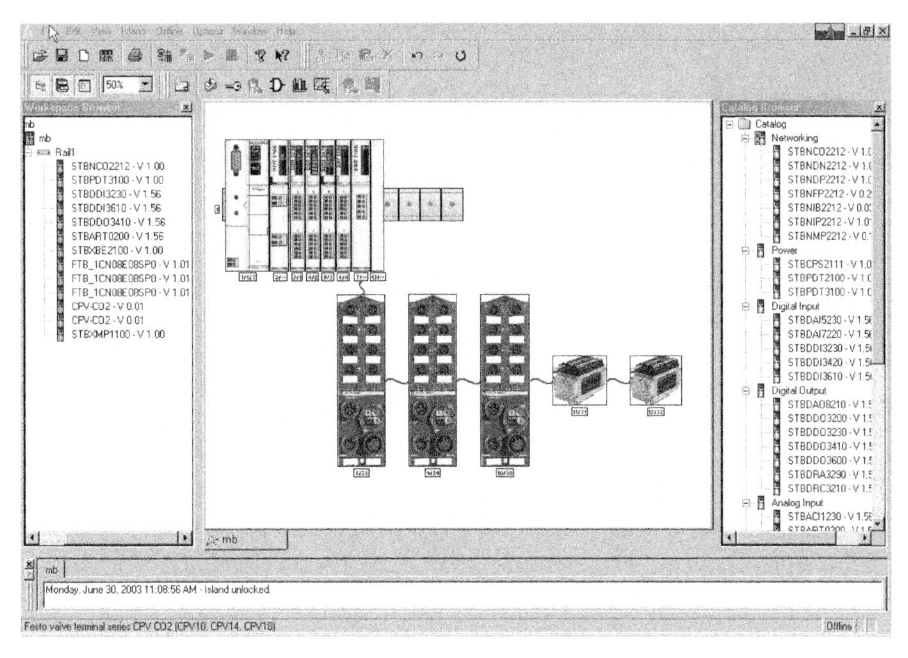

Figure 4.2 Schneider Advantys Configuration Software

All the process data of such an automation island is combined into one process image which is provided to the "upstream" CANopen network. The process image is simply mapped into the PDOs available to the upstream network.

Once the configuration is completed, an EDS can be generated. The EDS file is needed when integrating this particular automation island into a CANopen network.

4.3.2 CANopenIA Configuration

Another example for a device specific setup tool is that of Philips' CANopenIA. The "Middle Node" was chosen to be implemented based on the Philips CANopenIA-XA controller which also comes with an individual setup tool, CANopenIASetup.

The CANopenIA nodes do not support dynamic PDO mapping and thus the setup tool needs to be used to configure a CANopenIA chip. The configuration includes some hardware settings (like Node ID, bit rate used and which ports/pins of the controller are enabled) and software settings like the default heartbeat time or the PDO configuration (both communication and mapping parameters).

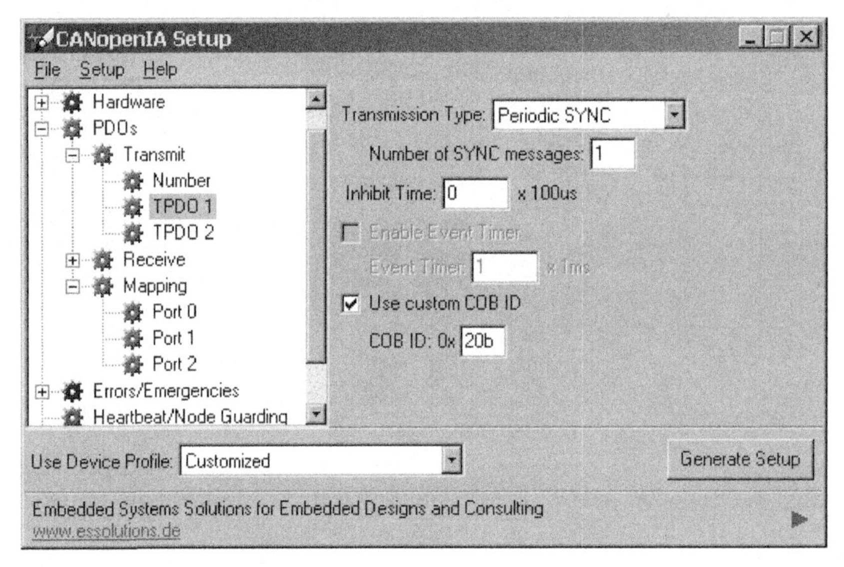

Figure 4.3 Philips' CANopenIA Setup Tool

The screenshot in Figure 4.3 shows the configuration of the first TPDO. The settings selected set a SYNC transmission for the TPDO and a CAN message ID of 20Bh to be used. Once the setup of a node is completed the setup information can be saved into a DCF (Device Configuration File) and a binary file that needs to be transferred to the CANopenIA chip. Although a download tool is provided, a regular CANopen configuration tool could be used for that purpose as well.

The reason why the setup tool provides a DCF instead of an EDS is that the DCF is already node specific. It contains the setup information for exactly one node with a specific node ID and not a generic setup for multiple nodes.

4.4 Overall Network Configuration

Objective

Although it is theoretically possible to configure an entire network on a "node-by-node" basis, it is not a very practical approach for bigger networks or for configuring multiple networks.

In this section we show how a configuration tool such as Vector's proCANopen can configure an entire network and automates the work of assigning process data variables to specific Process Data Objects (PDO).

4.4.1 Getting Started: Select Nodes

The configuration of an entire network using Vector's proCANopen starts by selecting the nodes of the network. For each node a name, Node ID number and an EDS file describing the node must be chosen. Figure 4.4 shows the selection of the EDS file for the master in the example application, and Figure 4.5 shows the node configuration options. Here the node is named "Schneider TSX Master," has the Node ID "1" and uses the EDS file "asbc259.eds."

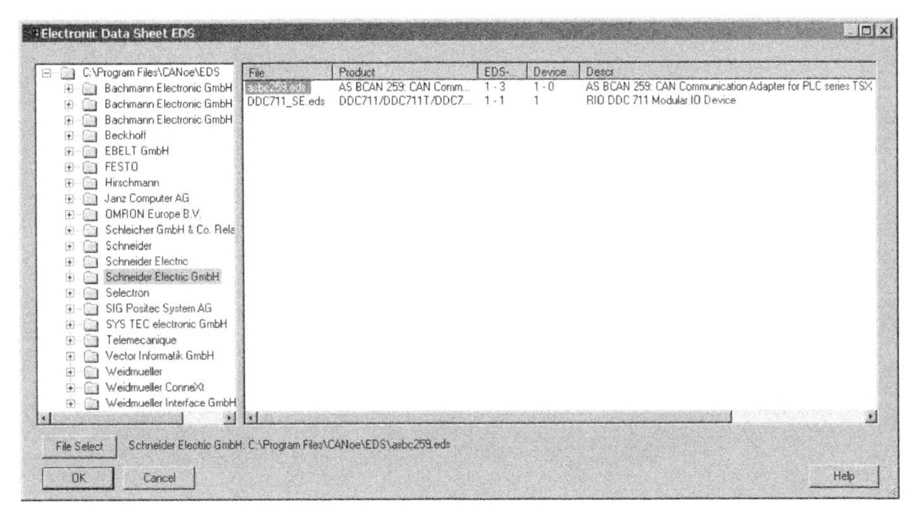

Figure 4.4 Selecting an EDS file for a Node in proCANopen

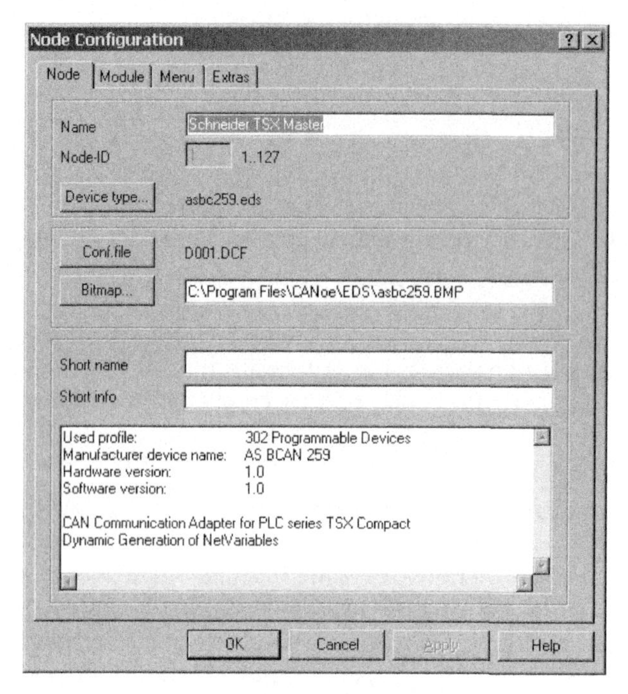

Figure 4.5 Configuration of a Node in proCANopen

Figure 4.6 below shows the PLC and the left and right node of the example system.

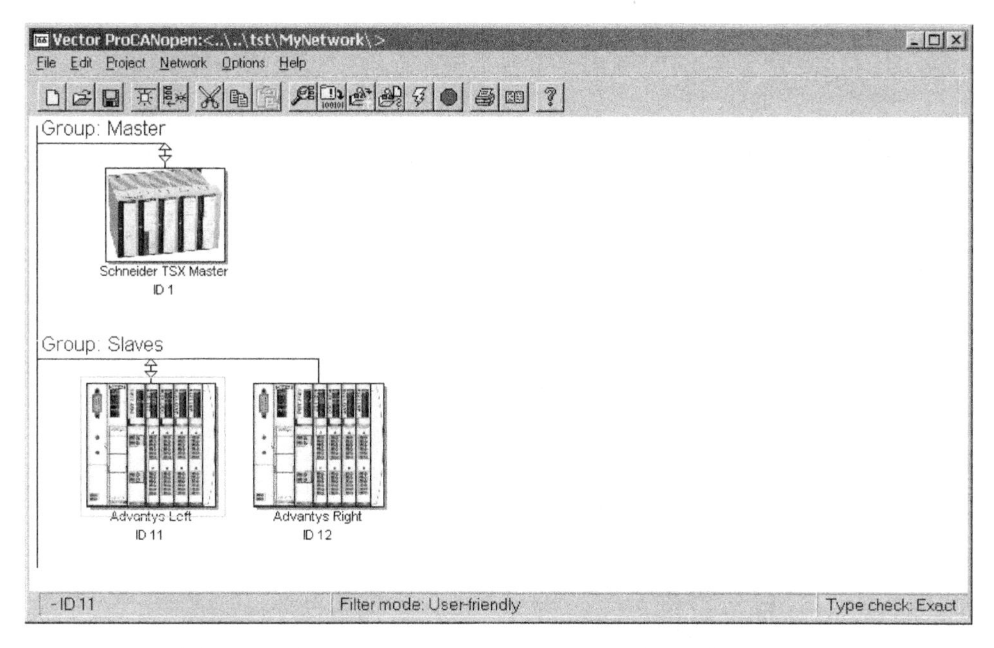

Figure 4.6 Main Overview Window of proCANopen

4.4.2 Establishing Connections

Once the nodes have been configured, connections need to be made. A connection is the link from one input variable to one output variable in the system.

It should be noted that the user only configures these connections (linking an input variable of one node to the output variable of another). The software in proCANopen then automatically assigns PDOs as required to implement these connections. So the tedious task of assigning individual variables to PDOs does not need to be executed by the user, it is automated.

A right mouse click on any of the nodes opens up a local menu, from which "Graphic Connection" can be chosen. For the configuration in this example, the left node is chosen with a right mouse click, then "Graphic Connection" is selected. Then the PLC is selected to be the communication partner. After clicking on the button "Insert PDOs" the window shown in Figure 4.7 pops up. It allows the specification of a prefix that is used for all variable names in the PLC. As all these variables come from the left Advantys node, the prefix entered is "AdvL_" which makes it easy to distinguish the variables from the left and the right node.

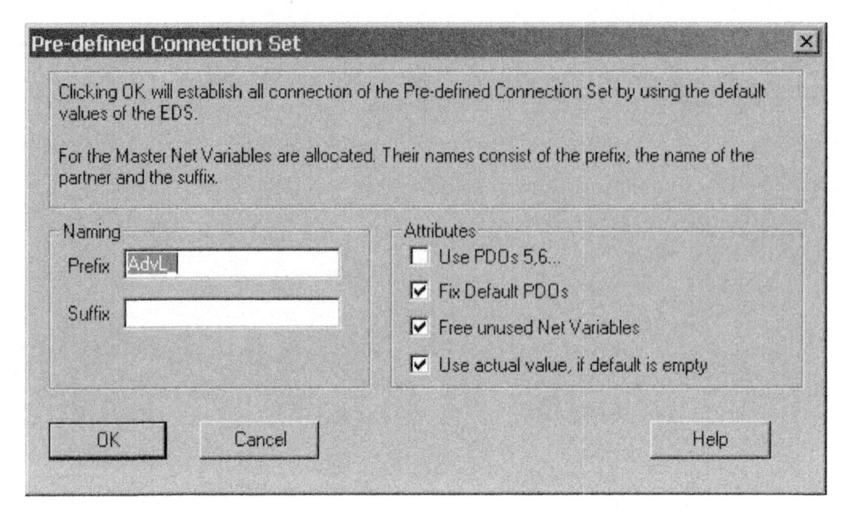

Figure 4.7 Step 1 of Making a Graphical Connection

After hitting "OK" proCANopen automatically connects all process variables from the left node to the PLC as shown in Figure 4.8. The names for the process variables in the Advantys node were directly extracted from the EDS file provided by the Advantys configuration software. The names generated for the corresponding variables in the master all have the specified prefix followed by an automatically generated name based on the process variable name used in the Advantys node.

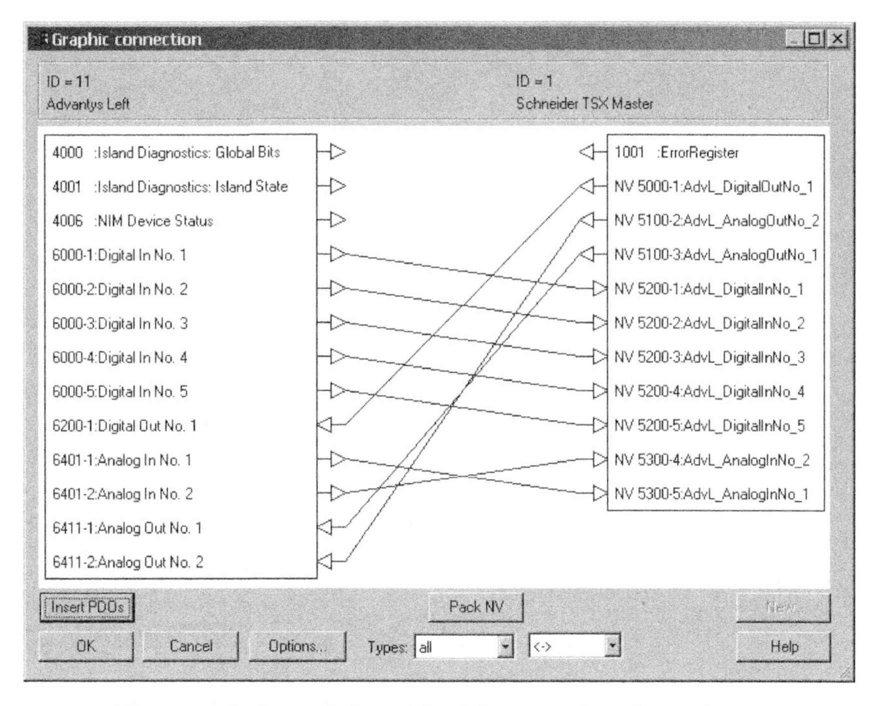

Figure 4.8 Step 2 Graphical Connection Completed

The importance of the prefix becomes more apparent in Figure 4.9 where both the process variables from the left and the right node are shown in the master.

As specified, the Advantys system in this example has only 2 bits of digital inputs and outputs and 2 words with analog inputs and outputs. The reason why a total of 5 bytes digital input are available is that Advantys also reports status and some echo information back.

The PDO transmission type used for the connections can be selected using the "Options" button. The default is "device profile specific," however any of the other possible transmission types may be chosen.

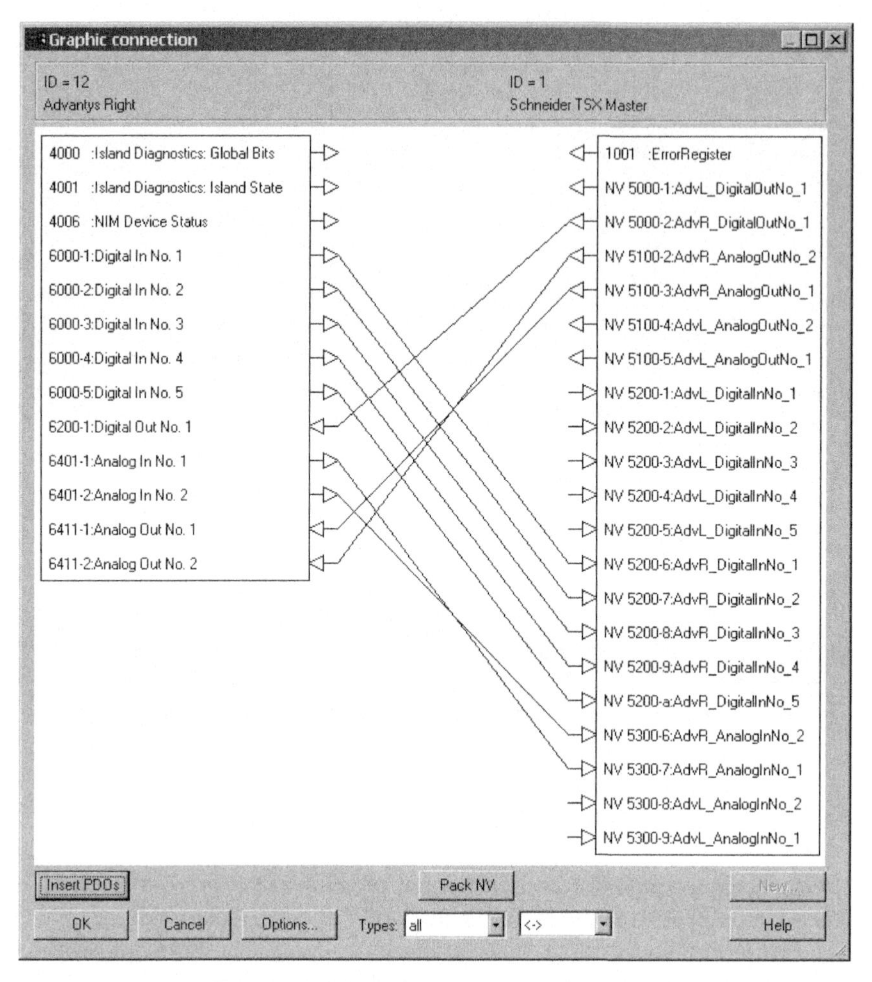

Figure 4.9 Step 3 Adding Connections for the Right Node

Once all the node specific configurations have been made, the list of all PDOs used in the network can be displayed. The left Advantys node with the Node ID 11 uses two transmit and 2 receive PDOs as shown in Figure 4.10. The first TPDO has 5 digital bytes mapped into it and uses the CAN message ID 18Bh which is the default from the pre-defined connection set.

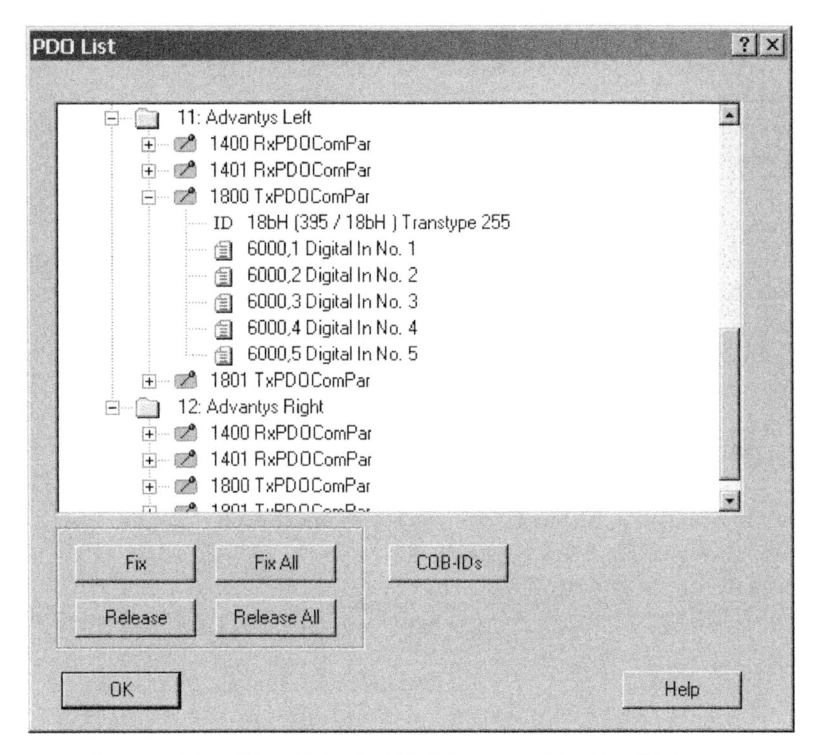

Figure 4.10 The list of all PDOs used in the Network

The final step of the configuration process is to start the "make" process that generates all the files needed to simulate this network using Vector's CANoe.

4.5 Network Simulation

> Objective
>
> This section shows how the network configured in the previous section can be simulated using Vector's CANoe.

One of the biggest advantages of a network simulation is that potentially crucial situations (like a high busload or how the overall system changes by adding or removing nodes) can be tested before all nodes are available physically. This allows tracking down potential danger zones where a system may reach its limits.

If you are not interested in the simulation of the network but would like to start "using" the real network, you can skip this section.

Once a CANopen network has been configured using Vector's proCANopen it can easily be simulated with CANoe. After starting a new CANoe configuration, the system configured by proCANopen can be imported. After ensuring that the CANoe configuration is saved and set to use the same CAN bit rate as the proCANopen setup, the simulation can be started by pressing the Tool button with the flash. The first messages that appear in the trace window are the boot-up messages of the individual nodes (see Figure 4.12, "CANoe Trace Recording of Simulation," on page 192, first 3 messages in trace window).

It should be noted that "simulation" in this sense actually means that the CAN traffic generated is actually sent on the network. As a minimum "network" CANoe expects a closed-loop system where both CAN ports provided by the Vector CAN hardware interface are connected together using a short CAN bus cable with two termination resistors.

As the PLC/Master is currently not simulated with the control algorithms it does not create the NMT startup message. As a result all nodes remain in the pre-operational state. However, all functions of proCANopen are fully usable – proCANopen does not realize that the network is simulated and thus can interact with all the simulated nodes.

To set all the nodes to operational, proCANopen can be used to generate the NMT message "Start all nodes." Once the nodes are in the operational state, the default communication mode is "change-of-state," which means a TPDO is transmitted whenever its input data changes. The inputs of each node are simulated by the automatically generated panel windows. By manually changing some of the inputs in the panels, the appropriate TPDO is transferred and the corresponding variable in the receiving node/panel is updated.

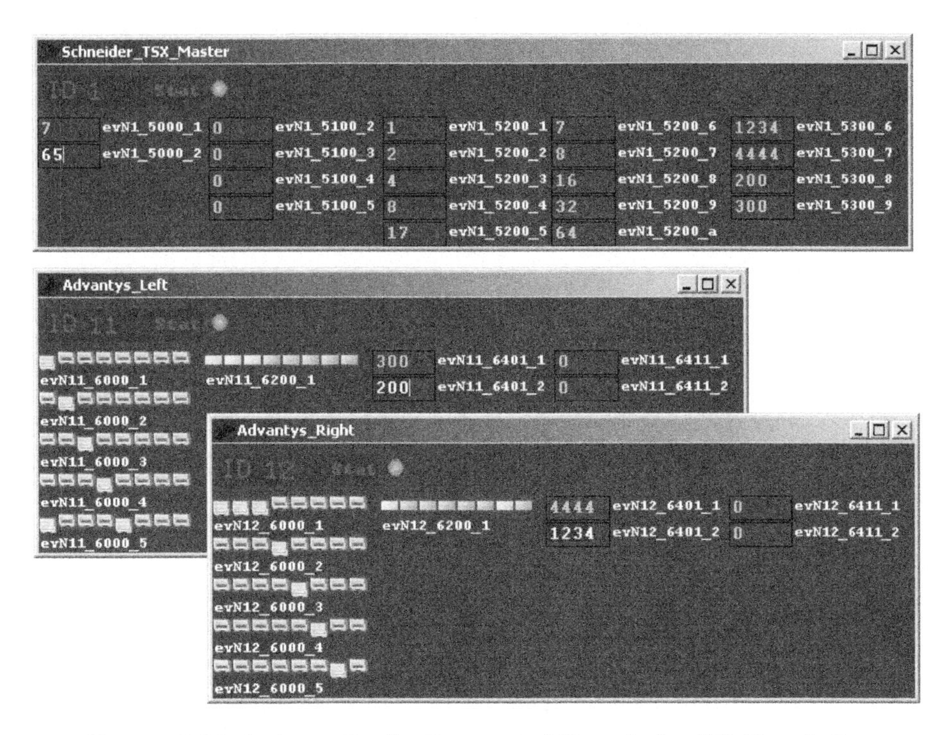

Figure 4.11 Automatically Generated Panels for I/O Simulation

The images in Figure 4.11 show the I/O panels for the nodes 1 (PLC), 11 (left Advantys system) and 12 (right Advantys system). The input data entered manually (for example the bits [6000h,01h-05h] of node 11) is transferred to the corresponding output data (here [5200h,01h-05h] of node 1).

Trace

Time	Chn	ID	Name	Dir	DLC	Data
0.055800	1	701	HBGuard_001	Tx	1	00
0.059520	1	70b	HBGuard_011	Tx	1	00
0.062920	1	70c	HBGuard_012	Tx	1	00
6.172770	1	77f	HBGuard_127	Rx	1	00
8.329680	1	6e0	DynSDOReq	Rx	0	
6.173770	1	00	NMTZeroMsg	Rx	2	01 00
91.551330	1	401	N12_T0	Tx	5	07 08 10 20 40
161.426510	1	501	N12_T1	Tx	4	5c 11 d2 04
108.141880	1	18b	N11_T0	Tx	5	01 02 04 08 11
Digital_In_No__5			17	[		11]
Digital_In_No__4			8	[		8]
Digital_In_No__3			4	[		4]
Digital_In_No__2			2	[		2]
Digital_In_No__1			1	[		1]
130.591970	1	28b	N11_T1	Tx	4	2c 01 c8 00
138.291000	1	20b	N1_T0	Tx	1	07
6.180810	1	30b	N1_T1	Tx	4	00 00 00 00
145.661850	1	381	N1_T2	Tx	1	41
6.182050	1	481	N1_T3	Tx	4	00 00 00 00

Figure 4.12 CANoe Trace Recording of Simulation

The trace window shown in Figure 4.12 lists the CAN messages that were generated so far. Each node generated its own boot-up message "HBGuard_xxx" with the data being 0. Node ID 127 is the default used by proCANopen which participates in any CANopen configuration as a true node of that system. The "DynSDOReq" was a dynamic SDO request of proCANopen to see if a CANopen SDO Manager is in the system. As there was no response, proCANopen took control and sent the "start all nodes" message "NMTZeroMsg." All following messages are TPDOs where the prefix of the message name indicates the Node ID number of the node transmitting the PDO and the suffix indicates the PDO number (starting at zero). Message "N11_T0" shows the same data that was visualized in the panels of Figure 4.11.

Because the CAN messages of this CANopen network are actually transmitted via the CAN interface, all network statistic analysis windows display the values for this network. These include information about the current and maximum bandwidth as well as statistics on how often which messages occur.

4.6 Network Commissioning

Objective

Whether the nodes of the network are simulated or physically existing on the network, the process for the final steps of the network configuration are the same. This section deals with these last steps of the configuration and how to apply them to a network.

4.6.1 Finalize the Configuration

In Section 4.4 some global settings and specifically the PDO mapping were already configured. What was not yet set up is some fine tuning concerning TPDO triggering methods, heartbeat or node guarding setup and error behavior. In addition, a method is needed to automate the configuration process.

The configuration of the nodes can be finalized using Vector's proCANopen and the "Device Access" functionality which allows read/write accesses to the Object Dictionary of each node. This can include things like using a different transmission type for specific PDOs and setting the heartbeat times used by individual nodes. Figure 4.13 below shows the "Device Access" window for the node 12, "Advantys Right." The Producer Heartbeat Time is in the process of being set to 300 milliseconds.

As an example, the TPDO communication parameters of the right and left Advantys node (ID 11 and 12) could be changed to use the SYNC signal instead of the default change-of-state transmission type. Whether the nodes are simulated by CANoe or physically present on the network, any changes made with proCANopen become immediately active and can be analyzed in the trace window of CANoe, CANalyzer or other analyzing tools.

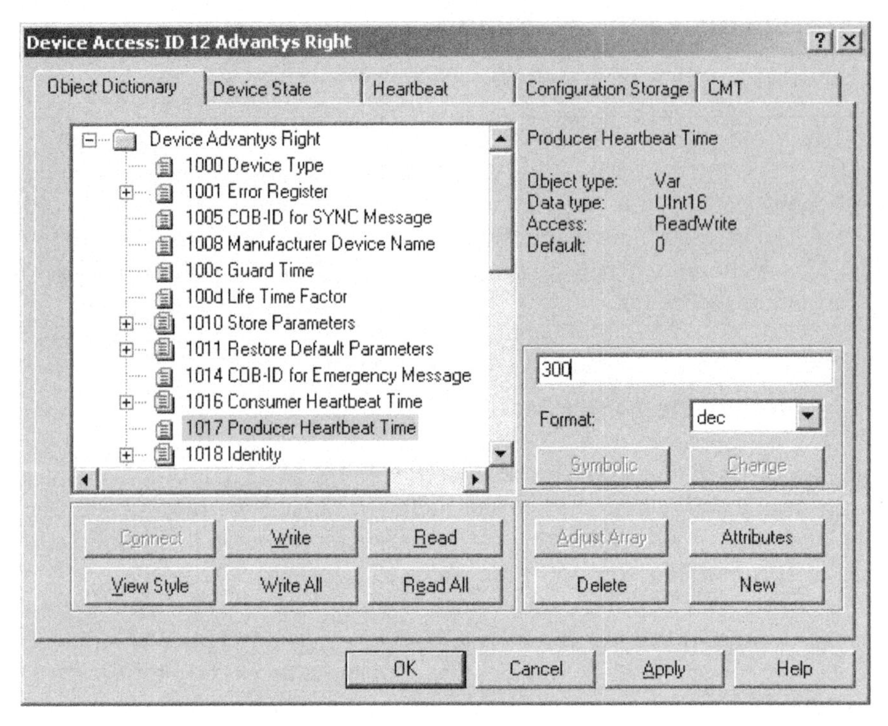

Figure 4.13 Setting the Heartbeat Producer Time of Node 12

The trace listing in Figure 4.14 uses relative timestamps, so the timestamp displayed is the time that expired since the last occurrence of a message with this message ID. The trace shows that the SYNC is sent about every 33 milliseconds (note that the actual value displayed here is 35.52 – the accuracy depends on the producer, in this case it comes from a regular PC) and the heartbeats of nodes 11 and 12 occur every 300 milliseconds.

Trace

Time	Chn	ID	Name	Dir	DLC	Data
0.035520	1	80	SYNC	Tx	0	
0.305800	1	70b	HBGuard_011	Tx	1	05
0.300280	1	70c	HBGuard_012	Tx	1	05
0.036300	1	18b	N11_T0	Tx	5	01 02 04 08 70
Digital_In_No__5		112		[	70]	
Digital_In_No__4		8		[	8]	
Digital_In_No__3		4		[	4]	
Digital_In_No__2		2		[	2]	
Digital_In_No__1		1		[	1]	
0.159780	1	28b	N11_T1	Tx	4	c8 00 b8 0b
Analog_In_No__2		3000		[	bb8]	
Analog_In_No__1		200		[	c8]	
0.034060	1	401	N12_T0	Tx	5	07 08 10 20 40
Digital_In_No__5		64		[	40]	
Digital_In_No__4		32		[	20]	
Digital_In_No__3		16		[	10]	
Digital_In_No__2		8		[	8]	
Digital_In_No__1		7		[	7]	
0.120700	1	501	N12_T1	Tx	4	d2 04 e0 15
Analog_In_No__2		5600		[	15e0]	
Analog_In_No__1		1234		[	4d2]	

Figure 4.14 Trace Listing of SYNC'd Communication

The timestamp can also be used to verify the triggering of each PDO. Here some other SYNC periods were chosen to add some diversity:

- N11_T0: Send with every SYNC
- N11_T1: Send on every 5th SYNC
- N12_T0: Send with every SYNC
- N12_T1: Send on every 4th SYNC

Once the network and all nodes have the desired configuration there are a number of methods to ensure that this configuration is used as the new default configuration right after startup of the network.

4.6.2 Downloading Configuration to Nodes

Vector's proCANopen stores any configuration change that is made to the Object Dictionary of a node into the corresponding DCF. This is true for all writes made to a live or simulated network as well for the changes made only in the input mask of a node's Object Dictionary.

At any time, the current configuration can be downloaded to the entire network or groups of nodes. This means that the configuration data stored in the DCF files is written to the appropriate Object Dictionary entries. This allows a quick configuration of an entire network or parts of a network after a power-up or reset cycle.

4.6.3 Storing the Current Network Configuration

In addition, proCANopen supports the setup of configuration management (CMT) for a CANopen Manager in the system. The entire network configuration can be stored in a concise format usable by CANopen Managers that are responsible for the configuration of the individual nodes.

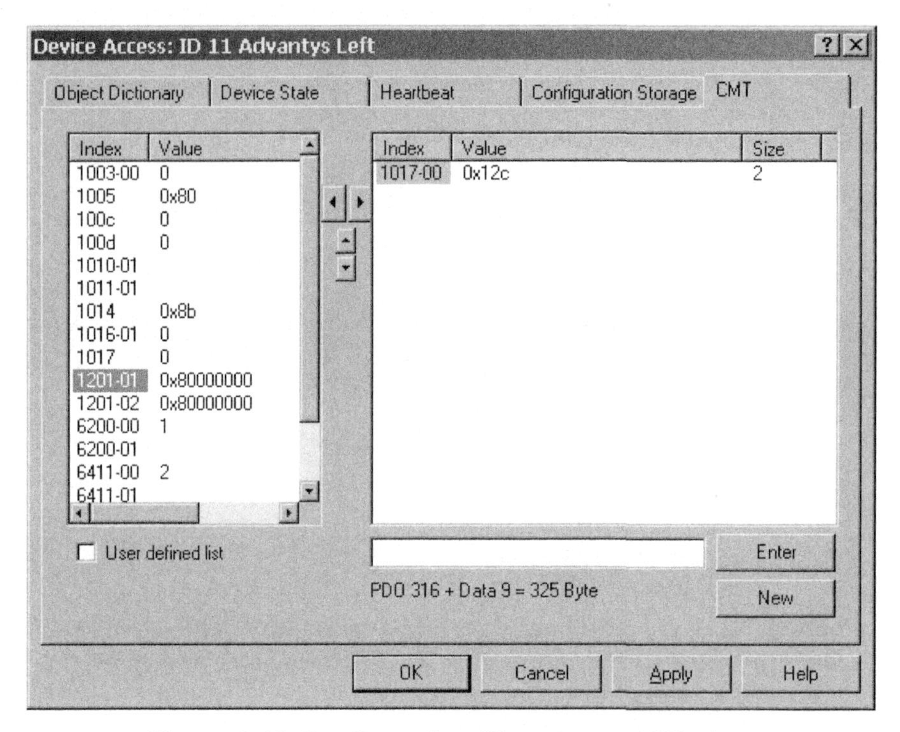

Figure 4.15 Configuration Management Window

The Object Dictionary entries used for these configuration cycles are either selected automatically or manually. For the manual selection each node has an individual list of selectable CMT parameters. In Figure 4.15 the list of CMT parameters is shown for the left Advantys node. The PDO parameters are always part of configuration management and are not listed. The only configurable parameter currently selected for CMT is the heartbeat producer time [1017h,00h]. It is set to 300 milliseconds. The total amount of configuration data (concise DCF format) is displayed at the bottom, currently 325 Bytes. This is the amount of storage needed in a CANopen Manager to store the configuration data for this node.

4.6.4 Alternatives with Store Parameters

If the individual nodes support the "Store Parameters" feature, an alternate configuration storage scenario is available. Each node can be instructed by proCANopen to save the current set of configurable parameters in its own non-volatile memory. This allows each node to store its own configuration and use it as the new default upon the next startup.

This feature can also be used if a Configuration Manager is not available in the network. However, it should be noted that without a Configuration Manager any new nodes added to the system (for example, repair replacements) must be configured before they are inserted into the system.

4.7 Advanced Features and Testing

Objective

This section summarizes extended features of the example setup. These features are not essential to the overall configuration, commissioning and maintenance cycle. However, they are useful for getting advanced and/or automated test results.

4.7.1 Advanced Node Simulation

So far only the communication between the nodes has been simulated, tested and analyzed. However, in CANoe it is possible to add algorithms to the simulated nodes, allowing them to produce certain data or interact with the data received.

The simulation of nodes is based on a programming language called CAPL (CAN Application Programming Language). Because the CAPL programs implementing the individual nodes are available in CANoe, they can be edited and enhanced to work with the data received or to be transmitted. An example of a CAPL program is shown in Figure 4.16.

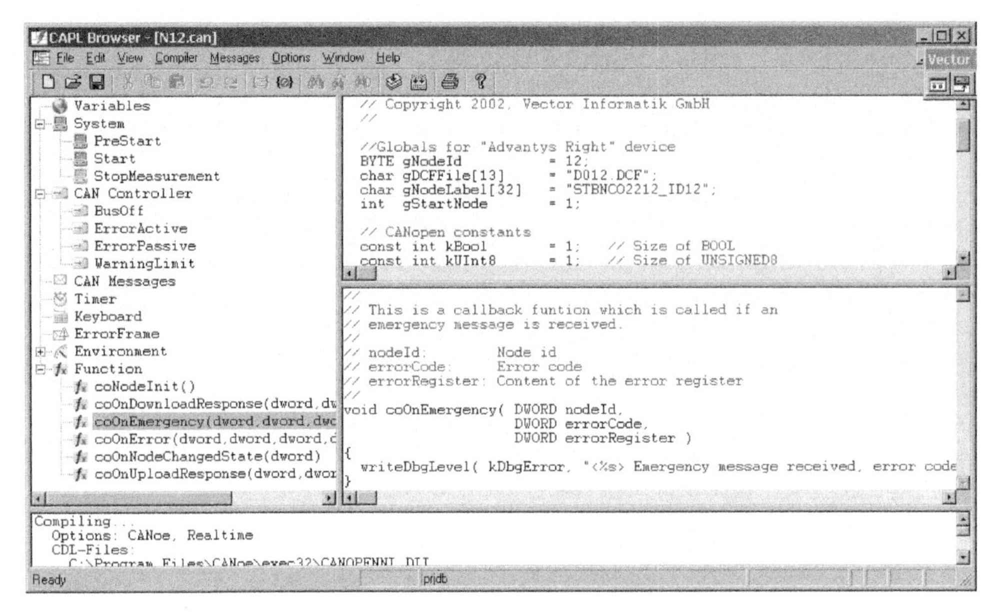

Figure 4.16 The CAPL Program for Node 12

4.7.2 Migration from Simulation to Physical Node

Another useful CANoe feature is that simulated nodes can be disabled individually. This allows for a variety of functions such as replacing the simulated nodes one-by-one with the physical nodes.

Another typical use is the test and configuration of a single node. Test and configuration would not need to be done on the live system, instead the CANoe simulation can be used with only that node disabled in the simulation that is currently connected physically and needs to be tested.

4.7.3 Advanced Panel Design

For applications that require extensive use of the I/O panels (for example if used for very thorough testing or as a test tool at the end of the production line), a more graphical representation of the data might be desirable.

The panels shown in Figure 4.11 are kept simple, as they are automatically generated. However, CANoe comes with a panel editor that allows the generation of customized I/O simulation panels based on standard graphical elements (dials and bars) and customized bitmaps. An advanced example for the panels is shown in Figure 4.17 where both the support feet and the ladder of a fire engine are animated.

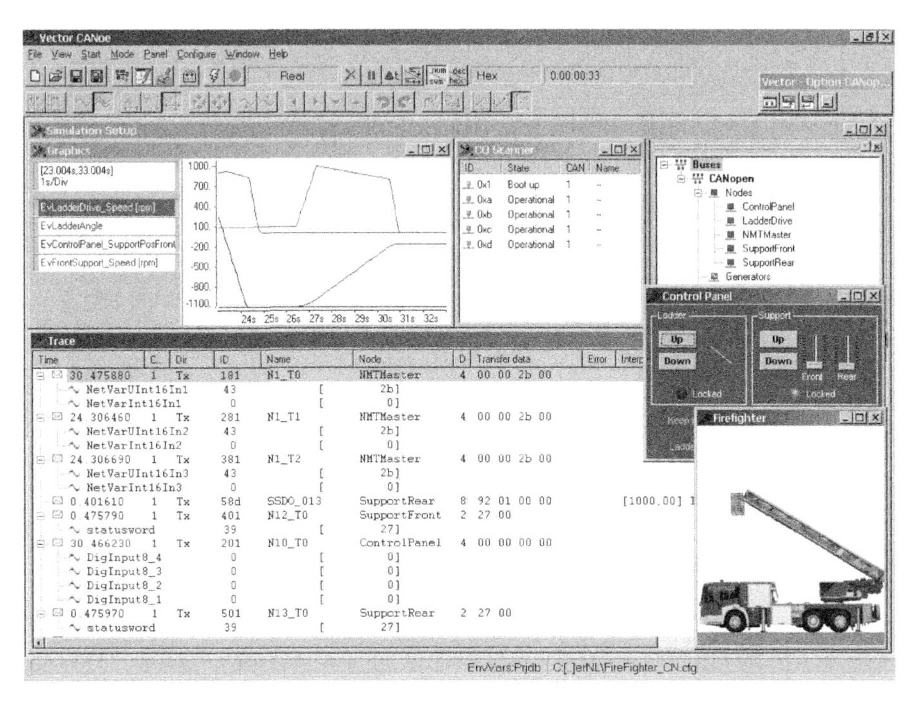

Figure 4.17 Example for an Advanced Panel Design

Part Two: CANopen Engineering

5 Underlying Technology: CAN

"The wireless telegraph is not difficult to understand.
The ordinary telegraph is like a very long cat.
You pull the tail in New York, and it meows in Los Angeles.
The wireless is the same, only without the cat."
Albert Einstein

CANopen was designed as a higher-layer protocol for CAN, operating with ISO 11898 compliant high-speed transceivers (line drivers). In general, CANopen is "open" enough that it can be operated with other CAN transceivers or even using completely different networking technologies. However, when choosing other transceivers or network technologies, one needs to understand that the result is a proprietary solution that is not compatible with anything else. Standard CANopen components only work if ISO11898 compliant high-speed transceivers are used.

About the Terminology
CAN – Controller Area Network – was invented in the late eighties. Since then some terms have been added and others replaced. As authors we had to make a choice on the terminology to use. Using the latest terms was one desire. However, in reality there are still many documents such as CAN controller datasheets, manuals and web pages using the established terminology, so just

ignoring the old terminology was not an option either. Here are the terms in question:

ISO 11898 vs. ISO 11898-X
This ISO standard specifies CAN physical layers. The original standard ISO11898 (which is still the only one published by ISO to date) only specifies one physical layer. However, a new version of the standard is in the works adding other physical layers, too. Assuming the current drafts are accepted, ISO 11898-2 will contain the specification for the high-speed transceivers as used by CANopen.

CAN 2.0A/B vs. Base Frame Format and Extended Frame Format
Many CAN related documents still make reference to CAN version 2.0A and 2.0B, which refers to the usage of 11-bit or 29-bit CAN message identifiers, respectively. The terminology used today is more intuitive. A CAN message frame with an 11-bit identifier is called base frame format, whereas the extended frame format corresponds to CAN message frames with a 29-bit identifier.

In order to truly understand CANopen and to be able to monitor, analyze and debug a CANopen system, some basic understanding about CAN is required. This chapter deals with CAN – the Controller Area Network – but with a focus on CANopen. CANopen does not use all of the features provided by CAN, and sometimes avoiding certain features is specifically recommended.

Features not used or not of direct concern (like synchronization mechanisms that shorten or lengthen individual bit times allowing all nodes to stay "in sync") are not covered in this book. Readers looking for additional details about CAN should consider reading [Etschberger01] or [Lawrenz97].

As an example of one of the unused features, the CiA recommends avoiding the use of the so-called extended frame format (which use 29-bit CAN message IDs instead of 11-bit IDs). Currently all CANopen functionality is entirely specified based on the 11-bit IDs of the base frame format. However, CANopen is open enough that it works with 29-bit identifiers, too.

So why shouldn't we use extended IDs?

A detailed answer is in the FAQ in Appendix A, which can be found in the Reference section. In summary, extended IDs "steal" from the data-bandwidth and are less secure because the checksum needs to cover additional bits.

5.1 CAN Overview

Objective

This is a brief summary of the most essential CAN features such as a high message rate per second, high reliability through extensive CRC and CRC checking by every node, typical physical media and arbitration by message priority. All of these features are explained in more detail in the following sections.

CAN was originally designed for automotive networks, where many small sensors need to report small values frequently. As a result, CAN features small message frames of only up to 8 data bytes but on the other hand can handle many message frames per second. At the highest bit rate of one Megabit per second, several thousands of messages could occur per second (see also Section 5.2.7).

The overhead per message includes an 11-bit message identifier and a 15-bit Cyclic Redundancy Checksum (CRC). A message can contain 50% overhead or more (doubling the length of a message) and makes CAN very secure and reliable, especially as the CRC is confirmed by *all* nodes. If a single node reports a CRC error, all other nodes discard the message and it automatically gets re-transmitted (see Section 5.2.9).

The typical physical medium is a twisted pair of wires and the maximum network length depends on the network speed chosen. At 1Mbps the maximum length is about 40m/120ft. Longer distances are achievable at lower speeds, for example about 500m/1500ft at a speed of 125kbps (see Section 5.2.1).

CAN is a multi-master network, so each node may send its data at any time. Collisions get resolved by priority. The message with the lowest message identifier wins the arbitration process and gets through. In order for this mechanism to work, all CAN message identifiers used in a network must be unique. Higher-layer protocols ensure the uniqueness of the CAN message IDs by assigning/reserving certain IDs for certain purposes (see also Section 5.2.8).

On the lowest level all message frames are broadcasts, meaning every single node receives every single message on the network. It is up to each individual node to decide if a particular message is needed by that node. To avoid a situation where each node must really examine every message on the network, most CAN controllers have filter techniques implemented in hardware (see Section 5.3).

5.2 An Introduction to CAN

Objective

This section explains the basic concepts of CAN. This includes physical signals, layout, speeds and limitations. After reading this section, you will know what kind of signals you will see when you hook up an oscilloscope to the CAN bus. In addition, you will understand how the error detection and re-transmission handling is implemented in CAN.

CAN is a very flexible communications network as it only implements parts of the physical layer and data link layer (see the end of Section 1.3.1 for more info on network layers and the 7-layer reference model). CAN may be implemented on different physical transmission media like twisted pair, power lines, optical and others. On top of the data link layer, a variety of higher-layer protocol standards are available, plus a countless number of in-house proprietary standards.

All major chip manufacturers provide microcontroller derivatives with CAN interfaces on-chip. The selection of an external transceiver determines the physical layer. All other CAN features described in this section are implemented on-chip as part of the CAN interface. This includes the entire error detection and message re-transmission mechanisms.

5.2.1 The Physical Layer based on ISO 11898

Although there are customized CANopen implementations using different physical implementations, the standard itself specifies the usage of "standard high-speed transceivers" in accordance with ISO11898. These transceivers (line drivers) are connected between the CAN controller and the physical medium: a pair of wires - preferably twisted with optional wires for shielding or additional customized signals.

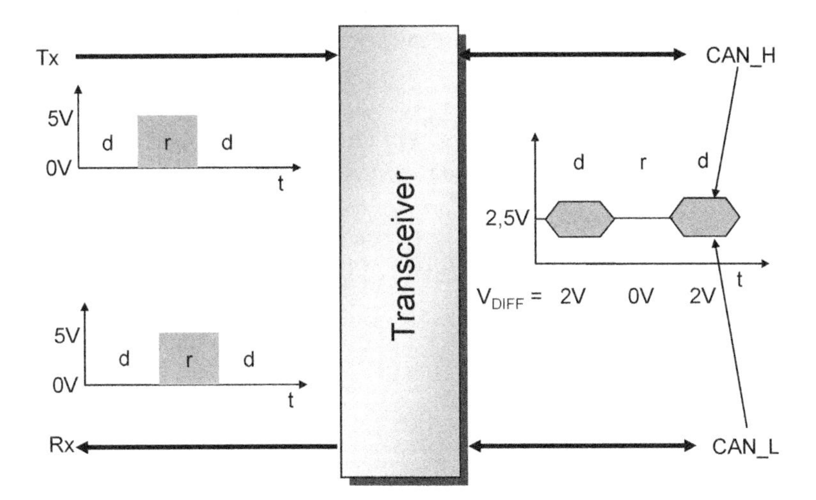

Figure 5.1 ISO 11898 Compliant High-speed Transceiver

As Figure 5.1 illustrates, the transceiver takes the TTL signal (on some parts this might be a 3V level) coming from the CAN controller's transmit pin and converts the signal to a differential signal between the two wires of the network cable (typically labeled CAN_L and CAN_H for low and high). In return, differential signals on the two wires of the network cable are converted back into TTL level and are fed back to the CAN controller's receive pin.

5.2.2 Signal States: Recessive versus Dominant

The letters "d" and "r" in Figure 5.1 stand for the so-called "dominant" and "recessive" signal states on CAN.

A logical 1 indicates the recessive state. It is represented by 5V (or 3V on 3V devices) on the TTL level and by a zero difference between the two network wires CAN_L and CAN_H. Both wires are at a level of about 2.5V.

A logical 0 indicates the dominant state. It is represented by close to 0 volt on the TTL level and by a 2 volt difference between the two network wires. CAN_L is driven about one volt lower and CAN_H is driven about 1 volt higher.

In CAN the dominant signal overwrites the recessive signal. If multiple nodes write to the network at the same time, the network will be in the dominant state if any single node writes the dominant signal. This mechanism is used to detect collisions: a node writing the recessive state and reading back the dominant state knows that there was a collision and can now start appropriate recovery actions.

Logically a CAN network behaves like a "wired AND gate." If any single node writes a dominant bit (zero), the entire network will be in the dominant state. Only if all nodes write a recessive bit (one) will the network be in the recessive state.

At this point it should be noted that the network only supports these two states. There is no third state for "idle." Extensive periods of the recessive state are used for "idling."

5.2.3 Signal Levels

The exact voltage levels used on the transmission wires are shown in Figure 5.2 below. In the recessive state, both CAN_L and CAN_H are at 2.5V. In the dominant state, CAN_L is pulled down by 1V and CAN_H is pulled up by 1V. So the levels are 1.5V for CAN_L and 3.5V for CAN_H with a 2V difference between them.

Many transceivers have some additional pins such as a "slope control" and a 2.5V "reference output." The slope control is only needed in applications that have hard limitations on the amount of electromagnetic interference they may produce. It allows softening the signal edges. However, if the CAN bus is used at higher bit rates, the bit time becomes so short that there is no room to play with the signal edges. Unless you really need this feature, we recommend not using the slope control and leaving the setting at "best signal quality."

The 2.5V reference output is not needed for the implementation of CAN or CANopen and can be ignored.

It would be nice to know how that discussion went when the designers of the transceiver had the choice: *What do we put on the extra pin that we have available?* and somebody probably came up with *We have a stable 2.5V level in the transceiver itself, why don't we put that out, maybe somebody else can use it?*

What a far-reaching decision that was: as a result, technical support engineers have one more frequently asked question to deal with that is heard over and over again: *What is that 2.5V reference for?* It would have been so much easier for technical support if the pin would have simply be left at NC – Not Connected.

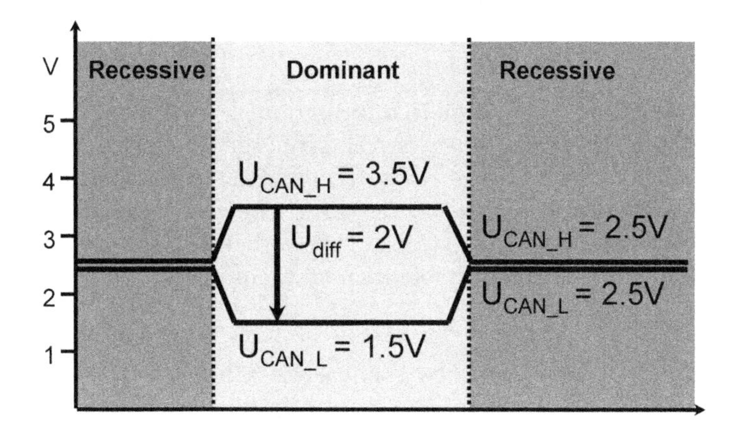

Figure 5.2 High-speed Signal Levels According to ISO-11898

The main benefit of a differential signal is its immunity to electromagnetic interference (EMI). If the signals are exposed to external EMI influence, that influence affects both wires as illustrated in Figure 5.3. In addition, the EMI generated by CAN itself is reduced by using a differential signal in wires close to each other.

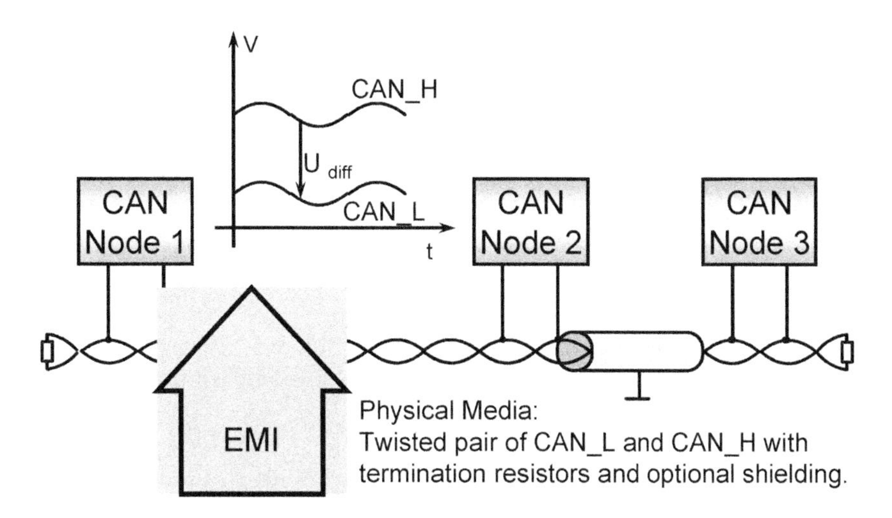

Figure 5.3 Using a Differential Signal

The figure shows how signals produced by node 1 can get altered during their transmission to nodes 2 and 3. However, the level of change is the same on CAN_L and

CAN_H and the transceivers of nodes 2 and 3 can still reliably detect the voltage difference between the two lines.

For those who are interested in more information about how CAN works we recommend the books [Lawrenz97] and [Etschberger01] that go into all the details. This section lists some details you will need to know when looking at low-level CAN signals and frames with oscilloscopes or logic analyzers. However, for CANopen development, monitoring and analysis, a CANopen monitor or analyzer is much more suitable than an oscilloscope!

Idling

A CAN bus can only be in one of two states: it is either recessive or dominant. So what about "idle" – does that mean we always have constant transmissions? This is true in a sense; by default every CAN controller constantly transmits the recessive state.

Between frames / messages, the bus defaults back to the recessive state, the logical 1. This is also called the "Inter-Frame Space."

Bit Coding

CAN uses NRZ (Non-Return to Zero) bit coding. This means that during an entire bit time the signal stays at the logical 0 or 1 level, without any edges or transitions within the bit time.

Bit Stuffing

Within data frames (messages), CAN uses a technique called bit stuffing. For synchronization purposes, edges are required within the communication data stream. To ensure that there are enough edges, the transmitter automatically inserts an opposite stuff bit into the communication stream after 5 consecutive bits of the same value.

The receiver counts consecutive bits received within a data frame and after 5 consecutive bits automatically removes (ignores) the following opposite stuff bit from the communication stream.

Note that the bit stuffing mechanism is only active within data frames and not during the inter-frame space (idling).

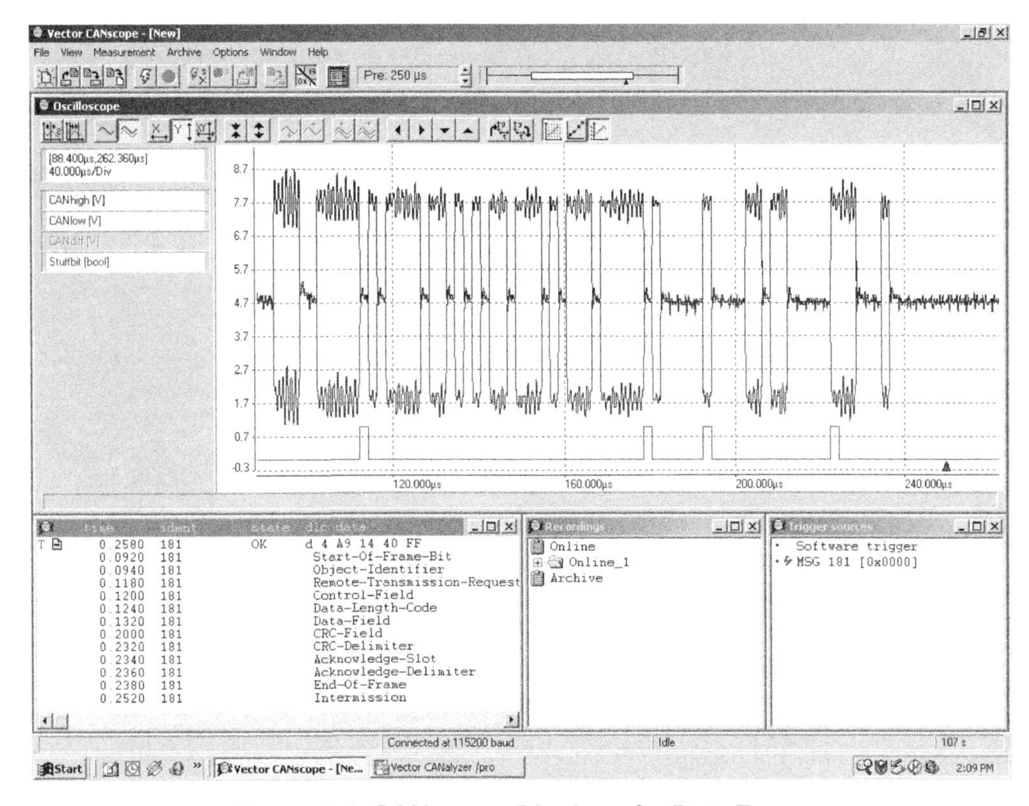

Figure 5.4 CANscope Display of a Data Frame

Figure 5.4 is from an oscilloscope-like tool with CAN awareness, the CANscope from Vector. Without the CAN awareness it would be very hard to recognize the different sections of the data frame displayed, especially the stuff bits. The message shown uses the identifier 181h (default ID of PDO1 from node 1) and has four data bytes: A9h, 14h, 40h and FFh. The bottom line indicates the four stuff bits of this message.

5.2.4 Wiring/Cabling

The differential signal used by ISO 11898 compliant transceivers already gives CAN a good level of EMI protection, and in some cases ordinary twisted-pair wiring without additional shielding can be used. However, for noisy environments using a shielded cable is still recommended.

The common perception is that CAN is a 2-wire network. However, an additional common ground is required for reliable operation. If the entire network is embedded in a machine or apparatus there typically is a common ground and it might not be necessary to actually use a third wire. If, on the other hand, the network spreads over a longer distance, the additional wire for ground should be part of the wiring.

Many applications use the same trunk of wiring for supplying the devices with power, which makes 4-wire CAN cabling one of the more popular variants. Two shielded twisted pairs are what many industrial automation applications use. One pair is used for the CAN signal and the other pair for the common ground and power supply.

[CiADRP3031] recommends the specific wiring, termination resistors and connectors to be used. In general, CAN is not very demanding in regards to the cabling, especially if only medium bit rates of 250kbs or below are used. Besides many variations of twisted pairs, there are applications that use flat ribbon cables, telephone cabling, PC serial cables, Ethernet cabling, Firewire cabling and others.

5.2.5 Connectors

Because CANopen is used in many very different applications, the wiring and the connectors are not part of the specification and are application specific. To allow a variety of connectors to still be compatible, the CiA recommends pin-outs for all major connector types typically used in CANopen environments. The recommendations are published in [CiADRP3031]: CiA Draft Recommendation Proposal 303-1.

The signals specified for connectors are:

- **CAN_L:**
  CAN_L is the bus line that is driven lower during the dominant bus state.
- **CAN_H:**
  CAN_H is the bus line that is driven higher during the dominant bus state.
- **CAN Ground:**
  This is the common ground used by the CAN nodes. If the nodes have a common ground anyway, this signal might not be needed.
- **CAN Shield (Optional):**
  An optional shield around the CAN_L and CAN_H signal is connected to this pin.
- **Positive Supply (Optional):**
  If a CAN node is supplied with its operating power via the cable, this pin

gets connected to the positive line of the supply power. The voltage levels are not specified. A commonly used voltage in industrial systems is 24VDC.

- **Ground (Optional):**
  This is an additional ground pin that in most applications will be identical to the CAN ground.

> In general, pins that are unused or reserved may be used for manufacturer specific purposes. However, you should keep in mind that the CiA reserves the right to change their recommendation for the reserved pins in the future.

5.2.5.1 9-Pin D-Sub

When using 9-pin D-Sub connectors, the male connector is expected to be on the device or network node. The female connector is used on the cable.

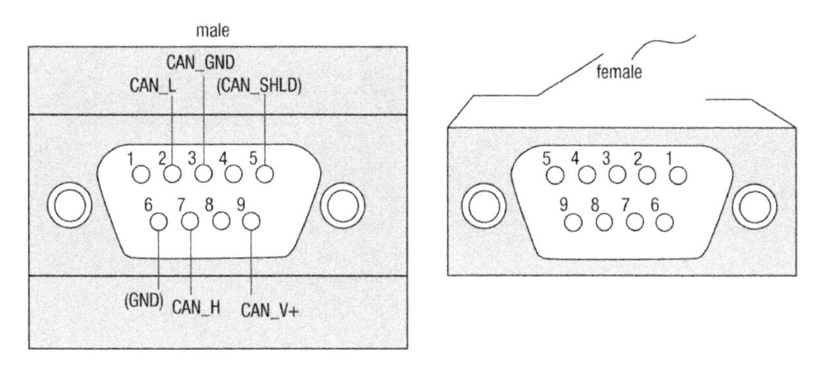

Figure 5.5 Pin Assignment for 9-pin D-Sub

Pin	Signal	Description
1	-	Reserved
2	CAN_L	CAN_L bus line (dominant low)
3	CAN_GND	CAN Ground
4	-	Reserved
5	(CAN_SHLD)	Optional CAN Shield

Table 5.1 Pin Assignment for 9-pin D-Sub

Pin	Signal	Description
6	(GND)	Optional Ground
7	CAN_H	CAN_H bus line (dominant high)
8	-	Reserved
9	(CAN_V+)	Optional CAN external positive supply (dedicated for supply of transceiver and opto-couplers, if galvanic isolation of the bus node applies)

Table 5.1 (Continued) Pin Assignment for 9-pin D-Sub

5.2.5.2 Muti-Pole or Dual Header Row

When connecting to a header row directly on a PCB, the layout is D-Sub compatible – meaning that if clamped 9-pin D-Sub connectors are used on a flat ribbon cable, it can be directly connected to a dual header row.

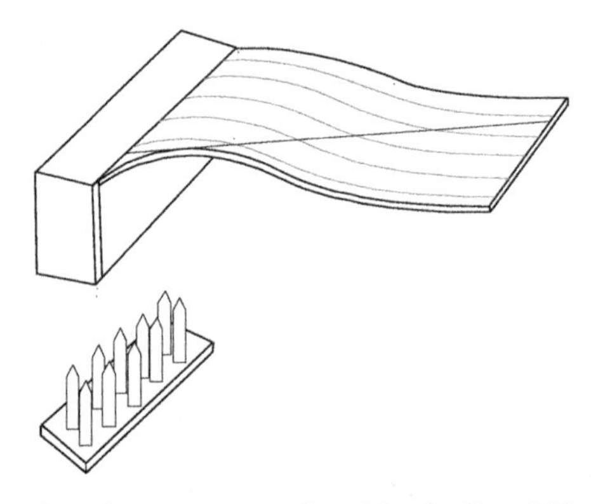

Figure 5.6 Pin Assignment for 10-pin Dual Header Row

Pin	Signal	Description
1	-	Reserved
2	(GND)	Optional Ground

Table 5.2 Pin Assignment for 10-pin Dual Header Row

Pin	Signal	Description
3	CAN_L	CAN_L bus line (dominant low)
4	CAN_H	CAN_H bus line (dominant high)
5	CAN_GND	CAN Ground
6	-	Reserved
7	-	Reserved
8	(CAN_V+)	Optional CAN external positive supply
9	-	Reserved
10	-	Reserved

Table 5.2 (Continued) Pin Assignment for 10-pin Dual Header Row

5.2.5.3 RJ10 – 4-pin

Some RJ10 sockets have a "connector inserted" detection built in. On devices with two sockets this can be used to implement auto-termination. As long as only one cable is inserted in any of the two sockets a termination resistor is connected. As soon as both sockets are used (two cables plugged in), the termination resistor is disconnected.

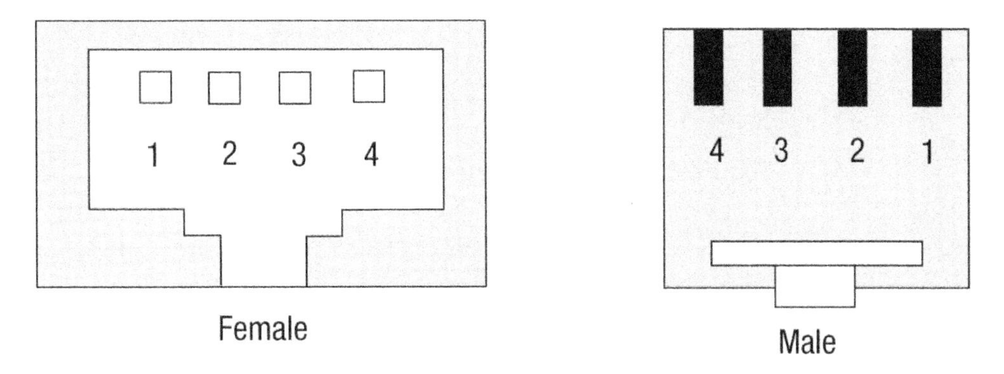

Figure 5.7 Pin Assignment for 4-pin RJ10

Pin	Signal	Description
1	(CAN_V+)	Optional CAN external positive supply (dedicated for supply of transceiver and optocouplers, if galvanic isolation of the bus node applies)
2	CAN_H	CAN_H bus line (dominant high)
3	CAN_L	CAN_L bus line (dominant low)
4	CAN_GND	Ground / 0 V / V-

Table 5.3 Pin Assignment for 4-pin RJ10

5.2.5.4 RJ45 – 8-pin

One of the benefits of RJ45 is that many cable configurations are readily available off-the-shelf as this connector is used by Ethernet. Therefore it is seldom necessary to manufacture customized cabling when using RJ45.

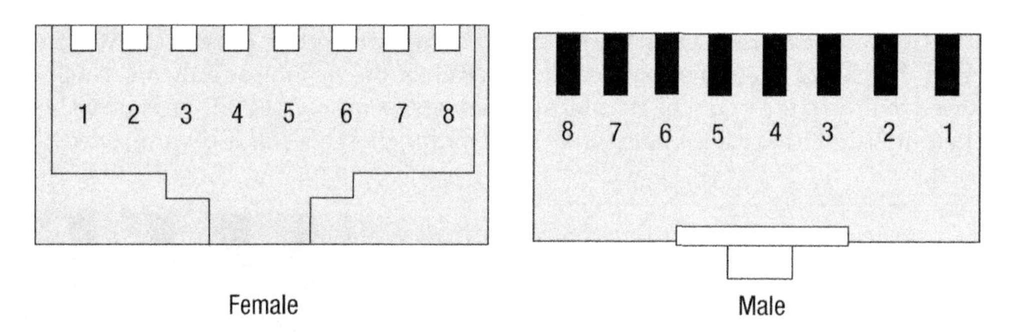

Figure 5.8 Pin Assignment for 8-pin RJ45

Pin	Signal	Description
1	CAN_H	CAN_H bus line (dominant high)
2	CAN_L	CAN_L bus line (dominant low)
3	CAN_GND	Ground / 0 V / V-

Table 5.4 Pin Assignment for 8-pin RJ45

Pin	Signal	Description
4	-	Reserved
5	-	Reserved
6	(CAN_SHLD)	Optional CAN Shield
7	CAN_GND	Ground / 0 V / V-
8	(CAN_V+)	Optional CAN external positive supply (dedicated for supply of transceiver and opto-couplers, if galvanic isolation of the bus node applies)

Table 5.4 (Continued) Pin Assignment for 8-pin RJ45

Additional recommended pin-outs for industrial connectors including "mini" and "micro" style connectors and for Firewire connectors are specified in [CiADRP3031].

5.2.6 Physical Layout

The physical layout of a CANopen network is that of a linear bus. The main trunk consisting of the CAN_L and CAN_H signals must have termination resistors at each end of the line. It is recommended that the termination resistors be 120 Ohm for buses running at a speed of 1Mbps. Slower and longer buses should use resistor values in the range of 150 Ohm to 300 Ohm.

If "junctions," "Y's" or "drops" are used, they may not exceed a maximum length. This length depends on the maximum speed used on the network. The higher the speed, the shorter the maximum drop length allowed. In a worst case scenario with a bit rate of 1Mbps the maximum drop length may not exceed 1 (one) foot.

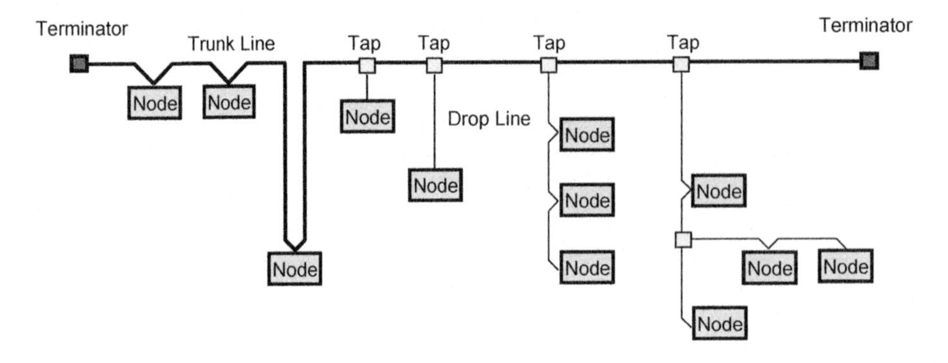

Figure 5.9 Physical Layout

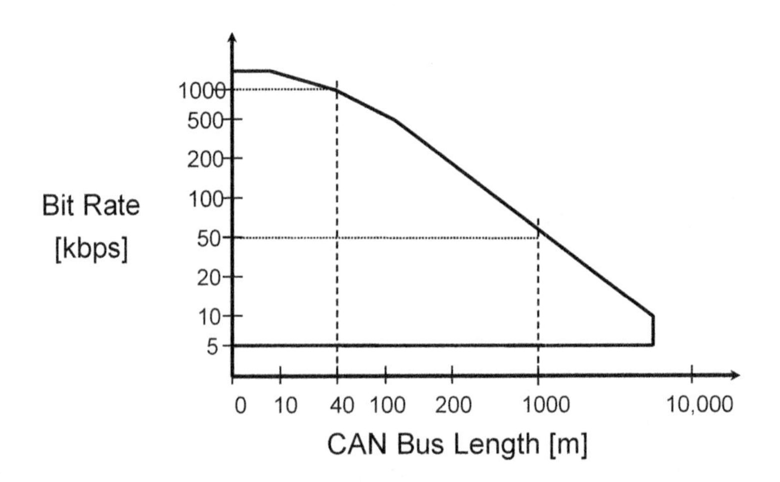

Figure 5.10 CANopen Bit Rate versus Bus Length

There are many factors involved when calculating maximum bus length and maximum drop length possible. Besides the conductivity factors of the cabling and connectors, one also needs to consider the number of nodes connected, the bus speed used, the delay time of the transceivers and the position of the sample point for read-backs (which specifies where in a bit time a node samples the bit for reading it – typical values are in the range of 70% to 87%).

The single most crucial value for the maximum bus length is the bus speed. Figure 5.10 shows how the bus speed influences the maximum bus length achievable. The physical background is quite simple. The way the arbitration and error detection system of CAN works, a single bit must be stable on the entire bus before the next bit can start. So as an estimate, the shortest bit time possible is the time it takes for a signal to travel from a node at one end of the bus to one on the other end – and back again. "And back" is required to ensure that any node still has a chance to overwrite the signal on the bus. So a node writing the recessive state could be overwritten by one writing the dominant state.

In summary, one bit time cannot be shorter than the time it requires for a signal to go through a transceiver onto the bus, roughly traveling at the speed of light to the other end, going through a transceiver again, and all the way back. At a bit rate of 1Mbps some 120 feet is about the best one can expect in terms of CAN bus length. If some of the other conditions are unfavorable, like lots of nodes connected to the bus or an early sample point for the read-back, the maximum length possible might not even reach 100 feet.

In many cases, a rough estimate for the maximum achievable bus length is sufficient. An estimate is shown in Figure 5.10.

5.2.7 The CAN Base Frame Format

Of the various messages/frames on CAN, the one used most often is the Data Frame containing process data. Another important frame is the Error Frame. Additional information about the frames listed here and other messages/frames such as remote frames and overload frames can be found in [Lawrenz97] and [Etschberger01].

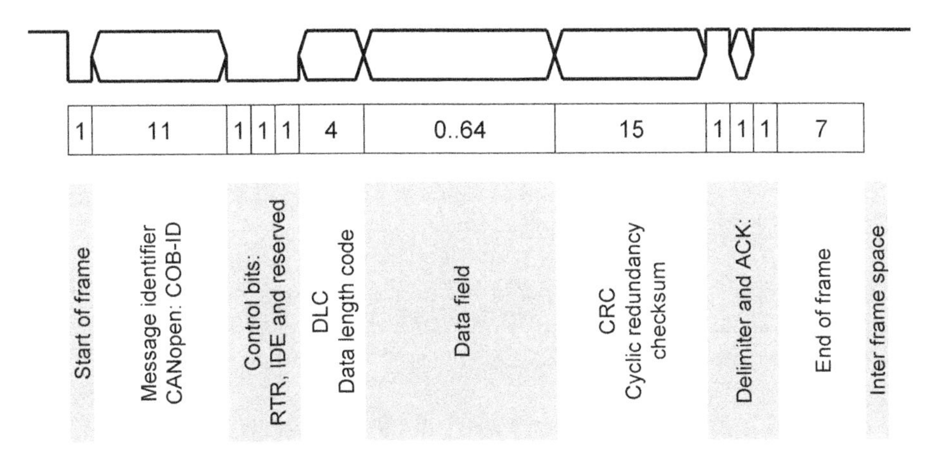

Figure 5.11 The CAN Base Frame Format

Consider Figure 5.11, reading from left to right. The CAN Data Frame begins with a dominant start bit. While idle, the bus is in the recessive state, so that any transition from idle to dominant is considered the start of a frame.

What follows is the 11-bit CAN message identifier. In CANopen this is usually part of the COB ID, the Connection Object ID. Because this field is part of the arbitration process (the resolving of collisions if multiple nodes transmit at the same time), it must be ensured that each message ID is unique in the network. This simply means that no CAN ID may be transmitted by more than one node at any time.

When using CANopen, the three control bits that follow should all be considered "reserved" and should be left at 0. The CiA recommends that developers not use RTR (remote request, used to poll a certain message) or IDE (used to enable a 29-bit identifier instead of 11-bit) in CANopen networks. See also the FAQ Appendix A for more information on RTR.

DLC stands for "Data Length Code" and specifies how many data bytes are in this frame. Although the DLC is a 4-bit value, the only values allowed for DLC are 0 through 8.

The data field contains as many data bytes as specified by the DLC. So the length in bits is either 0, 8, 16, 24, 32, 40, 48, 56 or 64. Right after the data field is the 15-bit CRC – Cyclic Redundancy Checksum.

The remaining control bits are the CRC delimiter, the ACK (acknowledgement) slot and the ACK delimiter. The delimiters are used to give all nodes some time to work on and react to the previous bits. Receiving nodes get one bit time (CRC delimiter) to compare the CRC calculated internally on the received data with the CRC received in the frame. They then have one bit time (ACK delimiter) to complete CRC calculation and one bit time (ACK slot) to acknowledge that they received the Data Frame. If the last delimiter is recessive, it confirms that all nodes received the frame and matched the CRC. See also Section 5.2.9.

The data frame ends with an end of frame sequence of 7 consecutive, recessive bits. A minimum period of 2 (new ISO 11898-1) or 3 (ISO 11898) recessive bits must follow as inter-frame space (idle between messages), before the next frame can begin.

Depending on the number of data bytes in a data frame, the length of a data frame varies from 44 to 108 bits. However, due to the bit stuffing the actual length of a data frame can be longer.

For more information on bandwidth and worst-case calculations such as the highest expected message rate (data frames per second) or the highest possible data bandwidth, see the FAQ Appendix A.

When the designers at Bosch who developed CAN had to decide how many data bytes would be allowed as minimum and maximum, several factors were evaluated.

One concern was that even the highest priority messages would have to wait until the message currently in progress was completed. So the maximum message length allowed directly determines the maximum delay even the highest priority message might have before it gets transmitted.

Another concern was the reliability of the checksum. The longer the message, the less effective the CRC.

The final compromise was to allow 0 to 8 data bytes, requiring a 4-bit DLC – Data Length Code. Some argue that a 4-bit DLC could also be used to allow a message length of up to 15 data bytes, However, that would entirely change the communication timing and reliability of CAN.

5.2.8 Collisions and Arbitration

The collision and arbitration process is one of the central features of CAN. It ensures that if a collision occurs (multiple nodes transmitting at the same time), they are resolved by priority and the highest priority message will get through. The entire process is implemented in such a way that no bandwidth is lost.

The process is similar to CSMA/CD (Carrier Sense Multiple Access with Collision Detection). "Carrier Sense" means that a node constantly "senses the carrier," listening to the communication on the network. It will not, however, interfere with a communication currently in progress. "Multiple Access" means multiple nodes have access to the carrier medium at the same time – if there is no communication on the network, multiple nodes may try to write to the network at the same time. The "Collision Detection" is the detection of a collision if indeed multiple nodes write at the same time.

The interesting part is what happens next. The traditional CSMA/CD as implemented on Ethernet would start a jamming sequence once a collision is detected. This way the communication is interrupted for everybody and aborted. All nodes will start over after a random delay. Unfortunately this process steals bandwidth, as nothing can be transmitted during the jamming sequence.

CAN uses a smarter approach. Instead of producing a jamming sequence, collisions are instantly resolved by priority. The message with the highest priority wins this arbitration and gets through. Not only does the highest priority message get through, the entire process is immediately repeated with all the messages that lost the arbitration cycle. So the message with the next highest priority is automatically transmitted next. Sometimes, this mechanism is called CA (Collision Avoidance) instead of CD (Collision Detection), arguing that collisions in the true sense do not happen in CAN. This entire process is implemented in hardware and does not require any software intervention.

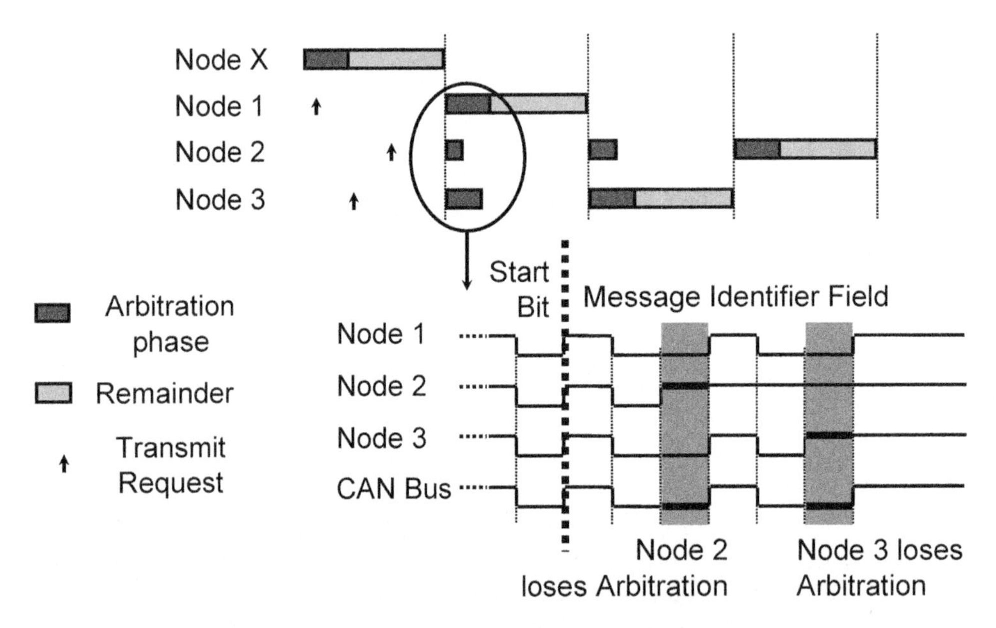

Figure 5.12 The CAN Arbitration Process

Figure 5.12 illustrates how the arbitration cycle of CAN works. The most likely scenario for a collision occurs when multiple modes internally get a request to send a frame while another transmission is currently in progress. In this case nodes 1, 2 and 3 get a request to send while node X is transmitting. Because nodes 1, 2 and 3 are constantly listening to the network, they know that the network is currently being used and they wait until node X completes its transmission.

Once node X completes its transmission, nodes 1, 2 and 3 will wait until the end of the minimum inter-frame space (three recessive bits) and then simultaneously start transmitting their messages. They all write the dominant start bit. Right after the start bit, the three nodes start the bit-by-bit arbitration cycle in which the 11-bit message identifier is used. The nodes write the message identifiers with the highest significant bit first.

There are three key points to the arbitration process that follows:

1. CAN message identifiers are unique in the network. An identifier is assigned to a node and only that node may transmit it.
2. Writing a 0 (the dominant state) overwrites a 1 (the recessive state).
3. Each node writing a bit also reads it back from the bus to confirm that transmission was successful, or if it was overwritten.

Because the identifiers are unique, there will be a collision somewhere within the 11-bit identifier field if multiple nodes try to transmit at the same time. Nodes that send a 1 and read back a 0 know that another node overwrote their 1 and that they lost the arbitration. They will step back from the bus and try again immediately after this frame ends. In this example, node 2 loses arbitration in the third bit of the identifier. It writes a 1 but reads back a 0, so it knows that it lost arbitration and will abort the current transmission. Instead, it will start receiving the message from the node which won the arbitration.

The algorithm started in each node with a request to transmit performs the following simplified steps:

1. Wait until the bus is available (wait until current message is completed).
2. Send the dominant start bit.
3. Send the next bit of the 11-bit CAN identifier.
4. Read back the bit from the bus lines.
5. If the bit from step 4 is different from the bit from step 3, receive the incoming message, then go back to step 1 and start over.
6. If number of bits arbitrated is not yet 11, go back to step 3.
7. Arbitration won – send data frame.

Although the arbitration process is one of the best features of CAN, it is also directly responsible for some of the limitations of CAN.

Because every node needs to be able to read back the bit that it just wrote, a bit must be stable on the entire bus before the next one can be transmitted. This limits CAN's maximum speed/distance compared to networks that operate as a pipeline – bits are pushed in as fast as possible without waiting. By the time the first bit reaches the end of the pipeline, several more bits might be already on the way.

The message identifier uniqueness requires that the 2048 (2^{11}) identifiers available need to be assigned to the network nodes to ensure that no single node uses an ID that is assigned to another node. All higher-layer protocols including CANopen implement some scheme to assign the identifiers. Most of these schemes require that each node be assigned a unique Node ID *before* it gets connected to the network. Typically nodes get pre-configured via some software setup tool that writes to non-volatile memory, or by switches or dials.

A true plug-and-play implementation would preferably not require such a setup from the user. In some applications, each setup step performed by the user is considered a "hazard" as there are always some users that will do wrong what they can do wrong.

There are some work-a-rounds for this problem, all with their own set of drawbacks. See the FAQ Appendix A for more information on possible work-a-rounds.

Note: In carefully controlled situations, multiple nodes may transmit the same message identifer to signal a request or condition, as long as the message does not contain any data. If more than one node attempts to transmit the message at the same time, a single message will appear on the bus as all nodes will think that they transmitted their message. It is then up to any listening node to determine what to do next, for example polling all nodes to see which one transmitted the message. An example of this is the Dynamic SDO Request message described in Section 3.2.2.

5.2.9 Error Detection Mechanisms

CAN has very sophisticated error handling implemented as part of the protocol. Most steps involved are implemented directly in the CAN controller hardware and are usually not influenced or controlled by the application program.

In low-level hardware, each node on the network actively monitors the network and checks the CRC of every message. Figure 5.13 illustrates a successful CRC comparison. In this case node 1 transmits a message and during transmission also calculates the cyclic redundancy checksum. The checksum covers all bits from the first bit of the message identifier up to the last bit of the last data byte transmitted. The calculated 15-bit checksum, in this case 4A2Fh, is transmitted right after the last data bit.

Any other node on the network which receives the data frame actively calculates the checksum. In this case node 2 receives the message, calculates the checksum and compares the one calculated with the one received via the network. If the two match, it pulls the acknowledgement bit to the dominant state as a confirmation that the reception process was completed successfully.

Because every node on the network performs this task, one might get the impression that the acknowledgement only confirms that at least one node received the message correctly. However, a look at the case for a mismatch illustrates that it actually means that *every* node on the network received it correctly, because if a single node detects a mismatch it will destroy the faulty message for *every* node.

It should be noted that the above is only true for nodes that are in error active mode. A node that is error passive or even "bus off" does not destroy messages with a CRC error.

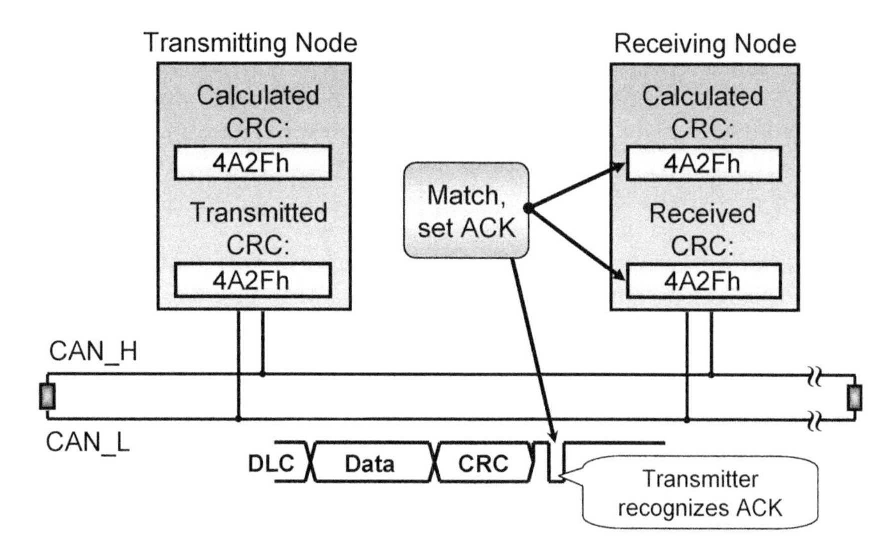

Figure 5.13 Cyclic Redundancy Check – Match

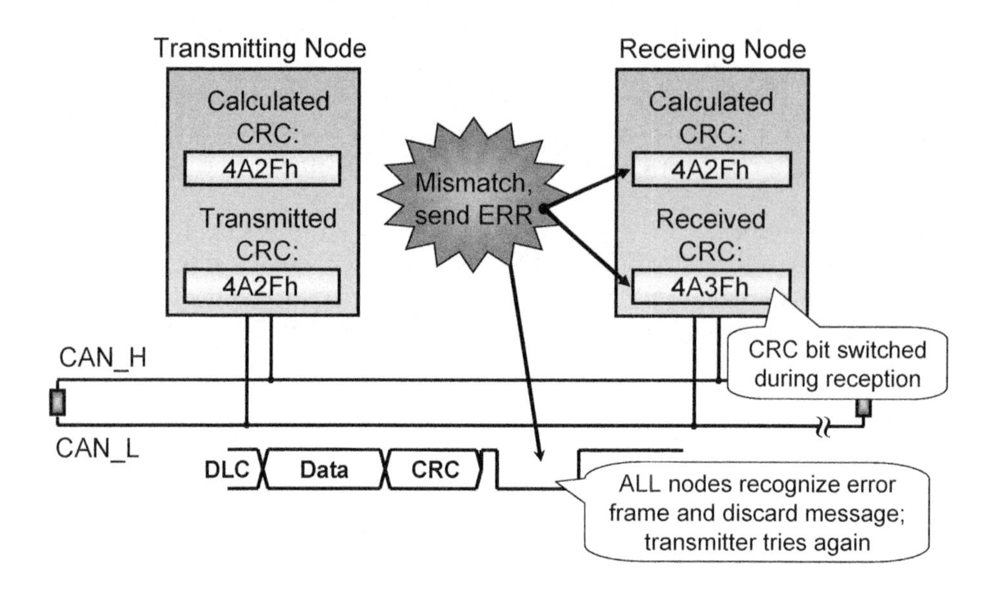

Figure 5.14 Cyclic Redundancy Check – Mismatch

In Figure 5.14 the scenario from Figure 5.13 is modified slightly. In this case, the CRC transmitted gets changed during reception.

> It should be noted that an error can also be caught by the transmitter, if it is a global change from a recessive bit (sending) to a dominant bit (detected by read-back) outside of the part of the message used for arbitration (mostly the identifier). In this case the transmitter will abort transmission and issue an Error Frame immediately.
>
> The Error Frame is a sequence of 6 or more consecutive dominant bits followed by an error delimiter of 8 recessive bits. Due to bit stuffing, there is no way that 6 consecutive dominant bits could occur during a regular data frame. An error frame detected within a data frame is an indication to all nodes that something is wrong with this data frame and that it should be discarded.

As a result of the change, a different checksum is received by node 2. Node 2 detects the mismatch and instead of setting the acknowledgement bit, it generates an Error Frame which is recognized by all other nodes on the network (including the one transmitting the message). Every node recognizes the error and discards the current message. After a specified timeout, the inter-frame space, the transmitting node will automatically re-try to transmit the message data frame again.

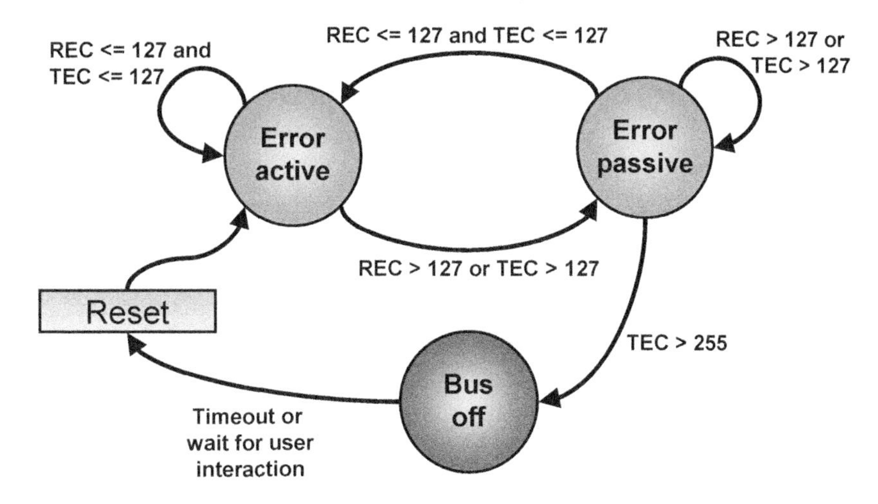

Figure 5.15 CAN Controller Error States and "Bus Off"

A question that typically arises at this point is *What happens if the check sum generator of a node is faulty*? It would destroy every communication attempt on the network.

To avoid such an error state, CAN nodes implement different error states. Figure 5.15 shows these error states and the transitions.

Per default all initialized nodes are "Error Active." In this state, a node actively performs all the previously described CRC comparisons as a result of a software initialization in the microcontrollers hosting the CAN controllers, typically executed after a power-up or reset.

There are two error counters, a receive error counter (REC) and a transmit error counter (TEC). These get incremented by a certain value with each error detected by the node. The more severe the error, the higher the increment value. However, they also get decremented with each message successfully received or transmitted.

If any of the two counters in a node reaches 127, the node goes into the "Error Passive" mode. In this mode, a node can still send and receive messages, but it does not actively destroy message frames on the bus. Because the error counters can get decremented, a self recovery from temporary network faults is possible.

If, on the other hand, the values of the error counters increment further and the transmit error counter overflows (greater than 255) the node goes into "Bus Off." Bus Off means that the CAN controller shuts down and stops transmitting or receiving CAN messages. Bus Off is a serious network error and self recovery is not possible. The Bus Off state can only be left by re-initializing the CAN controller.

How to treat a "Bus Off"

The Bus Off state is typically reported to the microcontroller by an error interrupt. What exactly happens in this interrupt service routine depends on the application software.

Because Bus Off is an indication of a serious network error, just re-initializing the CAN controller might not fix the problem. You should consider a controlled shut-down of the entire system if it is feasible.

However, if your application can not afford an immediate shut-down (like many automotive systems), the only thing you can do is to re-try. Reset the controllers involved and try again. To give other nodes a chance to catch-up or recover, the re-initialization should not be immediate (for example, by implementing an immediate reset upon Bus Off detection). Instead, there should be a minimal timeout before re-initializing. Typically the timeout should be in the hundreds of milliseconds.

5.2.10 The Safety of CAN: Error Statistics

The overall error statistics of a CAN network depend on several factors. As described earlier, any single node detecting an error will actively destroy the message for all nodes, so the more nodes that are connected and participating in the communication the more reliable the network becomes. Other factors include the level of electromagnetic interference along the network cable and the kind of shielding used. The differential signal provides more stable communication in noisy environments, but still CAN is not immune to interference.

CAN Error Statistics

It is often said that with statistics you can prove anything...

Nevertheless, the network reliability of CAN is very high. To those who are familiar with Cyclic Redundancy Checksums this should not come as a surprise. Just to give a comparison: on Ethernet protocols a 16 bit checksum is typically used to cover message blocks of up to 1,500 bytes. In CAN, a 15-bit checksum covers a maximum of 8 data bytes.

The guaranteed Hamming Distance for CAN is 6. The Hamming Distance is a measurement for the checksum reliability. In this case it means that up to 5 bit errors (bits randomly flipped in the data covered by the checksum and in the checksum itself) can be detected reliably.

One of the most stringent statistical requirements set by some automotive applications is fulfilled by CAN: if a network based on 250kbps operates for 2000 hours per year at an average busload of 25% an undetected error occurs only once per 1000 (one thousand) years.

An "undetected error" means that multiple bits in a message get distorted in such a way that the CRC does not detect it and a message with the wrong identifier or the wrong data contents gets through.

There are several research papers available online that examine the performance, reliability and vulnerability of CAN. See [Charzinski], [Nolte] and [Zuberi] in Appendix J. Interested readers should have no problem locating the papers using an online search engine.

5.3 Selecting a CAN Controller

"Just because something doesn't do what you planned it to do doesn't mean it's useless."
Thomas Edison

Objective

In this section we point out the main benefits and drawbacks of specific CAN implementations with a special focus on suitability for CANopen. Several of the major chip manufacturers producing CAN devices add their own twist to the CAN controllers by incorporating some additional hardware functionality. The main goals for these customizations are usually either to lower the burden on the host MCU (like filtering, buffering and/or queuing incoming messages) or providing extra safety and security (producing additional or more detailed error detection mechanisms).

The first CAN controllers implemented in the early eighties were the Intel 82526 and the Philips 82C200. Their features were quite different. The Philips 82C200 only provided a very basic set of communication functions, and thus was dubbed a "Basic CAN controller." The Intel 82526 (its successor 82527 is still used today) was referred to as a "Full CAN controller." Today there are so many variations of CAN controllers that the terms Basic or Full CAN often cannot be applied anymore and a more specific one-to-one comparison becomes necessary.

In general, all CAN controllers available today can be used to implement CANopen – even those that were designed with other protocol structures in mind.

5.3.1 Required Performance

The required communication performance, which of course depends on the specific application and implementation, is a crucial selection criterion. In any case, the worst-case scenario for CAN communication can be summarized as follows: The highest bit rate is 1Mbps. The longest possible message contains 8 data bytes. The shortest possible message (0 data bytes) takes about 50 bit times on the bus. At 1Mbps, 50 bit times correspond to 50 microseconds.

If the goal of an application is to handle CAN interrupts in real-time, the microcontroller would need to "digest" an incoming message with 8 data bytes in less than 50 microseconds. Potentially this is the shortest time the next receive interrupt could occur.

However, to leave enough MCU operating time for the real application (whatever is handled besides CAN communication), the "digesting" should take far less than 50 microseconds.

Experienced users of 8-bit microcontrollers will immediately see that such a worst-case scenario could be very challenging to some microcontrollers, and could easily keep them busy with nothing but CAN communication. However, it is seldom the case that a single node needs to receive and work on 100% of the messages on the bus. Typically a node only needs to listen to a certain percentage of the messages. While this helps to reduce the overall average MCU operating time required for handling the CAN communication, work-a-rounds are still needed to handle bursts of back-to-back messages that a node might need to receive.

Fortunately, modern CAN interfaces have hardware filtering and buffering features that help with the task of ignoring unwanted communication and buffering back-to-back messages.

5.3.2 Hardware Filtering with Match and/or Mask

The functionality of hardware filters is very similar on many CAN devices. While receiving a CAN message, the identifier (and sometimes even the data) can be compared to a configured filter. Only if the incoming message matches the filter does the message get stored into a receive buffer. The major differences in filters are usually the width of the filter, and whether or not it is a "match only" filter or also allows a mask to be used. A mask allows for the setting of individual bits as "don't cares," so a combination of match and mask registers can be used to select range filters, such as receiving all messages with an identifier in the range of 000h to 0FFh.

The filter width specifies how many bits of an incoming CAN message can be processed. For a standard CAN message identifier at least 11-bits are required. For an extended CAN message identifier it is 29-bits.

Where a "match only" filter looks for one exact match (for example, exactly one identifier), a combination of match and mask allows for filtering on message groups (for example, identifiers 100h to 11Fh). Usually, a bit set in the mask register means that the corresponding bit in the CAN message is a "don't care" value for the purposes of acceptance filtering. If a bit is cleared, it *must* match the value in the match register.

5.3.3 Different CAN Implementations

5.3.3.1 Traditional Basic CAN

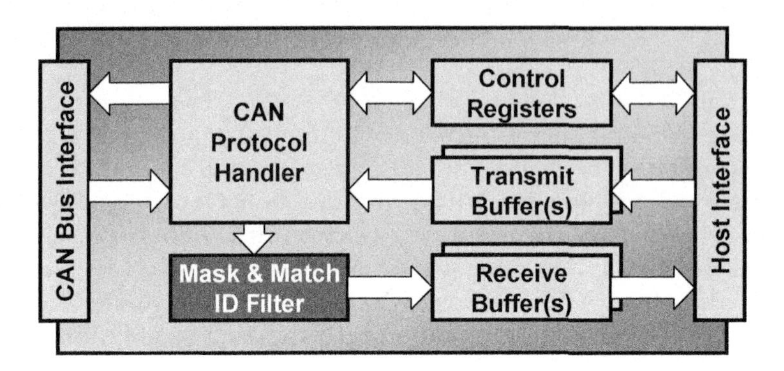

Figure 5.16 Block Diagram of a Basic CAN Controller

The first "Basic" CAN interface was implemented by the Philips 82C200. In comparison to the earlier Intel 82526 it only provided "basic" functionality. Basic CAN interfaces only offer a limited number of receive buffers and filters (typically one to three). If a node using such a controller needs to listen to a number of different messages (different CAN message identifiers), the filters usually have to be set "wide open" causing an interrupt with every single message on the bus. Obviously, the microcontroller will receive many CAN interrupts, as it has to check in software to see if a message can be ignored or needs to be worked on.

Today some CAN controllers have an "extended" Basic CAN interface that has additional buffers that can be used for either receive or transmit. However, using multiple buffers is the main idea of the Full CAN controller.

5.3.3.2 Traditional Full CAN

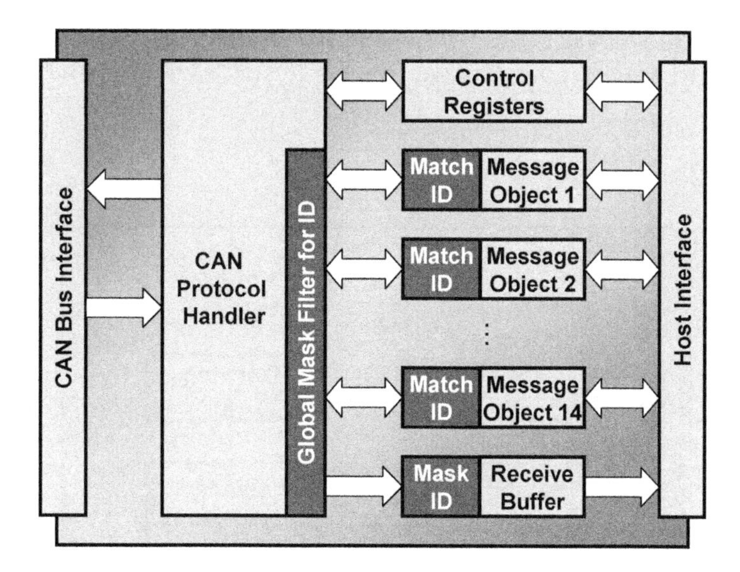

Figure 5.17 Block Diagram of a Full CAN controller

The very first CAN controller, the Intel 82526, used the so-called "Full-CAN" implementation. Full CAN controllers have a number of message objects (typically 15). Each message object is bi-directional (can be configured to either transmit or receive), each has its own transmit/receive buffer for one message, and each has one filter match register. This allows developers to set a message object to listen for exactly one message (one identifier).

As long as the total number of messages a node needs to listen to is smaller than the number of message objects available, these interfaces are very efficient. They will only cause an interrupt to the MCU if a "wanted" message is received. However, as soon as many different identifiers need to be received, there is no true benefit to a Full CAN interface over the Basic CAN interface. This is the case when implementing CANopen masters or managers that need to communicate with many or all nodes on the network.

In addition, the Full CAN mechanism does not offer any protection from a back-to-back worst-case scenario. Each message object has a single buffer and a matching incoming message will override the buffer's contents, so it is possible for messages to get lost. As long as a buffer is configured for a single message identifier, this scenario is not too problematic, as it is unlikely that the producer of that message will produce them back-to-back. But if any of the message objects are configured to receive multiple CAN identifiers (as required by most CANopen masters or managers), the microcontroller needs to be prepared for the possibility that these could come in back-to-back. On a 1Mbps CAN network that means about 50 microseconds from a receive interrupt occurrence to a potential overwrite of the message by the next incoming message.

5.3.3.3 FIFOs

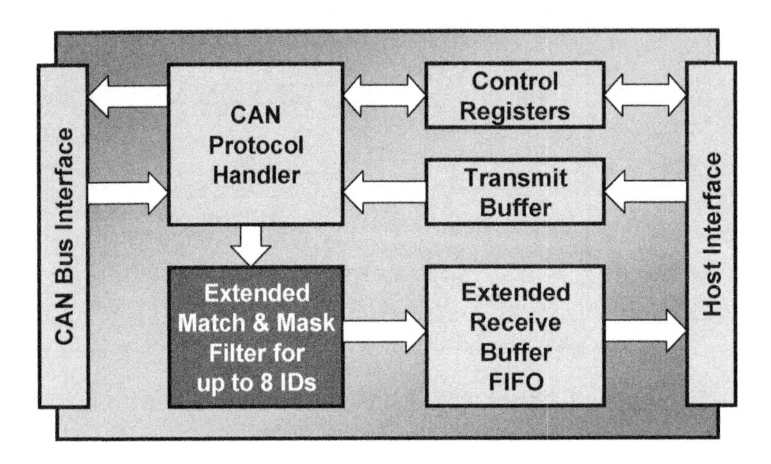

Figure 5.18 Block Diagram of Philips' PeliCAN (SJA1000, 87C591 and LPC99x)

The only way to get around the back-to-back message problem and the high performance and timing demands on the interrupt service routine is with a receive FIFO buffer (First In – First Out). A typical implementation features a number of filters that include both match and mask registers. Upon a filter match, the incoming message is moved into the FIFO buffer. An interrupt request to the MCU is made depending on configuration; either a certain fill-level is reached or a high priority filter received the last incoming message.

Even if such a FIFO can only hold 64 bytes it is still big enough to improve upon the back-to-back scenario mentioned earlier. If the FIFO is configured to cause an interrupt with every single incoming (matched) message, the MCU has at least 500 bit times until the FIFO will overflow. This is about 10 times more time available to the MCU than with Basic CAN or Full CAN implementations.

On the downside, messages in the FIFO cannot pass each other. So if the FIFO already contains several messages and an additional, but high priority message comes in, the MCU first needs to process all messages previously stored in the FIFO before it gets access to the high priority message. In a Full CAN interface it is up to the interrupt service routine to determine in what order the message objects are checked, and it may be possible for a higher priority message to pass previously received, lower priority messages.

5.3.3.4 Enhanced Full CAN with Receive FIFO

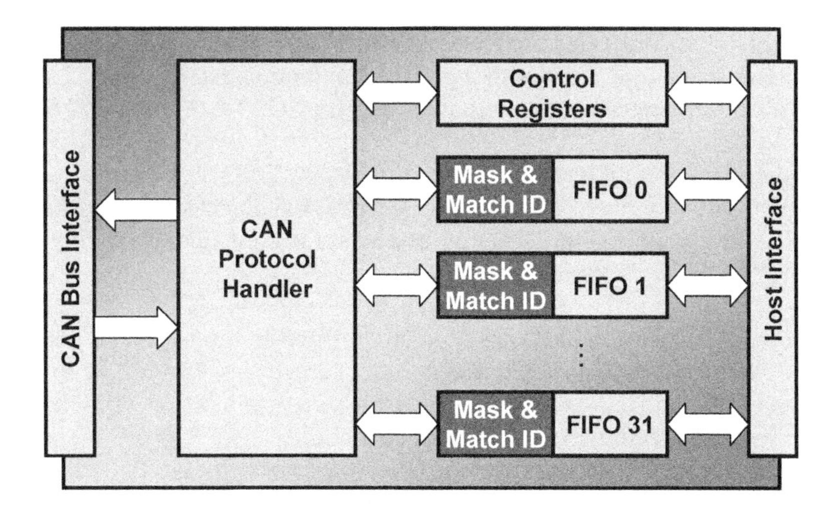

Figure 5.19 Philips' Enhanced CAN Interface of the XA-C37

The latest developments do not have standardized names because chip manufacturers have come up with their own customized improvements for the CAN interfaces. Several chip manufacturers now offer devices that combine the benefits of Full CAN and a FIFO.

The most powerful approach is a Full CAN implementation with a dedicated FIFO for each single message object. Although powerful, these are also the most complex controllers to configure, especially if each individual FIFO can be freely located in RAM and can be of individual lengths.

Another alternative is to be able to take a Full CAN implementation and concatenate message objects to a FIFO. So instead of one message object only having one buffer, a FIFO can be formed "borrowing" the buffers of other message objects. Although fairly flexible, the disadvantage is obvious - with each buffer added to a FIFO, one message object is lost. So the value of this feature increases with a high number of message objects, but decreases as the number of message objects decreases.

5.3.4 Physical Interfaces of CAN Controllers

A variety of interfaces have been implemented for the communication between the microcontroller and the CAN controller. The typical interface is "memory-mapped," meaning that the SFRs (the Special Function Registers) that control the CAN interface are mapped into the microcontroller's memory. Using memory read and write instructions the microcontroller can then access the registers; for example, those containing the bytes of a single CAN message. This method is used for both "stand-alone" (external to the microcontroller) and "on-chip" CAN controllers.

CAN controllers with many features also require many SFRs. For example a full-CAN style controller typically needs 16 registers per message object - four for the message ID, one for the length, up to eight data bytes and some control and/or filter mask registers.

On 8-bit microcontrollers the SFRs are typically located in internal data memory – which in the case of the 8051 is limited to a total of 256 bytes. Obviously there is not enough room in that address space to map all CAN registers into it, since that address space is also needed for non-CAN related registers.

Semiconductor manufacturers took different approaches to solve this problem on 8051 architectures. Some have chosen to place the CAN SFRs into the XDATA segment which is normally used for external memory. However, XDATA access is slower than internal memory access.

Philips Semiconductors chose a different path using just two SFRs: a selection or pointer register and a data register. By setting the selection/pointer register, the application can select which of 256 "hidden" CAN SFRs should be made available in the data register. A read or write instruction from or to the data register then performs a read or write to the selected CAN register. In addition, the read or write from or to the data register also auto-increments the selector/pointer. This allows the application to continuously read from or write to the data register if consecutive registers are to be accessed.

Sometimes a "stand-alone" CAN controller is equipped with another serial interface (I^2C or similar) towards the microcontroller. Before using one of these controllers, one should carefully examine the worst-case bus speed and bus load that could occur. The burden on the microcontroller can be quite high because the "regular" serial protocols require the microcontroller to react (typically via an interrupt) upon every byte transmitted. Using a memory-mapped CAN controller, the microcontroller only needs to react upon an entire message, so there are fewer interrupts that the microcontroller needs to deal with.

5.3.5 Code, Data Memory and CPU Performance Requirements

Another important selection criterion comes not from the CAN interface implemented on a specific microcontroller, but from the microcontroller or microprocessor itself. Some CANopen implementations can be quite demanding on memory and CPU performance requirements. Obviously CANopen masters/managers require more storage space for code and variables and more CPU power than average CANopen slave nodes. However, some CANopen slave nodes also might require more resources than an 8-bit microcontroller can provide.

While a specific implementation of a CAN controller can take workload from the microcontroller or microprocessor using it, the overall performance of a CANopen implementation also depends on the resources provided by the main processor. For example, a full-CAN style CAN controller can be setup to exclusively use one message object for each CAN message received or transmitted by a CANopen slave node (using the hardware filters provided, a message object can be assigned to one specific message ID). Typically the messages handled by a CANopen slave are one message object for receiving the NMT Master message, one for the heartbeat message produced, two for the SDO request and response channels and one for each PDO. For simple CANopen slaves this allows for a very efficient usage of the resources provided by the CAN controller.

However, if the nodes get more complex (increasing number of PDOs and/or monitoring of several heartbeats), it may be the case that the total number of CAN messages that need to be handled exceeds the number of message objects provided by the CAN controller. Once that point is reached the benefits of the "full-CAN" style CAN controller vanish. At this point one message object needs to be "opened" to receive *all* incoming messages (by removing all hardware filters). The result is that the CAN receive interrupt now gets far more messages to process and needs to run at a higher priority to deal with the additional network traffic. The interrupt service routine needs to look at the received message and decide in software if this message is needed by the local node or not. If it is needed, it needs to be removed from the CAN buffer before the next incoming CAN message overwrites the data. In order to achieve this, many implementations use additional memory buffers for the receive messages, requiring additional RAM.

In these cases the SJA1000-style CAN controllers from Philips Semiconductors are more suitable. They already have a receive buffer built-in, so the worst-case timing for a potential over-write of that buffer is much longer. As a general guideline, on an SJA1000 it takes about 10 times longer for a potential overwrite to occur, compared to a "Basic" CAN controller or a "Full" CAN controller with one message object set to receive everything.

When it comes to CPU performance requirements for CANopen implementations, it is unfortunately impossible to make exact comparisons as there are too many differences in the CAN controllers, the microcontroller architectures and the quality of the source codes and compilers.

The following is intended only as a rough estimate to give the reader some guidance as to whether or not the performance of the microcontroller chosen is sufficient, or if more detailed examinations have to be made.

To get an estimate of the CPU load required to handle CANopen, one should evaluate the CPU load available for handling the CAN communication (in comparison to how much CPU load is required for handling the application), the desired CAN bit rate, and the instruction execution rate of the microcontroller.

The question is *On average, how many instructions can the microcontroller use for work on the CANopen stack for each CAN bit time?* (i.e. InstructionsPerBitTime).

To get the number of InstructionsPerBitTime, one first calculates the CAN bit time. For example, at a bit rate of 1Mbps the bit time is 1 microsecond.

The next number needed is the number of instructions the microcontroller can execute during that time. A regular 8051 running at 12MHz would execute only one instruction per bit time. However a "6-clock" (double speed) part running at 24MHz would execute four.

Finally, it needs to be determined how much overall CPU load (as a percentage) is available for handling the CANopen communication, and this is applied to the number of instructions executed per CAN bit time. For example, if there are four instructions per bit time and the overall CPU load available for CANopen is 33% the result for "InstructionsPerBitTime" would be 1.333 instructions.

For 16-bit devices, multiply the result InstructionsPerBitTime by 1.5 and for 32-bit devices – don't even do this calculation, you should have enough performance.

If the resulting InstructionsPerBitTime is below 2, one would need to evaluate the entire system and scenario thoroughly and very carefully.

- If the resulting InstructionsPerBitTime is above 5, there should be ample performance for handling any CANopen implementation.
- For everything in between it might be necessary to guarantee a certain overall efficiency either by using an advanced CAN controller or by optimizing the CANopen software towards the specific microcontroller or a combination of both.

5.3.6 Controller Selection Summary

Selecting the CAN controller that is right for a particular application goes hand-in-hand with the selection of the microcontroller unit (MCU) used to run the application software.

Basic CAN controllers can dramatically increase the workload for the MCU and should only be considered for simple, minimal CANopen slave nodes in minimal networks. Because a lot of message filtering has to be done in software, the workload to handle the communication will greatly increase with the number of nodes connected to the network.

Full CAN controllers are ideal for CANopen slave nodes where the number of different CAN messages received does not exceed the number of "message object buffers" implemented in the controller. In this case, one message object can be configured to exclusively receive CAN messages with one particular CAN message identifier. If, however, the implemented node needs to receive more different CAN messages than message objects are available (for example, complex slaves or a CANopen master listening to many PDOs), the Full CAN interface does not have any advantage over the Basic CAN interface.

CAN controllers with one or multiple receive FIFO buffer(s) are suitable for any CANopen implementation, both slaves and masters or managers. Full CAN controllers that have the capacity to combine several message objects into a FIFO also fall into this category. CANopen nodes that need to receive many or all CAN messages on the network benefit greatly from these implementations, as these are the only ones that protect the application from high real-time demands. If a Basic or Full CAN controller is used to receive most or all CAN messages, the CAN interrupt service routine must often be implemented with the highest priority level. This is because after some 50 bit times on the bus (at 1Mbps about 50 microseconds) a data overrun could potentially occur. Using a hardware FIFO buffer, this time is multiplied by a factor of 10 or more, depending on the size of the FIFO.

5.4 CAN Development Tools

Objective

Unfortunately, there is no standard definition of what features a CAN monitor, CAN analyzer or CANopen configuration tool must have. As a result, there are many different products on the market with similar names but very different prices. It is not uncommon that a high-end tool costs ten times as much as a low-end tool.

In this section we do not recommend or compare any specific tools. Instead, we give the reader a list of functions that we found useful when using CAN monitors and analyzers. The reader can use this list when drawing comparisons between different commercial products.

There is a wide variety of development tools available that assist engineers in the development, debugging and testing process. When selecting tools like network monitors, analyzers, loggers, stimulators and simulators, one needs to evaluate what kind of functionality is available and which upgrade or options paths are available towards CANopen. In the same way an oscilloscope only offers limited visibility when looking at a CAN data frame, CAN monitors and analyzers lack functionality when the final application is CANopen.

The tool used most often is a network monitor or analyzer. These come in a very wide variety of both functionality and pricing. As with any development tool, additional functionality directly relates to development time saved. When calculating project budgets, the cost of the tool needs to be evaluated in comparison to the time needed for development, debugging and testing. It should also be considered that the tools can be re-used in future projects. However, engineering time spent on debugging is always a loss of time and money.

5.4.1 Functions Expected of a CAN Interface

5.4.1.1 Basic Functions

- Support a variety of CAN PC interfaces: ISA, PCI, PCMCIA/CardBus, "Dongle" for COM or LPT, USB and others
- 9-pin D-Sub male connector on CAN interface
- ISO-11898 compliant high-speed transceiver
- Support all CANopen bit rates: 1Mbps, 800kbps, 500kbps, 250kbps, 125kbps, 50kbps, 25kbps and 10kbps

5.4.1.2 Advanced Functions

- Able to use a variety of transceivers
- Support all CAN bit rates: from 1kbps to 1Mbps
- Support up to 100% busload on 1Mbps network
- Multiple CAN interfaces in one hardware
- Produce a high-resolution timestamp
- Able to generate Error Frames

5.4.2 Functions Expected of a CAN Monitor or Analyzer

5.4.2.1 Basic Functions

- Runs on all current Microsoft Windows operating systems
- Trace display (trace of all messages on the bus)
    - o Timestamp
    - o Chronological display or fixed position (new message with same ID overwrites previous display)
- Transmission of CAN messages
    - o Multiple messages configurable
    - o Transmit on key pressed
    - o Transmit in reply to a specific message received

5.4.2.2 Advanced Functions

- Symbolic display
    - o Replace identifiers with symbolic names
    - o Show single variables from CAN messages (for example display byte 2 and 3 of a CAN message as a word called "RPM")
    - o Support higher-layer protocols such as CANopen (knows and displays all the symbols known in CANopen)
    - o For CANopen support: recognizes EDS files, and can extract symbolic information from EDS files
- Trace display
    - o High-resolution timestamp
    - o Fully support "symbolic display" as described above
- Graphical Display
    - o Draws graphs showing how variables change over time
- Transmission of CAN messages
    - o Transmit periodically (every X milliseconds)
    - o Allow the data to be changed with each message sent
    - o Simulate specific nodes (script or program controlled reactions)
- Logging: record and replay messages
- Scripting language, usable to simulate nodes that are physically not yet available

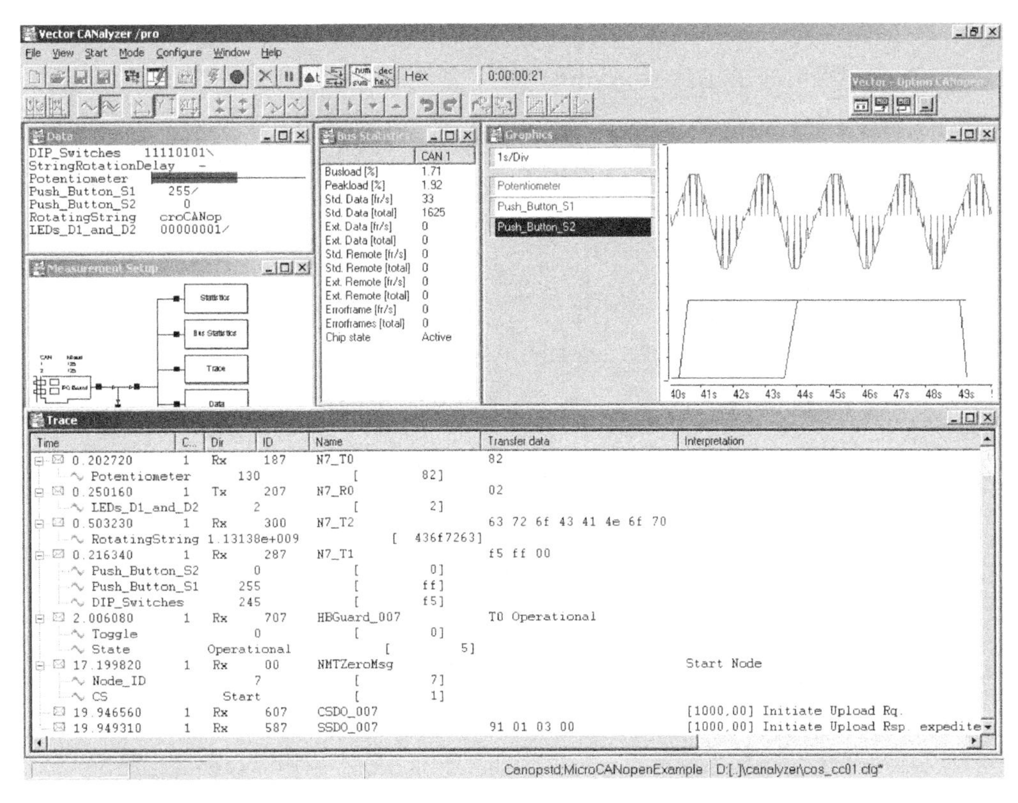

Figure 5.20 Screen Shot of Vector's CANalyzer with CANopen Option

5.4.3 Functions Expected of a CANopen Configuration Tool and Monitor

5.4.3.1 Basic Functions

- Runs on all current Microsoft Windows operating systems
- Send NMT message (start, stop nodes)
- Implements SDO write and read access
    - o Access any OD entry of any node on the network
    - o Support both expedited and segmented transfer
    - o Read/write from/to window or file
- CANopen-aware trace display

- Transmission of PDO or other CAN messages
    - o Multiple messages configurable
    - o Transmit on key pressed
    - o Transmit in reply to a specific message received

5.4.3.2 Advanced Functions

- Network Scan (find and display all nodes found)
- Implements SDO write and read access
    - o Support of block transfer mode
- Supports EDS files (extracts symbol information)
- Entire Network Configuration (versus node-oriented configuration)
    - o Uses dynamic mapping and linking to make point-to-point connections
    - o Graphical representation of network (graphic display of each node found)
- Can assume NMT and/or CANopen Manager functionality
    - o Produce SYNCs
    - o Monitor Heartbeats (or execute Node Guarding)
- Scripting language

6 Implementing CANopen

"I love deadlines. I love the whooshing sound they make as they fly by."

"A common mistake that people make when trying to design something completely foolproof is to underestimate the ingenuity of complete fools."

Douglas Adams

Objective

This chapter is for readers that need to develop and build a CANopen node. If you are integrating a CANopen network consisting of off-the-shelf products, you might want to skip this chapter.

When it comes to implementing CANopen, the "openness" of CANopen provides several benefits. However, there are also drawbacks involved.

In this chapter we show development engineers these benefits and drawbacks while comparing and contrasting different implementation approaches.

Making a decision on which of the available paths is the best for a specific product is a multiple-step process. In general, you will need to:

- Generate an overall communication layout
- Define the CANopen features required for each node
- Evaluate which implementation method fits best

For latest hints on implementation options and code examples visit this book's companion website at

www.CANopenBook.com

6.1 Communication Layout and Requirements

Objective

The first steps in any embedded network design should be to gather as much information as possible about the communication requirements. Only with that information can we make an estimate about the required bandwidth and response times.

Before one can start to make decisions about how to implement specific nodes of a CANopen network, it is essential to get an overview of the overall communication requirements. A reasonable understanding about the amount of communication that needs to be handled by individual nodes is required. This process also includes selecting the specific CANopen features that need to be implemented by each node. Are they following a specific device profile? And if yes, do all features of the device profile get implemented or can certain features be omitted?

To get a first impression of the required bandwidth, it is a good idea to start a table or spreadsheet with the maximum number of nodes and a list of all the process data variables that are produced by the nodes and transmitted in PDOs. For each process data variable produced, the table should have one line with the following columns: Node ID (or name), Name of the variable, data type and/or length of the variable and worst case transmission frequency. The transmission frequency should be in milliseconds. Typically it is the event time or inhibit time (whichever is smaller) to be used for the message that contains this variable.

An example is given in Figure 6.1. The column Ovr shows the number of bits of a PDO message that are not data bits, and the bps column indicates the bits per second. The overhead cannot be determined exactly, due to the bit stuffing done by CAN. However, 50 bits per message is an appropriate average for the purpose of getting an overview. For node 4 the overhead is only added once since the three variables Buttons_1 through Buttons_3 can all go into one PDO.

The formulas used are quite simple: "bps" is the total number of bits produced every second:

$$\text{bps} = \frac{1{,}000}{\text{Time (ms)}} \times (\text{Bytes} \times 8 + \text{Ovr})$$

The bandwidth used by an individual variable is:

$$\text{\% of total bandwidth} = \frac{\text{bps}}{\text{bus speed (kbps)}}$$

Produced Variables **Speed: 125 kbps**

Node	Variable	Bytes	Time (ms)	Ovr	bps	% of total
2	Temp_1	2	250	50	264	2.112%
3	RPM_1	2	25	50	2640	21.120%
4	Buttons_1	1	50	50	1160	9.280%
4	Buttons_2	1	50		160	1.280%
4	Buttons_3	1	50		160	1.280%

Total: 35.072%

Figure 6.1 Worksheet with Produced Process Variables

Having these values and formulas in a spreadsheet allows developers to quickly modify timing or speed values, or to add or remove variables and see what kind of impact the changes would have on total bandwidth usage.

The spreadsheet and several other examples can be downloaded from www.CANopenBook.com.

> How much bandwidth usage is acceptable?
>
> Please note that the given formulas are only rough calculations and do not account for all effects like bit stuffing or re-transmission of faulty messages. We also did not yet include heartbeats, node guarding and other potential messages.
>
> The bandwidth calculation method described here is good enough as long as your total bandwidth usage stays below 80%. If your usage is beyond 80% you should seriously consider choosing a higher bit rate for the CAN bus or reducing the amount of communication.
>
> If neither are possible, you would need to do a more detailed analysis of your worst case scenario. See publications such as [Lawrenz97] and [Etschberger01] for more hints on bandwidth calculations.

6.2 Comparison of Implementation Methods

> Objective
>
> In this section we introduce the major CANopen implementation options available and compare them with each other in regards to their individual benefits and drawbacks.

When it comes to implementing CANopen nodes, there are three primary implementation options available:

- Develop both hardware and software from scratch
- Develop the hardware from scratch and develop the software using a commercial CANopen stack (software library or source code)
- Design the hardware to use CANopen communication processors (peripheral chips or modules that implement CANopen)

None of the above can immediately be identified as the best method; the appropriateness of each depends on the application requirements. For some applications, portability to different microcontroller platforms might be important in order to build a variety of CANopen enabled products. Other applications might demand a certain performance, or require that a specific microcontroller be used.

More hints on selecting the appropriate method are listed in Section 6.7.

6.2.1 Develop Hardware and Software from Scratch

Although this route might sound tempting for many engineers it has several pitfalls, including the longest time for development, debug and test, as well as various "specification misinterpretation hazards."

However, the biggest pitfall is the incompleteness of the CANopen specifications, especially in regards to error behavior. On one hand, the specifications lack a detailed description of error behavior. For example, which errors should a CANopen compliant node report when a particular access sequence is wrongfully executed? On the other hand, this error behavior is checked by the CANopen conformance test.

As a result, developing a fully CANopen compliant software protocol stack is far more complex than a TCP/IP stack where good documentation and implementation examples are readily available. Yet many developers would consider buying a commercial TCP/IP stack implementation because they do not want to re-invent the wheel. In comparison, developing a CANopen stack from scratch is like re-inventing the wheel *and* the engine driving it. Before deciding on that path, engineers and managers should carefully evaluate all options.

However, the picture is quite different if 100% CANopen conformance is not really required. If all nodes of an application are designed and developed by the same engineering team and are never sold as "CANopen compliant off-the-shelf" products it is perfectly acceptable to deviate from the standard – indeed this is part of the openness of CANopen. The typical approach is to design the in-house CANopen nodes in a way that they have enough CANopen compliance to be able to communicate with other, third party, fully compliant CANopen nodes even if they are not 100% CANopen compliant themselves.

To assist engineers with the process of deciding if such an approach is feasible for their application, the authors developed MicroCANopen. MicroCANopen is a minimal CANopen implementation that can be downloaded for free from www.MicroCANopen.com. MicroCANopen is introduced in detail in Section 6.3.

MicroCANopen has some clear limitations that are a direct result of making the implementation "minimal" in terms of memory (both code and data) and CPU performance required:

- Object Dictionary entries are limited to 32-bit, no larger entries supported
- Entire CANopen configuration is "static": it is hard-coded and cannot change during operation
- Not all PDO triggering methods are supported
- Only the newer heartbeat is supported, not the original node guarding

6.2.2 Using Commercial CANopen Software

A very common approach to developing CANopen nodes is to buy CANopen software in the form of libraries or source code that implement CANopen. Such commercial solutions are available for a wide variety of microcontrollers and microprocessors from various companies. For a current listing of companies offering such products, see www.canopen.org and www.canopen.us.

There are several benefits to using commercial CANopen software. The vendor typically guarantees CANopen conformance of their stack and the examples delivered with the product. In addition, these codes are highly adaptable and portable, supporting a wide variety of microcontrollers.

Engineers using a commercial CANopen solution can get quick results on the CANopen side of the project and can thus concentrate on the application side. Indeed, experienced engineers and consultants can create the prototype of a new CANopen node within a week if it is based on a commercial software stack with which they are familiar.

Like all CANopen implementation paths, the commercial CANopen software solutions have some pitfalls, too. The primary development goal for most commercial solutions is portability. The manufacturers want the code to be executable on the widest variety of processors possible. That literally includes the high-end PC as well as the low-end 8-bit microcontroller. Obviously there must be performance drawbacks somewhere, as the price for portability is "software overhead" in the form of additional function interfaces or process queues.

Usually the 8-bit microcontrollers with both limited memory space and CPU performance are the ones that do not get "optimal" support from commercial CANopen software. Depending on the configuration, portable CANopen stacks for 8-bit micro-controllers tend to require at least 12 kbytes to 48 kbytes of code and 500 to 1000 bytes of RAM. However, this is more than some of the smallest microcontrollers with CAN interface have available on-chip.

Besides using additional off-chip resources, the only solutions for such devices are highly customized implementations that typically are less portable, but strongly optimized towards the microcontroller used. There are several companies offering consulting services specializing in such optimized CANopen implementations. On some processor architectures the gain can be a factor of 2 or 3 – an optimized CANopen implementation is about 2-3 times faster than the "generic portable" implementation and requires only 1/3 to 1/2 of the memory.

6.2.3 Using CANopen Processors or Modules

Several companies offer CANopen chips, co-processors, modules or dongles that implement the CANopen protocol "in hardware." Typically the protocol is not handled in hardware, but by a regular microcontroller that is pre-programmed with software to handle the CANopen communication.

Usually there are two types of such CANopen implementations; one that is a stand-alone CANopen I/O node by itself, and one that requires a host processor.

The first provides direct access to digital and analog inputs and outputs. One needs to simply design the chip or module into the hardware and directly connect the inputs and outputs to the chip or module. No further software development is required. Obviously this is one of the fastest implementation methods available, as it completely skips the software development process.

The second type, where a host processor is required, is more of a CANopen peripheral or communication coprocessor. It provides a communication channel to the host processor, like a serial interface or a shared, external memory area. The host processor and the CANopen coprocessor primarily exchange process data variables. When and how those are transmitted via CANopen is entirely handled by the coprocessor. This keeps the software interface between host and coprocessor minimal, as only a few commands need to be supported and the host does not need to know much about CANopen other than how to identify process data variables.

In summary, the main benefit of these hardware solutions is the need for little or no software development, resulting in the shortest development times.

One of the drawbacks is the limited flexibility. Because the developer does not have access to the code within the coprocessor, customized CANopen enhancements are difficult or impossible to implement. An application would need to stick with the exact features provided by the chip or module.

In addition, all "common sense hardware purchasing guidelines" apply: What's the cost and availability of the parts in the volumes required? What kind of long-term supply guarantees are there?

6.3 Simple Do-It-Yourself Implementation: MicroCANopen

"Beware of bugs in the above code; I have only proved it correct, not tried it."
Donald Knuth

> Objective
>
> For those engineers that are not yet ready to make the step towards a full-grown, higher-layer CANopen implementation, "MicroCANopen" might be an alternative. MicroCANopen is not an existing standard, just a concept which suggests that rather than shooting for full CANopen compliance it might be better to just adapt the basic ideas of CANopen to your own network layout to get a quick start. If the design requires more complex CANopen features (or full compliance) in the future, the system would not need to be re-invented, since the communication basics are already in place and compatible.

MicroCANopen was introduced by the Embedded Systems Academy as an "entry-level" alternative to CANopen for deeply embedded applications with limited resources. Code and data sizes required depend on the microcontroller used and functionality desired. On an 8051 with on-chip CAN interface MicroCANopen requires as little as 4 kbytes of code and some 200 bytes of RAM, compared to the 50+ kbytes of code and 1 kbyte of RAM for some "full-featured" CANopen implementations. The small size of MicroCANopen makes it especially suitable for some of the smallest CAN microcontrollers around, such as the Philips LPC99x microcontroller family. Table 6.1 shows a feature comparison between MicroCANopen and CANopen. Code examples for MicroCANopen are available at www.MicroCANopen.com.

	CANopen	**MicroCANopen**
CAN bit rates (in kbps)	10, 20, 50, 125, 250, 500, 800, 1000	10, 20, 50, 125, 250, 500, 800, 1000
Max. nodes per segment	127	127
Network Management	Originally designed to use a Master or Manager, but can operate without	Originally designed to operate *without* a Master or Manager, but can use one
Node guarding / heartbeat	Node guarding done by master or heartbeat monitoring by any node	All nodes produce heartbeat, can be monitored by communication partners
Configuration of nodes	Nodes can typically be configured via the network	Nodes are pre-configured, configuration cannot change during operation
Object Dictionary: ID entries	Available, optional with ASCII string	Available, 32-bit IDs only
Object Dictionary: Process data variables	Available, often with multiple access (8/16 bit)	Not available in OD, only in process data messages
Object Dictionary: Process data configuration	Available	Not available
Object Dictionary: Support of long variables	Supports variables and data fields of any length	A single OD entry may not be longer than 4 bytes
Mapping of multiple variables into one CAN message	Supports dynamic re-mapping of variables into CAN messages	One fixed, pre-configured mapping
Triggering methods of CAN messages with process data	Any combination of time-based, polled, change-of-state, synchronized or manufacturer specific; Inhibit time supported	Time-based and/or change-of-state only; Inhibit time supported

Table 6.1 Comparison of CANopen and MicroCANopen

6.3.1 Basic Concepts of MicroCANopen

To be able to implement a minimal CANopen-like system, a few basics and limitations need to be addressed. Please note that some of these are not limitations of CANopen, but limitations necessary to achieve this "minimal" version of CANopen.

6.3.1.1 Bit Rate/Bit Timing

All network nodes start up with the same CAN bus bit rate. The bit rate used in one system/application may be 10 kbps, 20 kbps, 50 kbps, 125 kbps, 250 kbps, 500 kbps, 800 kbps or 1 Mbps.

6.3.1.2 Node ID

Each network node has a unique Node ID, and this ID is in the range of 1 to 127, allowing for a total of 127 nodes in the system. This ID must be assigned and known to the node before it gets onto the live network.

6.3.1.3 Byte Ordering

In multi-byte variables the bytes are ordered by significance, the lowest significant byte coming first.

6.3.1.4 Process Variables

Any *single* variable shared via the network is 1, 2, 3 or 4 bytes in length. Note that as in CANopen, *multiple* variables can go into a single CAN message. For ease of use, MicroCANopen supports a process image where all process variables communicated via the network are stored in one array of bytes.

6.3.1.5 Network Management Master (NMT)

A typical CANopen network would expect the presence of a CANopen NMT Master to actually start and monitor the nodes. In deeply embedded applications where all nodes are pre-configured and know what they need to do, a master might not be required.

Beginning with DSP302 Version 3.21 [CiADSP302], CANopen provides a standardized method of operating without a master by allowing the slave nodes to start autonomously. In MicroCANopen it is assumed that this is the default operation mode: there is no master present and that all nodes automatically startup after power-up.

6.3.2 Functionality of a Single MicroCANopen Node

Each MicroCANopen node implements a minimal CANopen Object Dictionary. The Object Dictionary (OD) of a regular CANopen implementation holds all process variables that a node needs to receive or transmit. Each entry implemented has a unique 16-bit Index and 8-bit Subindex value that identifies one process variable in this CANopen node. Note: This is different from the CAN ID, which identifies a unique message on the bus.

In MicroCANopen, only the OD entries listed in Table 6.2 are implemented. The process data variables are not implemented in the OD, they are only available via the process data messages listed in Table 6.3 and Table 6.4.

Index	Subindex	Description
1000h	0	32-bit Device Type, typically set to represent generic I/O
1001h	0	8-bit Error Register
1018h	0	8-bit entry of 4 – number of Subindexes in this record
1018h	1	32-bit Vendor ID
1018h	2	32-bit Product Code
1018h	3	32-bit Revision Number
1018h	4	32-bit Serial Number (optional)

Table 6.2 CANopen Object Dictionary Entries Implemented by MicroCANopen

The main functionality of any CANopen implementation is to receive and transmit messages. MicroCANopen nodes produce and consume the following CAN messages:

- Upon startup, a MicroCANopen node transmits a boot-up message and continues to regularly transmit a heartbeat message in the specified heartbeat time interval. Other nodes can take this as an indication about the current status of this MicroCANopen node.
- Read accesses to the Object Dictionary are accepted and replied to. This allows standardized CANopen configuration tools or Network Management Masters to recognize a MicroCANopen node.

- Up to 4 separate transmit messages with process data can be triggered individually by a timer (every x milliseconds) or automatically by a detected change-of-state (COS) in the data to be transmitted.
- Up to 4 separate receive messages with process data can be received.

6.3.3 Assigning CAN Message Identifiers

MicroCANopen and CANopen both use CAN base frames (11-bit identifier field). The CAN message identifiers used in the system are assigned in accordance with the CANopen pre-defined connection set, which embeds the Node ID number into the identifier field. For transmitting data a MicroCANopen node uses the CAN IDs specified in Table 6.3. As an example, the node with ID 3 uses CAN ID 703h to transmit the boot-up message.

Table 6.4 shows the CAN IDs a MicroCANopen node listens for. The entries marked with "*" can be customized to ensure that a node directly listens to the process data message it needs to receive. So if we want the RPDO2 of node number 5 to directly consume TPDO1 of node number 8, the CAN ID listened to would need to be changed from 305h (default receive ID for RPDO2 of node 5) to 188h (transmit ID for TPDO1 of node 8).

CAN ID	Used for transmitting
080h + Node ID	Emergency Message (optional)
180h + Node ID	Transmit Process Data Message 1 (TPDO1)
280h + Node ID	Transmit Process Data Message 2 (TPDO2)
380h + Node ID	Transmit Process Data Message 3 (TPDO3)
480h + Node ID	Transmit Process Data Message 4 (TPDO4)
580h + Node ID	Service Data Response (SDO tx)
700h + Node ID	Boot-up message and heartbeat

Table 6.3 CAN Identifiers for Transmitting Data

CAN ID	Used for receiving
000h	Network Management Master Message (NMT)
* 200h + Node ID	Receive Process Data Message 1 (RPDO1)
* 300h + Node ID	Receive Process Data Message 2 (RPDO2)
* 400h + Node ID	Receive Process Data Message 3 (RPDO3)
* 500h + Node ID	Receive Process Data Message 4 (RPDO4)
600h + Node ID	Service Data Request (SDO rx)

Table 6.4 CAN Idenifiers for Receiving Data

A CANopen Network Management Master or generic configuration tool can access a single node by using the appropriate SDO "channel." A channel consists of two messages, one used for the SDO request from the configuration tool to the node and one used for the SDO response from the node back to the requester. Thus to access node number 3, a configuration tool would use CAN ID 603h and would expect a response coming back using CAN ID 583h.

Figure 6.2 shows a screenshot of a trace window with CANopen messages. The trace recoding was made with Vector's CANalyzer and shows the power-up cycle of a MicroCANopen node with the node ID 3. After transmitting the boot-up message, node 3 starts transmitting its heartbeat about every 2.5 seconds. Using a CANopen configuration tool such as Vector's CANsetter (see Figure 6.3) a read request is made to the Object Dictionary entry at Index 1018h, Subindex 1. That location contains the Vendor ID, in this case 00455341h.

After receipt of the NMT Master message "Start Node" the MicroCANopen node starts transmitting the process data messages with the CAN IDs 183h and 283h.

Trace

Time	Chn	Dir	ID	Node	Transfer data	Interpretation
6.617570	1	Rx	703	Node3	Boot-up	
9.116820	1	Rx	703	Node3	TO Preoperational	
11.615930	1	Rx	703	Node3	TO Preoperational	
12.566720	1	Rx	603	Node3		[1018,01] Initiate Upload Rq.
12.568670	1	Rx	583	Node3	41 53 45 00	[1018,01] Initiate Upload Rsp.
14.115160	1	Rx	703	Node3	TO Preoperational	
16.614410	1	Rx	703	Node3	TO Preoperational	
19.113500	1	Rx	703	Node3	TO Preoperational	
20.113020	1	Rx	00	Node3		Start Node
20.114700	1	Rx	183		ff 00 00 00	
20.115510	1	Rx	283		00 d1 00 00	
20.216920	1	Rx	183		ff 00 00 00	

Figure 6.2 Vector CANalyzer Trace Recording of Power-up Cycle

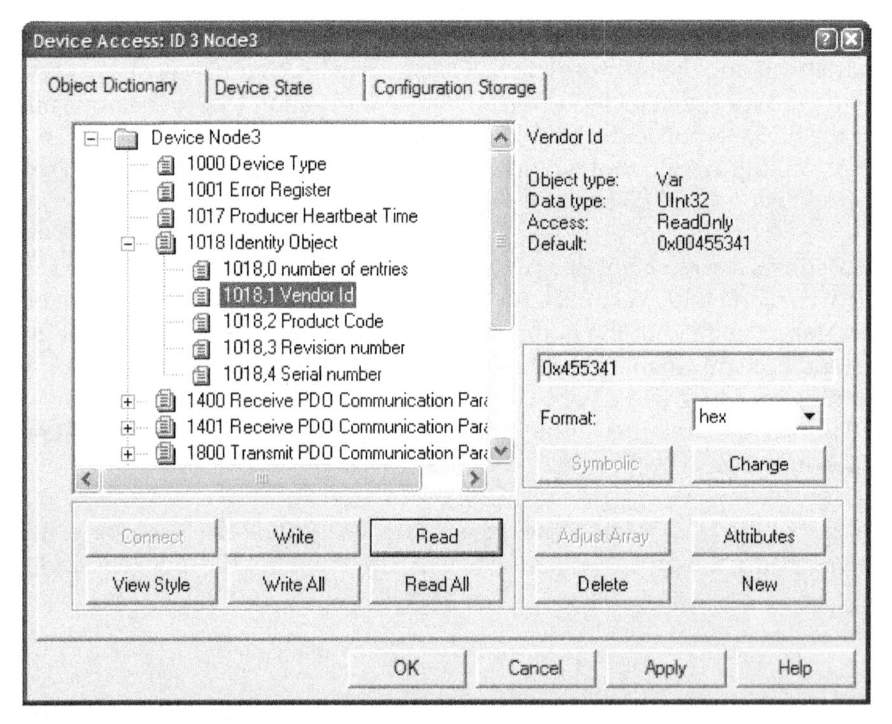

Figure 6.3 Reading Object Dictionary Entry [1018h,01h] (Vendor ID) from Node 3

6.3.4 Message Contents

Now that the CAN identifiers are assigned, it is time to examine the required message contents. The process data messages (PDOs or Process Data Objects), are the easiest.

They may simply be filled with one or more variables (as mentioned before, only use 1, 2, 3 or 4 byte variables and ensure the byte ordering for multiple byte variables). Figure 6.4 illustrates an example of how the 8 bytes of a CAN message could be used in a PDO; bytes 1 and 2 contain the 8-bit variables A and B. Bytes 3 and 4 are used for the 16-bit variable C and bytes 5 and 6 are used for variable D. Bytes 7 and 8 remain unused and are not transmitted.

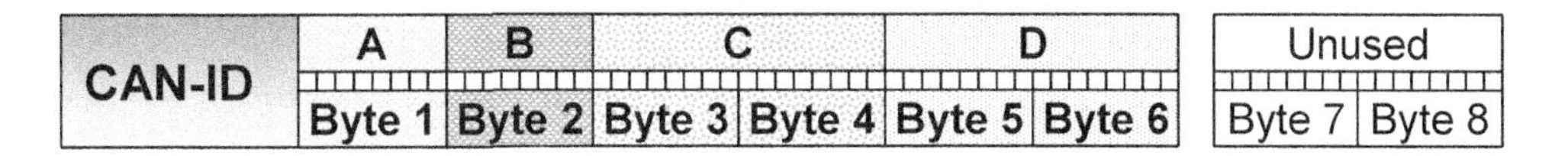

Figure 6.4 Example for PDO Mapping of Variables

6.3.5 Message Triggering

In CANopen there are several conditions that can trigger the transmission of a message. For a minimal implementation we will focus on the following triggering methods:

6.3.5.1 Boot-up and Heartbeat Message

As soon as the internal initialization is completed, MicroCANopen nodes transmit their boot-up message. If the node is configured to auto-start, it is followed by a continuous heartbeat. The heartbeat frequency is configurable and is typically in the area of hundreds to thousands of milliseconds, depending on application requirements. The timer resolution is a multiple of milliseconds.

6.3.5.2 Service Data

A master or configuration tool initiates any SDO communication. The MicroCANopen nodes may not trigger SDO communication by themselves, only in response to a request. Any SDO request directed at a node requires a response, which can either be the correct SDO response or an appropriate abort message.

6.3.5.3 Process Data

CANopen supports several triggering mechanisms for PDO messages containing the process data. Of the available mechanisms MicroCANopen nodes support event timer transmission and event change (COS, or change-of-state) transmission. In event timer mode a PDO is transmitted every *n* milliseconds. In event change mode the PDO is transmitted whenever a change-of-state is detected in any of the data to be transmitted.

To avoid situations where frequently changing data continuously triggers messages, the MicroCANopen nodes support the "inhibit time" feature of CANopen. This timer prohibits event change PDOs from being transmitted back-to-back. With each transmit an "inhibit timer" is started and the next transmission will not occur until the timer expires.

6.3.6 Implementing MicroCANopen

In MicroCANopen there are two software communication interfaces to deal with. To the bottom, an interface to a CAN driver is needed that provides some minimal functionality to receive and transmit CAN messages. To the top, a user's interface for the application is needed.

6.3.6.1 The MicroCANopen hardware driver interface

Portability to different microcontrollers is not easy to achieve due to the major differences in the implementations of CAN controllers. For this implementation we simply assume that the drivers are taking care of all real-time issues including the CAN interrupt service routine and providing both a receive and a transmit queue. In order to avoid making the CANopen communication a high priority within the embedded system, MicroCANopen is implemented as a background task that can be called within the main while(1) loop.

Listing 6.1 shows the "mcohw.h" file specifying the driver functions required by MicroCANopen. The first functions are needed for initialization of the hardware. In addition, the driver must implement some sort of receive and transmit queue for messages. These queues are accessed by the "PullMessage" and "PushMessage" functions.

```
/**************************************************************************
MODULE:    MCOHW
CONTAINS:  Hardware driver specification for MicroCANopen implementation
           The specific implementations are named mcohwXXX.c, where
           XXX represents the CAN hardware used.
COPYRIGHT: Embedded Systems Academy, Inc. 2002.
           All rights reserved. www.microcanopen.com
           This software was written in accordance to the guidelines at
           www.esacademy.com/software/softwarestyleguide.pdf
DISCLAIM:  Read and understand our disclaimer before using this code!
           www.esacademy.com/disclaim.htm
```

```
LICENSE:    Users that have purchased a license for PCANopenMagic
            (www.esacademy.com/software/pcanopenmagic)
            may use this code in commercial projects.
            Otherwise only educational use is acceptable.
VERSION:    1.00, Pf/Aa/Ck 07-OCT-02
---------------------------------------------------------------------------
HISTORY:    1.00, Pf 07-OCT-02, First Published Version
---------------------------------------------------------------------------

Implementation recommendations:

1.) CAN interrupt
The CAN interrupt should check all the possible error flags and set the
global variable gMCOHW_status accordingly. Fatal errors must result in
a call to MCOUSER_FatalError with an error code in the range of 0x8000
to 0x87FF.

If a transmit queue is implemented, the transmit interrupt should be used
to trigger transmission of the next message in the transmit queue.
On "Basic CAN" controllers the receive interrupt copies the incoming message
into a receive queue. CAN controllers with "Full CAN" style capabilities
or internal receive queue might not need to maintain a software queue.
In case a hardware queue or buffers are used, the interrupt should still
check for a potential overrun and set bit RXOR in gMCOHW_status in case
of an overrun.

2.) Timer interrupt
A 1ms timer interrupt needs to implement a local 1ms WORD timer tick.
The timer tick is only accessible via the functions MCOHW_GetTime and
MCOHW_IsTimeExpired to avoid data inconsistency.

In case only a multiple of 1ms is available on a system, the timer tick
would need to be incremented in each interrupt in a way that the timer
tick is still accurate (for example increment by 4 all 4ms).
****************************************************************************/

#include "mco.h"

// Status bits for function MCOHW_GetStatus
#define INIT 0x01
#define CERR 0x02
#define ERPA 0x04
#define RXOR 0x08
#define TXOR 0x10
#define BOFF 0x80

/**************************************************************************
DOES:     This function returns the global status variable.
CHANGES:  The status can be changed anytime by this module, for example from
          within an interrupt service routine or by any of the other
```

```
        functions in this module.
BITS:   0: INIT - set to 1 after a completed initialization
                  left 0 if not yet inited or init failed
        1: CERR - set to 1 if a CAN bit or frame error occurred
        2: ERPA - set to 1 if a CAN "error passive" occurred
        3: RXOR - set to 1 if a receive queue overrun occurred
        4: TXOR - set to 1 if a transmit queue overrun occurred
        5: Reserved
        6: Reserved
        7: BOFF - set to 1 if a CAN "bus off" error occurred
***************************************************************************/
BYTE MCOHW_GetStatus
  (
  void
  );

/**************************************************************************
DOES:    This function implements the initialization of the CAN interface.
RETURNS: 1 if init is completed
        0 if init failed, bit INIT of MCOHW_GetStatus stays 0
***************************************************************************/
BYTE MCOHW_Init
  (
  WORD BaudRate   // Allowed values: 1000, 800, 500, 250, 125, 50, 25, 10
  );

/**************************************************************************
DOES:    This function implements the initialization of a CAN ID hardware
        filter as supported by many CAN controllers.
RETURNS: 1 if filter was set
        2 if this HW does not support filters
          (in this case HW will receive EVERY CAN message)
        0 if no more filter is available
***************************************************************************/
BYTE MCOHW_SetCANFilter
  (
  WORD CANID      // CAN-ID to be received by filter
  );

/**************************************************************************
DOES:    This function implements a CAN transmit queue. With each
        function call is added to the queue.
RETURNS: 1 Message was added to the transmit queue
        0 If queue is full, message was not added,
          bit TXOR in MCOHW_GetStatus set
NOTES:   The MicroCANopen stack will not try to add messages to the queue
        "back-to-back". With each call to MCO_ProcessStack, a maximum
        of one message is added to the queue. For many applications
        a queue with length "1" will be sufficient. Only applications
        with a high busload or very slow bus speed might need a queue
```

```
        of length "3" or more.
***************************************************************************/
BYTE MCOHW_PushMessage
  (
  CAN_MSG *pTransmitBuf // Data structure with message to be send
  );

/**************************************************************************
DOES:    This function implements a CAN receive queue. With each
         function call a message is pulled from the queue.
RETURNS: 1 Message was pulled from receive queue
         0 Queue empty, no message received
NOTES:   Implementation of this function greatly varies with CAN
         controller used. In an SJA1000 style controller, the hardware
         queue inside the controller can be used as the queue. Controllers
         with just one receive buffer need a bigger software queue.
         "Full CAN" style controllers might just implement multiple
         message objects, one each for each ID received (using function
         MCOHW_SetCANFilter).
***************************************************************************/
BYTE MCOHW_PullMessage
  (
  CAN_MSG *pTransmitBuf // Data structure with message received
  );

/**************************************************************************
DOES:    This function reads a 1 millisecond timer tick. The timer tick
         must be a WORD and must be incremented once per millisecond.
RETURNS: 1 millisecond timer tick
NOTES:   Data consistency must be insured by this implementation.
         (On 8-bit systems, disable the timer interrupt incrementing
         the timer tick while executing this function)
         Systems that cannot provide a 1ms tick may consider incrementing
         the timer tick only once every "x" ms, if the increment is by "x".
***************************************************************************/
WORD MCOHW_GetTime
  (
  void
  );

/**************************************************************************
DOES:    This function compares a WORD timestamp to the internal timer tick
         and returns 1 if the timestamp expired/passed.
RETURNS: 1 if timestamp expired/passed
         0 if timestamp is not yet reached
NOTES:   The maximum timer runtime measurable is 0x8000 (about 32 seconds).
         For the usage in MicroCANopen that is sufficient.
***************************************************************************/
BYTE MCOHW_IsTimeExpired
  (
```

```
  WORD timestamp // Timestamp to be checked for expiration
  );
/***************************************************************************
// Recommended implementation for this function (8051 version):
{
WORD time_now;

  EA = 0; // Disable Interrupts
  time_now = gTimCnt;
  EA = 1; // Enable Interrupts
  timestamp++; // To ensure the minimum runtime
  if (time_now > timestamp)
  {
    if ((time_now - timestamp) < 0x8000)
      return 1;
    else
      return 0;
  }
  else
  {
    if ((timestamp - time_now) > 0x8000)
      return 1;
    else
      return 0;
  }
}
```

Listing 6.1 Driver Functions Required by MicroCANopen

6.3.6.2 The MicroCANopen user interface

Listing 6.2 shows parts of the "mco.h" file which specifies the functions provided for the application interface. The function "ProcessStack" must be called frequently as a background task (for instance calling it from the while(1) loop in main). The only messages made transparent to the application are those with process data. All service messages are handled within "ProcessStack" without additional interfacing to the application. The "ReceivedPDO" function is a call-back function, which means it must be implemented by the user/application. It gets called from within the stack whenever new process data arrives. The default is that this function is called from within "ProcessStack" – meaning that there is an unpredictable delay from the time a message arrives until this function actually gets called. Experienced users facing tougher real-time requirements can modify the code and call this function from within the interrupt service routine (ISR) receiving the CAN messages. However, in this case it must be ensured that the code executed within the callback function is absolutely minimal, as it will add to the execution time of the ISR.

```
/**************************************************************************
MODULE:    MCO
CONTAINS:  MicroCANopen implementation
COPYRIGHT: Embedded Systems Academy, Inc. 2002-2003.
           All rights reserved. www.microcanopen.com
           This software was written in accordance to the guidelines at
           www.esacademy.com/software/softwarestyleguide.pdf
DISCLAIM:  Read and understand our disclaimer before using this code!
           www.esacademy.com/disclaim.htm
LICENSE:   Users that have purchased a license for PCANopenMagic
           (www.esacademy.com/software/pcanopenmagic)
           may use this code in commercial projects.
           Otherwise only educational use is acceptable.
VERSION:   1.20, Pf/Aa/Ck 19-AUG-03
---------------------------------------------------------------------------
HISTORY:   1.20, Pf 19-AUG-03, Code changed to use process image
           1.10, Pf 27-MAY-03, Bug fixes in OD (hi byte was corrupted)
                 OD changed to indefinite length
                 Support of define controled MEMORY types
           1.01, Pf 17-DEC-02, Made Object Dictionary more readable
           1.00, Pf 07-OCT-02, First Published Version
***************************************************************************/

/**************************************************************************
GLOBAL TYPE DEFINITIONS
**************************************************************************/

// Standard data types
#define BYTE  unsigned char
#define WORD  unsigned int
#define DWORD unsigned long

// Boolean expressions
#define BOOLEAN unsigned char
#define TRUE 0xFF
#define FALSE 0

// Data structure for a single CAN message
typedef struct
{
  WORD ID;                  // Message Identifier
  BYTE LEN;                 // Data length (0-8)
  BYTE BUF[8];              // Data buffer
} CAN_MSG;

// This structure holds all node specific configuration
typedef struct
{
```

```
  BYTE Node_ID;               // Current Node ID (1-126)
  BYTE error_code;            // Bits: 0=RxQueue 1=TxQueue 3=CAN
  WORD Baudrate;              // Current Baud rate in kbps
  WORD heartbeat_time;        // Heartbeat time in ms
  WORD heartbeat_timestamp;   // Timestamp of last heartbeat
  CAN_MSG heartbeat_msg;      // Heartbeat message contents
  BYTE error_register;        // Error regiter for OD entry [1001,00]
} MCO_CONFIG;

// This structure holds all the TPDO configuration data for one TPDO
typedef struct
{
#ifdef USE_EVENT_TIME
  WORD event_time;            // Event timer in ms (0 for COS only operation)
  WORD event_timestamp;       // If event timer is used, this is the
                              // timestamp for the next transmission
#endif
#ifdef USE_INHIBIT_TIME
  WORD inhibit_time;          // Inhibit timer in ms (0 if COS not used)
  WORD inhibit_timestamp;     // If inhibit timer is used, this is the
                              // timestamp for the next transmission
  BYTE inhibit_status;        // 0: Inhibit timer not started or expired
                              // 1: Inhibit timer started
                              // 2: Transmit msg waiting for expiration of inhibit
#endif
  BYTE offset;                // Address of data in process image
  CAN_MSG CAN;                // Current/last CAN message to be transmitted
} TPDO_CONFIG;

// This structure holds all the RPDO configuration data for one RPDO
typedef struct
{
  WORD CANID;                 // Message Identifier
  BYTE LEN;                   // Data length (0-8)
  BYTE offset;                // Address of data in process image
} RPDO_CONFIG;

/**************************************************************************
GLOBAL FUNCTIONS
**************************************************************************/

/**************************************************************************
DOES: This function initializes the CANopen protocol stack.
      It must be called from within MCOUSER_ResetApplication.
**************************************************************************/
void MCO_Init
  (
  WORD Baudrate,   // CAN baudrate in kbps(1000,800,500,250,125,50,25 or 10)
  BYTE Node_ID,    // CANopen node ID (1-126)
  WORD Heartbeat   // Heartbeat time in ms (0 for none)
```

```
  );

/***************************************************************************
DOES: This function initializes a transmit PDO. Once initialized, the
      MicroCANopen stack automatically handles transmitting the PDO.
      The application can directly change the data at any time.
NOTE: For data consistency, the application should not write to the data
      while function MCO_ProcessStack executes.
***************************************************************************/
void MCO_InitTPDO
  (
  BYTE PDO_NR,        // TPDO number (1-4)
  WORD CAN_ID,        // CAN identifier to be used (set to 0 to use default)
  WORD event_tim,     // Transmitted every event_tim ms
                      // (set to 0 if ONLY inhibit_tim should be used)
  WORD inhibit_tim,// Inhibit time in ms for change-of-state transmit
                      // (set to 0 if ONLY event_tim should be used)
  BYTE len,           // Number of data bytes in TPDO
  BYTE offset         // Offset to data location in process image
  );
/***************************************************************************
DOES: This function initializes a receive PDO. Once initialized, the
      MicroCANopen stack automatically updates the data at offset.
NOTE: For data consistency, the application should not read the data
      while function MCO_ProcessStack executes.
***************************************************************************/
void MCO_InitRPDO
  (
  BYTE PDO_NR,        // RPDO number (1-4)
  WORD CAN_ID,        // CAN identifier to be used (set to 0 to use default)
  BYTE len,           // Number of data bytes in RPDO
  BYTE offset         // Offset to data location in process image
  );

/***************************************************************************
DOES: This function implements the main MicroCANopen protocol stack.
      It must be called frequently to ensure proper operation of the
      communication stack.
      Typically it is called from the while(1) loop in main.
***************************************************************************/
BYTE MCO_ProcessStack
  ( // Returns 0 if nothing needed to be done
    // Returns 1 if a CAN message was received or sent
  void
  );

/***************************************************************************
USER CALL-BACK FUNCTIONS
These must be implemented by the application.
***************************************************************************/
```

```
/**************************************************************************
DOES: This function resets the application. It is called from within the
      CANopen protocol stack, if a NMT master message was received that
      demanded "Reset Application".
**************************************************************************/
void MCOUSER_ResetApplication
  (
  void
  );

/**************************************************************************
DOES: This function both resets and initializes both the CAN interface
      and the CANopen protocol stack. It is called from within the
      CANopen protocol stack, if a NMT master message was received that
      demanded "Reset Communication".
      This function should call MCO_Init and MCO_InitTPDO/MCO_InitRPDO.
**************************************************************************/
void MCOUSER_ResetCommunication
  (
  void
  );

/**************************************************************************
DOES: This function is called if a fatal error occurred.
      Error codes of mcohwxxx.c are in the range of 0x8000 to 0x87FF.
      Error codes of mco.c are in the range of 0x8800 to 0x8FFF.
      All other error codes may be used by the application.
**************************************************************************/
void MCOUSER_FatalError
  (
  WORD ErrCode // To debug, search source code for the ErrCode encountered
  );
```

Listing 6.2 Application Interface Functions Provided by MicroCANopen

6.3.6.3 Debugging and Testing

Several tools are available for debugging and testing CANopen nodes. Besides monitor and analysis tools, CANopen-specific tools to maintain and access the Object Dictionary are very helpful. Figure 6.5 shows Vector's CANeds, an editor for CANopen electronic data sheets and device configuration files (EDS, DCF). These files specify which OD entries are implemented by a particular node. Configuration tools, analysis tools or NMT Master implementations can use this information to directly access the OD entries of that particular node.

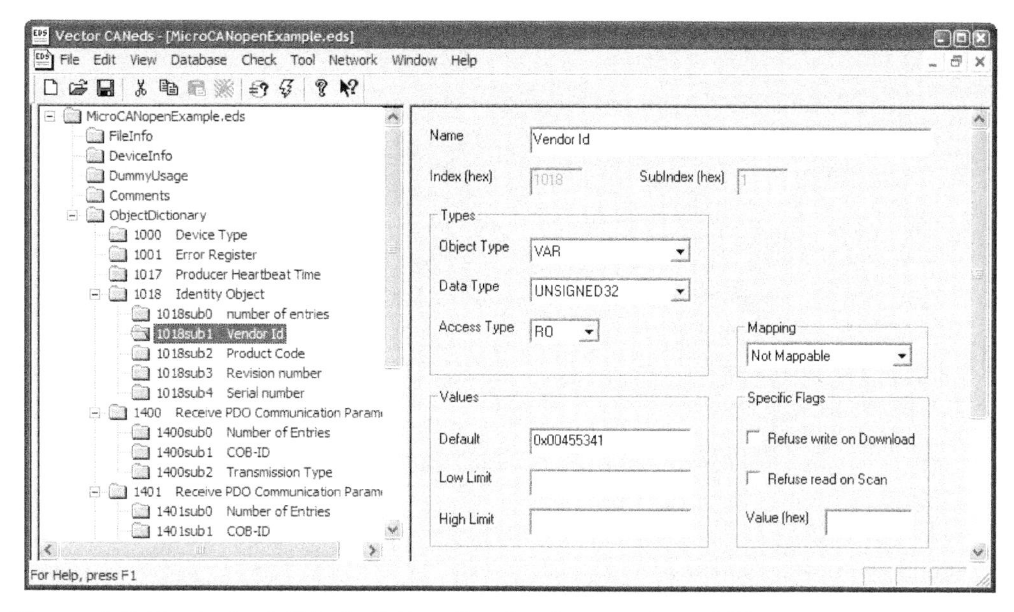

Figure 6.5 Vector's CANeds Editor for the Generation of Electronic Data Sheets

6.3.7 Summary: What Does it Do?

The communication methods implemented in MicroCANopen allow the sharing of process data among several CAN nodes in a CANopen-style manner. Data can be transmitted with a pre-set frequency (every n milliseconds) or on COS (change-of-state). MicroCANopen is sufficiently CANopen compatible that regular CANopen configuration tools, monitors/analyzers and master implementations such as those available from Vector CANtech can be used with it.

Listing 6.3 shows an example of the main function implementing a MicroCANopen node, in this case a temperature sensor with some digital outputs. The CAN bit rate is 125 kbps, the Node ID is 3, the heartbeat time is 2.5 seconds and the temperature value is transmitted every 500 milliseconds.

The process data is organized in a process image, the variable gProcImg, an array of bytes. The offset VAR_temp and VAR_digiout are defined to provide symbolic access to the offsets.

```
// Process Image to hold all process data, in total room for 16 bytes
BYTE gProcImg[16];

#define VAR_temp   0 // offset for temperature variable
```

```
#define VAR_digout 2 // offset for a digital output

void main (void)
{

  InitIO();

  // 125kbps, Node ID 3, 2.5s heartbeat
  MCO_Init(125,3,2500);

  // TPDO1, default CAN ID, 500ms timer, no inhibit, 2 bytes of data
  MCO_InitTPDO(1,0,500,0,2,VAR_temp);

  // RPDO1, default CAN ID, 1 byte of data
  MCO_InitRPDO(1,0,1,VAR_digout);

  EA = 1; // End of initialization, Enable all interrupts

  while(1)
  {
    MCO_ProcessStack();
    // Process the data
    gProcImg[VAR_temp+1] = GetHiByteFromADConverter();
    gProcImg[VAR_temp]   = GetLoByteFromADConverter();
    ApplyDigitalOutput(gProcImg[VAR_digout]);
  } // end of while(1)
}
```

Listing 6.3 Implementing a Temperature Sensor with MicroCANopen

6.3.8 *Flow Charts for the Main Function Blocks*

The following five flow charts illustrate how the MicroCANopen implementation operates. Some of the charts are broken into two parts for readability. Larger copies of all the flowcharts are available at www.CANopenBook.com. The main function executed is illustrated in Flow Chart 6.1. After checking whether this is the first time the function is called, it polls the next receive message from the driver. If a message was received and it is an NMT Master message, an RPDO message or an SDO request, the associated code sections get executed - see Flow Chart 6.2, Flow Chart 6.3 and Flow Chart 6.4.

If no message was received or the message received was not for the local node to handle, ProcessStack continues with potential transmissions that are due. First, the TPDO transmissions are checked (Flow Chart 6.5). If no TPDO transmission is due, the heartbeat producer time is verified. If it is expired, a heartbeat message is generated.

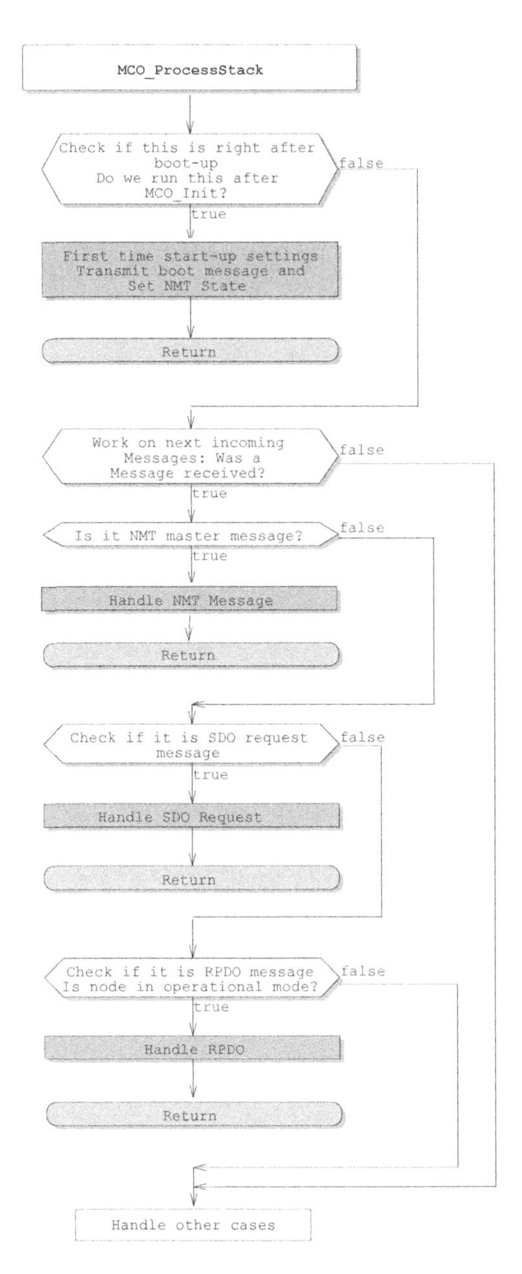

Flow Chart 6.1 Process Stack (Continued Next Page)

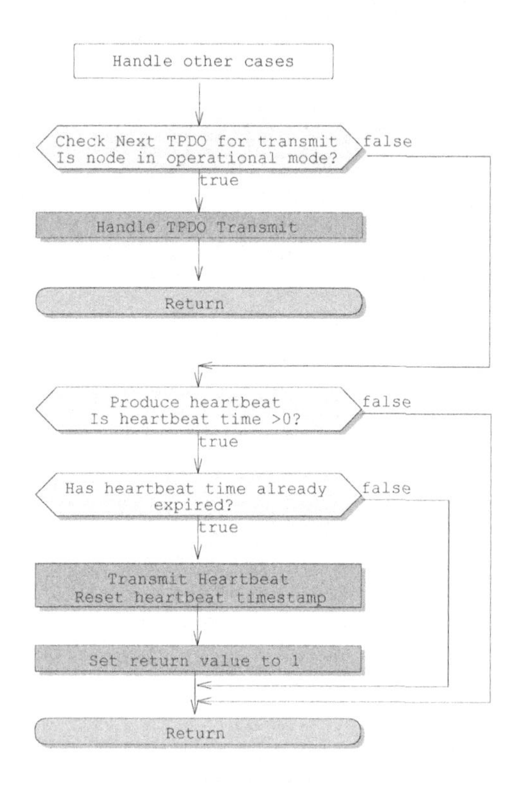

Flow Chart 6.1 Process Stack (Continued)

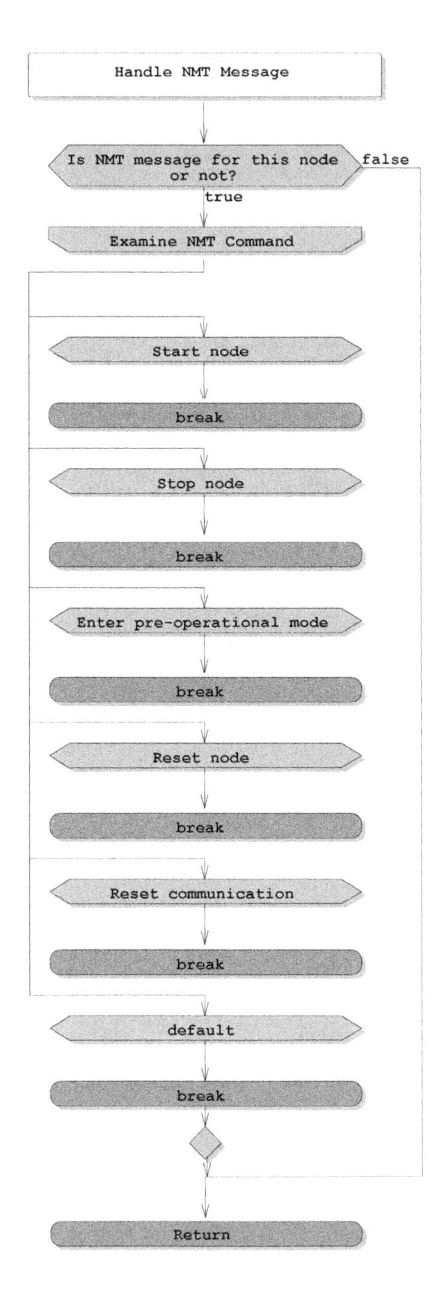

Flow Chart 6.2 Handle NMT Message

Upon reception of an NMT message the current state of the node must be switched according to the command received.

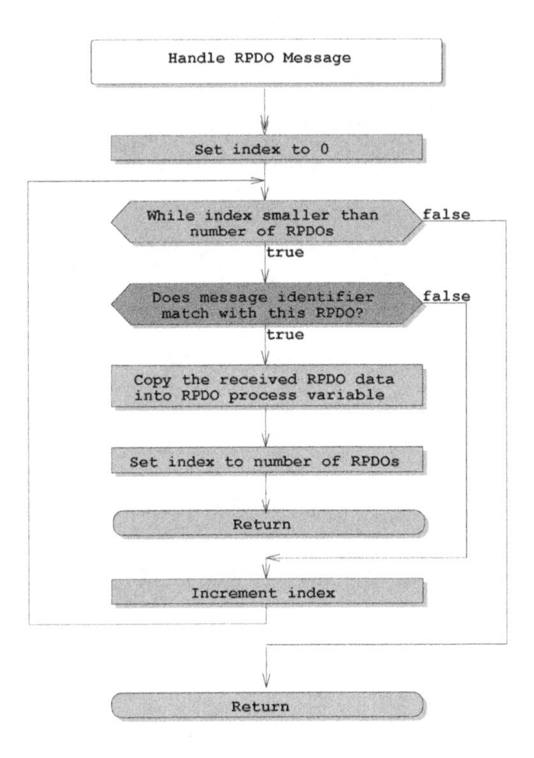

Flow Chart 6.3 Handle RPDO Message

When handling a Receive PDO, MicroCANopen runs through a loop checking all configured RPDOs to see if the identifier of the message received matches any of the identifiers used for the RPDOs. If a match is found, the data received is copied to the appropriate process variable.

Handling an SDO Request is simplified to the point where only two OD entries are treated directly: [1001h,00h] the Error Register and [1017h,00h] Producer Heartbeat Time. For all other entries it is assumed that they are constant read-only values coming from the lookup table, accessed by function Search_OD().

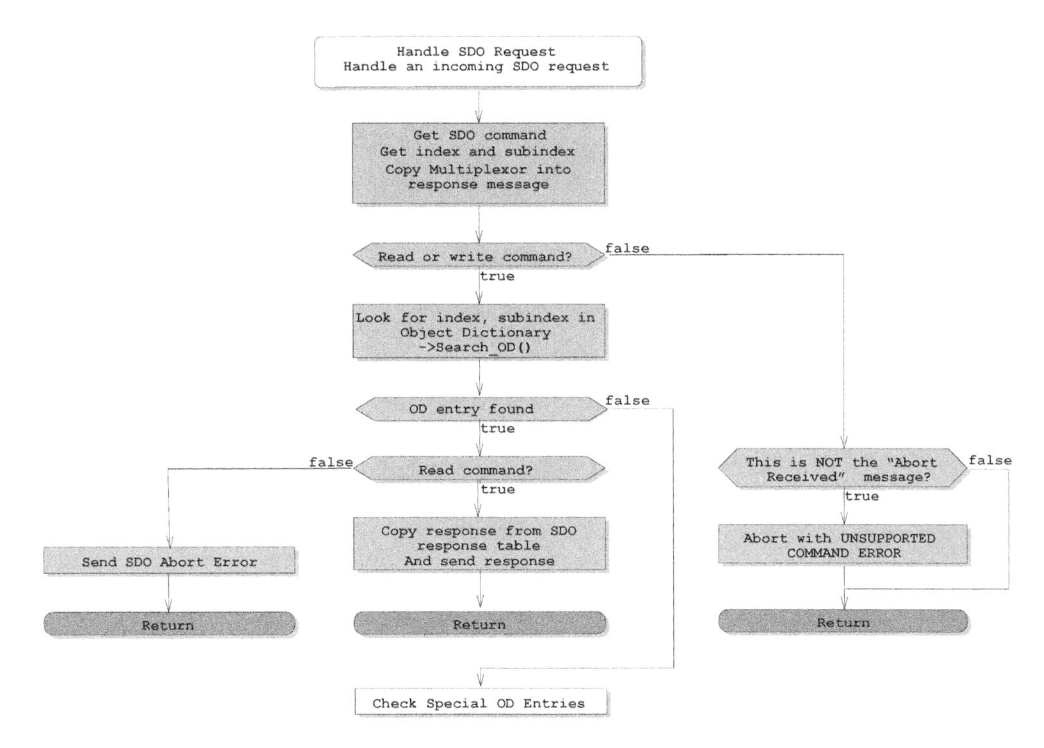

Flow Chart 6.4 Handle SDO Request (Continued Next Page)

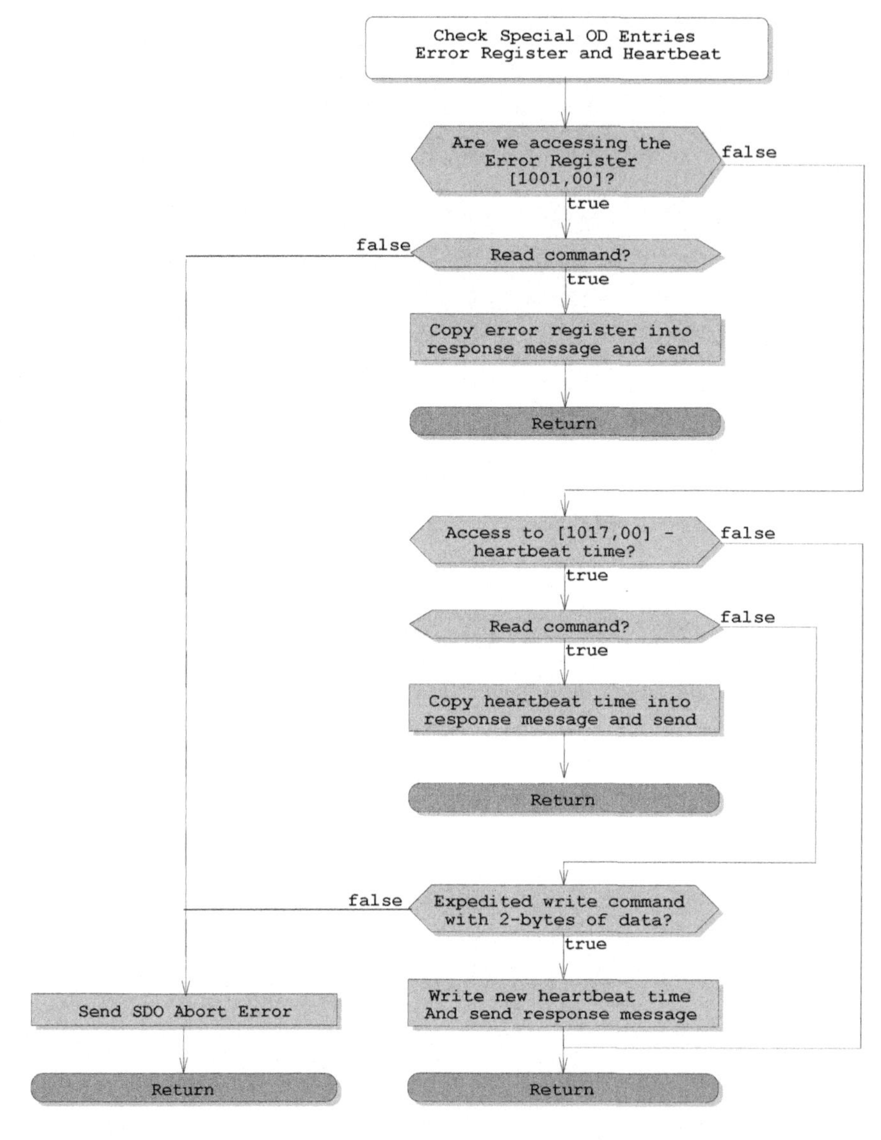

Flow Chart 6.4 Handle SDO Request (Continued)

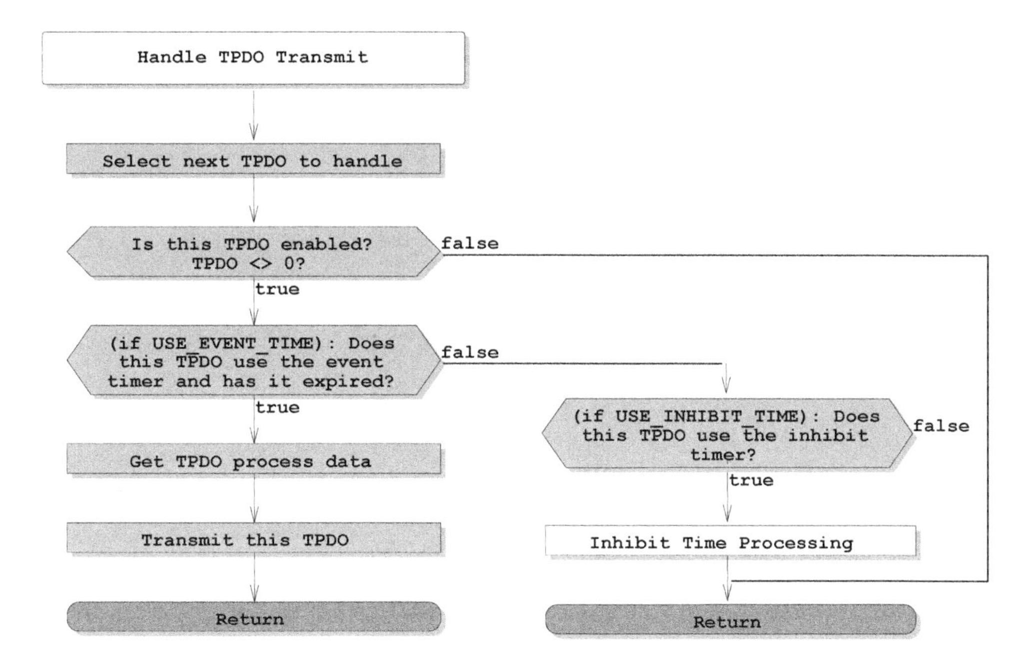

Flow Chart 6.5 Handle TPDO Transmit (Continued Next Page)

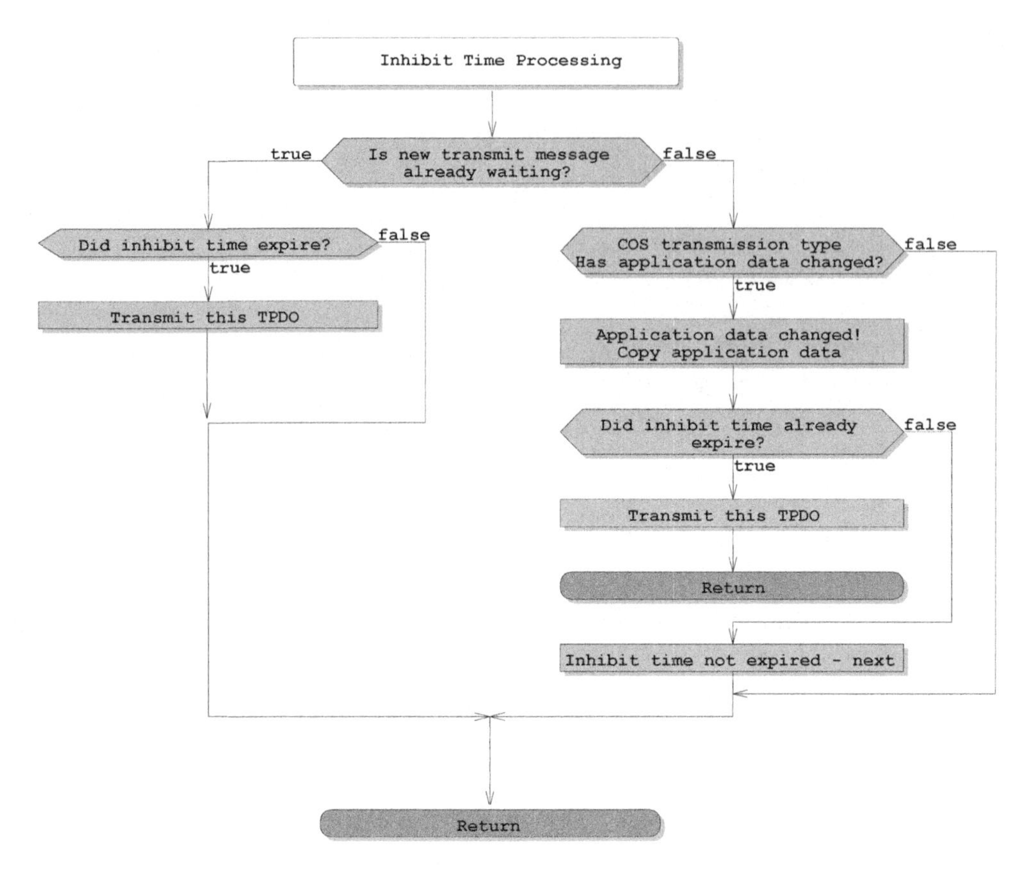

Flow Chart 6.5 Handle TPDO Transmit (Continued)

Only one TPDO per call to ProcessStack is checked for transmission. This is done to avoid bursts of back-to-back TPDOs that may be due for transmission. When using an Event Time (TPDO is transmitted every X milliseconds), the handling of the TPDO is simple. If the time is expired, re-set the timer and transmit the TPDO. However, if change-of-state detection is used with an Inhibit Time, the first step is to check if the TPDO is already due for transmission and is just waiting for the Inhibit Time to expire. If that is not the case, the last transmitted data needs to be compared with the current data (trying to detect a change-of-state). If the data changed, it needs to be copied to the transmit buffer. However, it can only be transmitted if the Inhibit Time has already expired – otherwise transmission needs to wait. The Inhibit Time is reset with every transmission of the TPDO.

6.4 Using CANopen Hardware Modules or Chips

Objective

This section explains how CANopen modules or chips can be used to develop a CANopen node. It also discusses the typical operation modes; "stand-alone" usage or usage as a "communication co-processor." Examples given are for the Philips CANopenIA technology.

One of the fastest ways to design and implement a CANopen node is to use existing CANopen hardware in the form of CANopen chips or modules. What all these solutions have in common is that the CANopen protocol stack is pre-programmed into a microcontroller which can be incorporated into a hardware design either directly or in the form of a module (daughter-board). Typically at least two operation modes are supported: some sort of "stand-alone" operation and a "communication co-processor" operating mode.

6.4.1 Stand-Alone Operation

In stand-alone operation a CANopen hardware solution directly implements a specific Device Profile and provides the inputs and outputs required for a particular application. This allows the chip to be used for this particular application without the requirement of an additional microcontroller. A simple example would be that of a Generic I/O node, where the CANopen chip or module directly provides the digital and analog inputs and outputs on pins of the chip or module. Figure 6.6 shows how Philips' CANopenIA solution works in the stand-alone mode. It provides a total of 20 digital signals (configurable in groups of 4 bits to be used either as inputs or outputs). Analog signals are provided using external D/A or A/D chips with SPI interface.

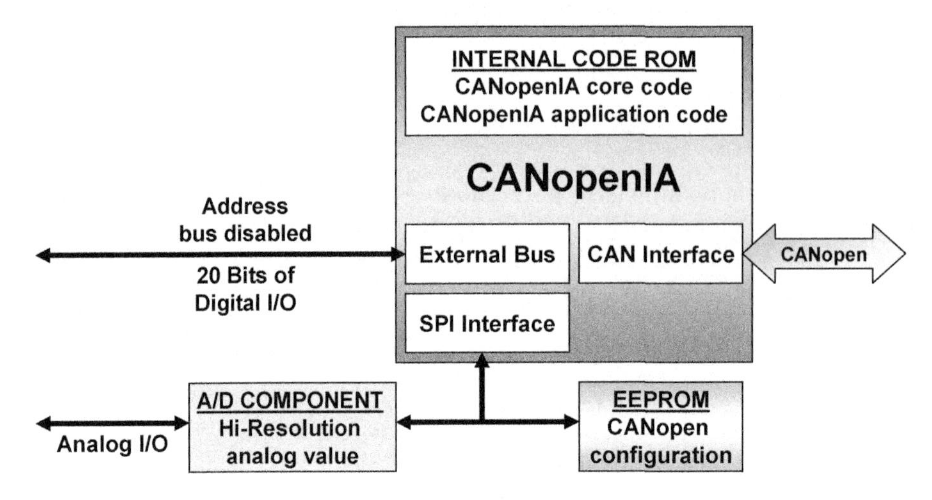

Figure 6.6 CANopenIA in Stand-Alone Mode

This mode is most suitable when a specific device profile needs to be implemented, such as an encoder, joystick or battery. It allows for the direct connection of a standardized device to a CANopen network using the CANopen chip or module.

6.4.2 Co-Processor Operation

In co-processor mode, a CANopen chip or module acts as a peripheral chip to another host controller. The interface between the host and the CANopen hardware is often implemented using a serial port or a shared memory area. Figure 6.7 shows how Philip's CANopenIA chip operates in the "co-processor" mode. In order to simplify the software requirements on the host, the only data exchanged is the process data. The entire CANopen side of what happens with the data (and when) is handled by the CANopenIA chip.

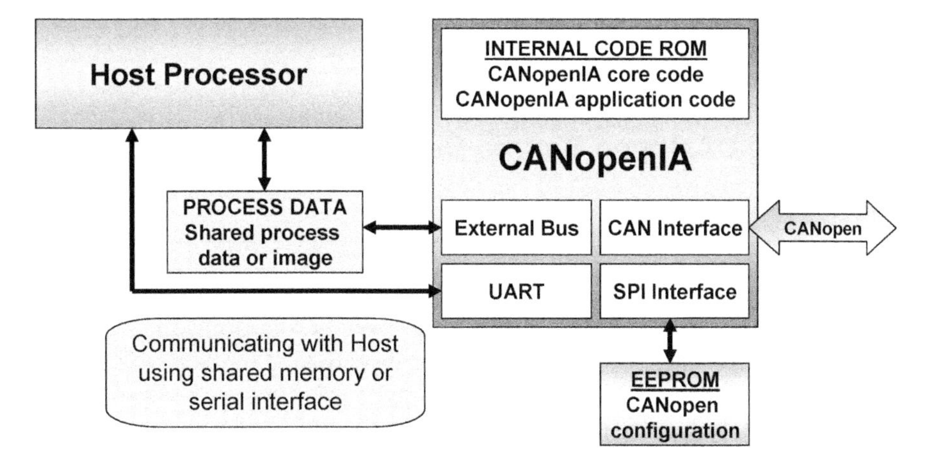

Figure 6.7 CANopenIA in Co-Processor Mode

This mode is most suitable for applications that require a certain amount of customization (as opposed to implementing a standardized Device Profile) and which will benefit from off-loading the burden of processing a communication protocol stack to a co-processor. An example would be adding a CANopen interface to an existing design, where the current main controller does not have enough memory and/or performance left to implement an entire CANopen protocol stack.

6.4.3 Setup and Configuration

Different CANopen chip or module solutions use a variety of configuration options. Typically the CANopen related configuration (for example, the PDO communication and mapping parameters) is stored in non-volatile memory and can be changed with standard or customized CANopen configuration tools. In the case of the CANopenIA this includes the hardware setup of the pins, establishing which pins are used, and how they are used (for example, input or output).

The other essential settings of any CANopen node are the bit rate used on the CAN bus and its Node ID. A total of seven bits are required to set the Node ID (1-127) and three bits are required to select one of the eight bit rates supported by CANopen. Depending on the application, these parameters are either set in software (using setup software, and stored in non-volatile memory) or by hardware (switches or dials). When set in hardware a total of 10 bits need to be provided to the CANopen chip or module. In order to keep the number of pins used for these settings to a minimum, the CANopenIA uses shift registers. A total of 10 switches, dials or jumpers are connected to the shift registers and can be used to set the Node ID and the bit rate used. Upon boot-up this information is shifted into the CANopenIA serially, only requiring a total of three pins.

6.5 Using CANopen Source Code

> Objective
>
> In this section we summarize the typical configuration options one has when using purchased CANopen source code and how to best make usage of these options. All examples given apply to the Vector CANopen slave source code. However, other source codes typically have very similar configuration features.

One of the most common approaches in designing and implementing a CANopen node is to purchase a code library or the source code for a CANopen slave node. The biggest benefits of this CANopen implementation approach are portability and customizability. In addition, the providers of the source code typically guarantee that their source code passes the CANopen conformance test. Although the final responsibility for conformance lies with the engineering team using the source code, it helps to know that all essential CANopen functions as delivered have passed the conformance test.

- **Portability**
  The providers of CANopen libraries or source code ensure good portability simply because they want to be able to sell their product, no matter which microcontroller is used. As a result, most commercial solutions directly support a wide variety of microcontrollers and can be easily adapted to "exotic" or legacy systems, such as a Z80 with external CAN controller.

- **Customizability**
  Many CANopen-based systems take advantage of its "openness" by customizing and optimizing certain aspects of the CANopen communication. This can include special message trigger methods, customized emergencies or customized Object Dictionaries (such as password protected access). Customizations like these are only possible if the engineers and programmers have access to the CANopen source code.

6.5.1 Code Configuration through Conditional Compilation

In order to further support the customization, source codes typically use #define statements to control which CANopen functionality is included when compiling and building the code. Changing any of these defines can have severe consequences since completely different code gets compiled and generated, and some #define combinations might not work with each other. Thus, one should carefully document all changes that are done along with the reasons and the persons who did them. Even in cases where a setting is switched back to a previous value, it should be documented with an additional comment so that the complete history of the changes is traceable.

The following examples are taken from the CANopen slave source code from Vector CANTech. Although the names of the #define statements are those used by Vector in their "cos_main.h" file, similar statements can be found in almost all commercial CANopen source codes.

6.5.1.1 Buffering or Queuing

Due to differences in CAN controllers, most CANopen implementations provide at least two different operation modes on the driver level. One of them, termed "buffered" mode, tries to make best usage of the CAN buffers provided by Full CAN controllers. The other, termed "queued" mode, implements one or more software message queues. Incoming messages are copied to the queue by the CAN receive interrupt service routine for later processing. The queued mode is enabled by setting QUEUED_MODE to 1 (else 0).

The idea of buffered mode is to use one message object of the Full CAN controllers for each CAN message received or transmitted. Obviously this can only work if the total number of unique message IDs transmitted or received does not exceed the number of message objects provided in the CAN controller. The buffered mode is enabled by setting FULLCAN_BUFFER_MODE to 1 (else 0). In addition, NUM_CAN_BUFFER needs to be set to the number of CAN message objects supported by the CAN controller.

In general, the queued mode requires more code and data memory than the buffered mode, since additional code and storage is required for the handling of the queue. Due to the extra overhead in handling the queue, this mode also requires more CPU processing time.

6.5.1.2 SDO Transfers

CANopen defines a total of three different SDO transfer modes: expedited, segmented and block transfer. The expedited transfer is a basic requirement and must be supported by all CANopen nodes. It allows read and write accesses to the slave's Object Dictionary entries (which cannot exceed 4 data bytes).

As soon as any Object Dictionary entry in a node is longer than 4 bytes that node must support segmented SDO transfer (where up to 7 bytes are transferred in each segment). In comparison to the expedited SDO transfer, the segmented transfer requires significantly more code and data memory. Both communication partners need to track each segment of the entire transfer. The receiver typically needs an extra memory buffer big enough to hold the entire data block transferred. A transfer is only considered successful after the entire data block was received – so only at that point may it be copied to its final destination. The segmented transfer for write accesses can be enabled by setting SDO_WRITE_SEG_ALLOWED to 1 (else 0), the segmented transfer for read accesses can be enabled by setting SDO_READ_SEG_ALLOWED to 1 (else 0).

The block transfer mode not only requires additional memory, it also requires that the main MCU have a certain level of performance. Typically the block transfer mode is not suitable for 8-bit microcontrollers. Using the block transfer mode, up to 127 segments (with up to 7 bytes each) are transferred back-to-back with only one acknowledgment message for the entire block. This means the receiver needs to be powerful enough to receive these back-to-back messages. The SDO Block Transfer mode can be enabled by setting SDO_BLOCK_ALLOWED to 1 (else 0).

In addition, the block transfer mode uses an optional CRC checksum over the entire data block transferred. If implemented, both communication partners must calculate this CRC in software. Sometimes there are options as to which method to use to calculate the CRC; either a lookup table or dynamically. Dynamically means that the microcontroller performs the required checksum calculation with every data byte transferred. This uses less code memory, however, it requires greater CPU performance. When using a lookup table, the CPU has less to calculate but additional memory to store the lookup table is required. To enable CRC calculation, SDO_BLOCK_USE_CRC must be set to 1 (else 0). CRC calculation is done using a lookup table if CRC_LOOKUP_TABLE is set to 1. If set to 0 it is dynamically calculated.

6.5.1.3 SDO Clients and Servers

Per default, each CANopen slave node implements one SDO server that serves the data of the local Object Dictionary to the network using the default SDO channels.

If an additional SDO server is required (to allow other nodes to send SDO requests to this node at the same time), MULTI_SDO_SERVER must be set to 1 (else 0) and NUM_SDO_SERVERS must to be set to 2. In this case SDO_SERVER2 must also be set: to a second CANopen Node ID. The second server requires that the SDO channels used are "stolen" from another Node ID. In other words, the physical network may not have a CANopen node on it with the Node ID that is used for the second server.

Some applications might require that a CANopen slave node also become an SDO client, allowing it to send SDO requests to other nodes. This feature can be enabled by setting CLIENT_ENABLE to 1 (else 0).

6.5.1.4 PDOs

The communication and mapping parameters for each PDO are configurable as either dynamic or static. The communication parameters determine which CAN message ID is used by the PDO, and when and how transmission is triggered. The mapping parameters determine which variables (in the form of Object Dictionary entries) go into the PDO. Dynamic means that the parameters can be changed during run-time, static means that they are fixed, frozen, hard-coded and cannot change during run-time.

Many deeply embedded applications use static PDOs if the network configuration is always the same. However, if the network configuration can change (typical for off-the-shelf industrial I/O components), dynamic PDOs are required.

RPDO_PAR_READONLY must be set to 1 if the RPDO communication parameters are static, or to 0 if they are dynamic.

TPDO_PAR_READONLY must be set to 1 if the TPDO communication parameters are static, or to 0 if they are dynamic.

RPDO_MAP_READONLY must be set to 1 if the RPDO mapping parameters are static, or to 0 if they are dynamic.

TPDO_MAP_READONLY must be set to 1 if the TPDO mapping parameters are static, or to 0 if they are dynamic.

6.5.1.5 NMT Startup, Heartbeat and Emergencies

Per default, CANopen slave nodes boot-up and stay in the "pre-operational" state until a master sets them to "operational." Sometimes it is desirable that nodes go into operational by themselves without waiting for a master (for example, in applications without a master). If STARTUP_AUTONOMOUSLY is set to 1 (else 0), the node goes into operational by itself, without waiting for a master.

If the node should also start to produce a heartbeat message right after boot-up, START_HEARTBEAT_PRODUCER must be set to 1 (else 0). START_HEARTBEAT_TIME defines the heartbeat time to be used in milliseconds. After boot-up, the node will transmit a heartbeat message every START_HEARTBEAT_TIME milliseconds.

A node can be configured to generate or not generate emergency messages. If ENABLE_EMCYMSG is set to 1 (else 0) the node produces emergency messages if communication faults are detected. A typical emergency would be to receive a PDO that is of a different length than expected.

6.5.1.6 Signals

There are several #define statements to configure SIGNAL_XXX. If enabled, these use callback functions to signal certain events back to the application. If a certain signal is enabled, the corresponding call-back function must be implemented by the application.

One example of such a signal is SIGNAL_BOOTUPMSG. The corresponding call-back function is called once the boot-up message is transmitted. This is an indication to the application that the initialization of the CAN bus interface was successful and that at least one other node is out there using the same bit rate.

6.5.2 The Object Dictionary

The Object Dictionary is implemented as an array of structures in the file "objdict.c". The structure for a single entry needs to hold information such as the type and/or status for the entry, the Index and Subindex, the length of the data in this entry in bytes, a pointer to the data and information about the access and mapping options (read-only, write-only, etc.) The access options defined are listed in Table 6.5.

Access Type	How Object Dictionary Entry Accessed
RO	Read-Only, cannot be mapped to a PDO
WO	Write-Only, cannot be mapped to a PDO
RW	Read-Write, cannot be mapped to a PDO
ROMAP	Read-Only, can be mapped into a TPDO
WOMAP	Write-Only, can be mapped into a RPDO
RWRMAP	Read-Write, can be mapped into a TPDO, but not RPDO
RWWMAP	Read-Write, can be mapped into a RPDO, but not TPDO

Table 6.5 Access Types for Object Dictionary Entries

To simplify the way entries are made, macros for the different types of entries are defined.

Example: SINGLE_OBJ (index, length, address, access)

This macro is used if the entry is a single object (only Subindex 0 is implemented and the entry is stored at Subindex 0). The parameters are:

- Index: 16-bit Index of the Object Dictionary entry
- length: the length of the Object Dictionary entry in bytes
- address: a pointer to the data of this Object Dictionary entry (typically a pointer to a variable or to a location in the process image)
- access: the access type for this entry (see Table 6.5)

This macro could be used as follows to specify that the 1-byte variable "gMyStatus" is made available at Index [2100h,00h] for read and write accesses:

```
SINGLE_OBJ(0x2100, 1, &gMyStatus, RW)
```

6.5.3 PDO Mapping

The PDO mapping parameters are implemented in two arrays of structures in the file "mapping.c" – one for RPDOs and one for TPDOs. There is one structure for each PDO and each structure contains the number of entries mapped for the PDO and a total of 8 mapping entries (default). There are always 8 entries, even if fewer entries are actually used. However, this parameter is configurable, so if all PDOs use less than 8 entries it may be modified to the maximum number of PDO map entries used. Several macros are provided to make the entries more readable.

Sample mapping entry for one PDO:

```
NUM_OF_MAP_ENTRIES(2),
MAP_ENTRY(0x2110, 0x00, 16, &gMyVar1),
MAP_ENTRY(0x2120, 0x00, 8, &gMyVar2),
VOID_MAP_ENTRY,
VOID_MAP_ENTRY,
VOID_MAP_ENTRY,
VOID_MAP_ENTRY,
VOID_MAP_ENTRY,
VOID_MAP_ENTRY,
```

The example above is for a PDO with a total of 2 mapping entries (NUM_OF_MAP_ENTRIES is set to 2). What follows are two mapping entries, the first one maps the Object Dictionary entry [2110h,00h]. There are 16 bits to map and the address for the data is the address of the variable "gMyVar1". The second entry maps the Object Dictionary entry [2120h,00h]. There are 8 bits to map and the address for the data is the address of the variable "gMyVar2". A total of 6 entries of VOID_MAP_ENTRY are used for the unused mapping entries (these can be avoided if the maximum number of map entries allowed per PDO is reduced).

6.6 CANopen Conformance Test

"Error, no keyboard – press F1 to continue"
PC BIOS

Objective

In this section we will not try to explain all the technical details of the CANopen conformance test nor try to find an answer as to why some specific access sequences that are not documented in the CANopen specification are tested. Sometimes mysteries are better accepted as such.

Instead we will give some guidelines on when the conformance test should be used and when it can be replaced by other test procedures.

The official CANopen conformance test was developed with significant involvement from the engineers of the CiA and is available through National Instruments and the CiA. When a device gets conformance tested by the CiA, they use exactly this software and simply confirm that the device passed or failed.

6.6.1 What Does it Do?

The conformance test not only tests a physical device, it also tests the Electronic Data Sheet (EDS) associated with it. It is important to confirm that an EDS is syntactically correct and that a device perfectly matches the EDS – in other words, it has exactly those communication parameters implemented as specified in the EDS file.

In part, the CANopen conformance test not only checks to see if all Object Dictionary entries specified are available in the device, it also scans for hidden entries that a device might have and that are not mentioned in the EDS. Due to this scanning process, the entire run-time of the CANopen conformance test is several hours.

It should be noted that the CANopen conformance test only checks the CANopen communication behavior. It cannot test PDO data contents or reactions to certain data.

6.6.2 Who Should Use It?

As experienced consultants and tutors who also teach classes on software quality, our first thought concerning the CANopen conformance test was to recommend to all our students that every CANopen node they develop should pass the CANopen test. The idea is that even if a network is completely embedded into a machine, it would still give everybody participating in the design of CANopen nodes a measurement of how well a specific node is implemented. If, for example, something does not work in the communication between two nodes, the CANopen conformance test could be used to test if a specific node is really behaving as it should.

Well, so much for the theory. As usual the real-world works slightly differently. In recent years, the CANopen specification has been enhanced and updated, but not all of these updates found their way into the conformance test. So some test failures are actually acceptable. Acceptable failures, for example, include a lack of support for node guarding if heartbeat is supported, or exceeding unspecified time-outs for SDO transfers.

Another issue is that the entire set of conformance requirements, tests and procedures is not well documented. So if a certain test fails there is a lot of guesswork left to the engineers with regards to fixing the problem.

If a CANopen device is designed and developed for the open market with the intention to sell it as stand-alone "CANopen Gizmo," then this node should pass the CANopen conformance test and get a certificate to that effect. There are several instances where end-users had bad experiences with uncertified products, and as a result more and more end-users are demanding that only certified products are used in their systems.

If a CANopen design is based on a self-developed CANopen implementation, passing the conformance test is more likely to become a significant hurdle. This can be avoided if a design is based on commercial CANopen source code or a CANopen chip or module; passing the conformance test should not be a problem since the manufacturers of these products ensure that their products can pass.

If a CANopen device is intended for "internal use only" there is no real need or requirement to pass the CANopen conformance test. This includes all developments where the CANopen network is truly "embedded" – hidden within a machine, virtually invisible to the end-user of that machine.

This can even be extended to scenarios where third party CANopen devices get integrated into such a machine. As seen in Section 6.9 on page 299, there are applications where CANopen nodes were intentionally designed in such a way that 100% CANopen conformance is not a requirement. Instead the requirement is that they just need to have as much conformity as required in order to work with "regular" conforming CANopen devices.

> Note, however, that we definitely recommend that any CANopen device designed be tested. However, using the official CANopen conformance test to do so might not always be required or even the best choice.

6.6.3 Other Test Options

There are a variety of tools besides the CANopen conformance test that can be used to test CANopen devices. Most of them even offer the opportunity to do data dependent testing, which is something the CANopen conformance test does not do. Data dependent testing includes sending specific process data to a device as well as verifying the process data received from a device.

In general, any CAN monitor or analyzer with a scripting, DLL or batch interface can be used to send test messages to a device and analyze the responses. This path requires a deep knowledge about the CANopen messaging system, since all messages would need to be generated manually.

Things become a little simpler if the monitor or analyzer is CANopen aware and if that awareness also finds its way through to the scripting interface, allowing for direct execution of SDO and PDO accesses. This path simplifies the generation of CANopen specific test sequences, such as a segmented SDO transfer.

If a CAN interface with CANopen DLL is used, standard test software like LabView® can be used to execute and log test sequences. This path is very attractive for test engineers that already have test software in place and would like to continue to use the same software for their CANopen devices.

6.7 Choosing an Implementation Path

"Experience is the name everyone gives to their mistakes."
Oscar Wilde

> Objective
>
> In this section we give some general guidelines on when to choose which of the implementation paths listed in the previous sections.
>
> Note that applications differ and a more detailed evaluation of a particular scenario might be required. This is especially true when it comes to "reaching the limits," such as getting above an 80% bus load, or single nodes requiring a specific response time (or other real-time behavior).
>
> Simply ask yourself the questions listed in this section and read the recommendations to get some clues about which implementation path is best for you and your application.

CAN is right for this application, but is CANopen?
If CAN is definitely going to be used, but it is not yet certain if CANopen or a custom, proprietary higher-layer CAN protocol should be used, then MicroCANopen might be the perfect match. MicroCANopen can provide a basic communication structure that is upward compatible to CANopen. So the final decision to be fully CANopen compatible does not need to be made at this point.

Are the CANopen nodes used internally only, or will they be sold individually on the open market as CANopen devices?
If a developed CANopen node is to be sold on the open market as a CANopen device, the implementation should either be based on commercial CANopen source code or a CANopen hardware module or chip. If the CANopen protocol is developed from scratch internally, passing the CANopen conformance test becomes a challenge.

What is the expected volume of the device planned?
The general rule is that the higher the expected volume of the device, the higher the demand for an optimized solution that takes best advantage of the microcontroller resources provided. Typically that means a commercial CANopen software solution with many target-specific optimizations or a customized implementation, for example based on MicroCANopen.

How much software development expertise does the engineering team have?
If the software development expertise of the engineering team is limited, using CANopen hardware modules or chips minimizes the software development required. If that is not an option, a commercial source code or library should be used to minimize the software development for the CANopen protocol stack.

How much hardware development expertise does the engineering team have?
If the hardware development expertise of the engineering team is limited, using CANopen hardware modules can minimize the hardware development required.

What are the time-to-market requirements?
If the team is under pressure because the CANopen device must be available in a very short time, both hardware and software development should be minimized. This can be achieved by using CANopen hardware modules or chips. If a solution using source code or libraries is required, hiring an experienced consultant/tutor should be considered to kick-start the project.

How will CANopen be used – strictly as CANopen or is it likely that the system will need to be tweaked or optimized?
If a CANopen node needs to support some specific communication features not standardized in CANopen, software solutions such as a commercial CANopen source code or MicroCANopen are preferred over CANopen hardware modules and chips.

Are there real-time requirements?
If the CANopen node must fulfill certain real-time requirements, CANopen hardware modules and chips have the advantage that their timing behavior is already known and typically published in their data sheets. If the application requires a software solution, the performance of the implementations for the target microcontroller needs to be carefully evaluated. If a lower-end microcontroller is used, a customized implementation, for example based on MicroCANopen, might be required.

6.8 Implementing CANopen Compliant Bootloaders

Objective

This section discusses the implementation of a CANopen compliant bootloader. Such a bootloader allows for the use of a standard CANopen configuration tool to load a hex file with new code via the CANopen network into the embedded device.

Note: There is no specification in the CANopen drafts and standards for the implementation of a CANopen compliant bootloader. The following is a recommendation by Embedded Systems Academy. The bootloader functionality described here is already used in multiple applications.

Embedded systems often use flash memory in order to simplify the process of updating the software/firmware running in embedded devices. Typically microcontrollers used with flash memory also provide ISP (In-System Programming) functionality - the microcontroller can communicate via one of its communication channels (typically a serial interface) and accept new code that is programmed into the flash memory.

In order to provide this functionality, a "bootloader" is required, a minimal piece of software that implements the communication and flash programming functions. The bootloader is often located in a protected memory area to prevent its accidental erasure. It can be activated during the boot-up of the processor (hence "bootloader") by setting a switch, button or jumper during reset.

If such a device is connected to a CANopen network, it would make sense to make the bootloader CANopen compliant. This frees the "other" communication channels from the bootloader task, as well as allowing the use of standard CANopen configuration tools as the communication partner providing the new code (hex file) to be loaded into the flash memory.

6.8.1 Minimal Functionality Required

A CANopen node whose only purpose is to accept a hex-file for loading into flash memory does not really have to be 100% CANopen compliant. It just needs to provide enough CANopen compatibility so that it does not interfere with any other communication on the network and provides a fully functional SDO server. This ensures that

SDO clients (like Masters, Managers or Configuration Tools) can make read and write accesses to the Object Dictionary in the node.

Thus the only CANopen features and communication channels that truly need to be implemented are the SDO server and the SDO request and response channels.

Sometimes it is desirable for the bootloader to be activated without having to physically touch the device (like setting a jumper, switch or button). Assuming the device in question is a CANopen node that also has a CANopen bootloader, a mechanism is needed that switches the device from its regular operation mode into its bootloader mode. In CANopen the straight-forward method would be to use a selected write sequence to an Object Dictionary entry as a command to switch the node into the bootloader mode.

An additional safety level can be added by adding checksum verification to all downloadable program segments. A bootloader should only jump to a code piece if it has been verified that the code piece in question is "real code." One method to implement this is by ensuring that any code piece downloaded must always have a pre-defined checksum – if a code piece in flash memory does not have that checksum, it is considered "trash" and should not be executed.

Segmented Transfer vs. Block Mode Transfer

Recall from Section 6.5.1 that when it comes to transferring larger blocks of data or code, there are two SDO transfer types that could be used: the segmented transfer (up to 7 bytes of data for each segment with one response message for each segment) or block transfer (up to 127 messages of 7 bytes each with one response message for the entire block).

Although the block transfer mode is more efficient for large transfers, it is not suitable for implementation on 8-bit devices. In addition, flash programming typically requires some timeouts after every byte (or group of bytes) so a communication model with only 7 new bytes at a time is quite welcome for the purpose of programming flash memory.

For the purpose of this section, we will simply assume that segmented transfer is used. However, in the spirit of the "openness" of CANopen, we leave it up to the individual engineers building the nodes to make the final decision.

6.8.2 Object Dictionary Entries Suggested for a Bootloader

The Object Dictionary entries that should be supported are:

- **OD entry [1000h,00h]: Device type information, read-only**
  Because there is no device type number standardized for a bootloader, a manufacturer specific value can be used. The Embedded Systems Academy uses 746F6F62h (ASCII representation is "boot") in their bootloader implementations.

- **OD entry [1001h,00h]: Error register, read-only**
  The bootloader can use this register to signal flash erase or programming failures. As an example, setting the voltage error could indicate that the flash erase or programming failed (often a specific voltage needs to be set to a pin of the flash memory to enable the erasing or programming functionality). Setting the manufacturer specific error bit could indicate an out-of-range error, seen if an attempt is made to program a memory location that is either protected or at which there is no flash memory.

- **OD entry [1018h,00h-04h]: Identity Object**
  The standard Identity Object as specified in CANopen [CiADS301]. At a minimum the Vendor ID, Product code and Revision number should be provided.

- **OD entry [1F50h,xxh]: Download program data**
  This Object Dictionary entry is described in [CiADSP302] and is used to directly accept the code programmed into the target memory. Subindex 0 is used to quantify how many different program or flash memory areas are available. The following Subindexes can each handle the download to one program or memory area. For many applications it is sufficient to implement one area (Subindex 1).

  The data type for the OD entry is "Domain" which indicates variable data size. The SDO download process itself uses segmented SDO transfer which supports such variable data sizes. A download is considered successful if no SDO Abort message occurs at any time during the download.

  Although not specified by the standard, the Embedded Systems Academy recommends using standard ASCII hex files as the files containing the program or data. Using a hex file has two benefits: the file contains the target address for the programmed data and the file also contains checksums making the downloading process more secure.

Because flash memory often needs to be erased before it can be re-programmed, Embedded Systems Academy further recommends implementing a specific erase command. For example, an erase could be initiated by sending the value 66726C63h (ASCII representation is "clrf") to the Object Dictionary entry [1F50h,01h], or other Subindexes to differentiate between different blocks or segments of flash memory.

- **OD entry [1F51h,xxh]: Program Control Object**
  This Object Dictionary entry is described in [CiADSP302] and is used to control the program(s) downloaded to [1F50h,xxh]. The essential command to implement is "Start program" which requires writing a 1 to the Object Dictionary entry. For example, if a program is downloaded to [1F50h,01h] it can be started by writing a 1 to [1F51h,01h].

  This Object Dictionary entry could also be used to activate the bootloader itself. If the regular CANopen application running on this node supports this entry, it should activate the bootloader upon receiving a 0 (zero = Stop Program).

6.8.3 Bootloader Flow Chart

Flow Chart 6.6 shows the basic operation steps that a CANopen bootloader should follow. It should be the first piece of code that is executed after the reset. It then needs to make a decision if it should stay in the bootloader mode or try to execute the application program. Typically the decision is made by reading some hardware and/or software settings that may or may not enable the bootloader. This could include a hardware switch or jumper, as well as a software flag that may have been set by the application before the reset.

Before calling the application code, the bootloader must ensure that valid code is in the code memory. This is typically accomplished by doing a checksum test. The bootloader should only jump to the application code if it is sure that valid application code is present at that location.

Once the bootloader is activated, it sends the CANopen boot-up message to inform other nodes on the network that it is initialized, and then waits for configuration and/or start-up. It will then activate the SDO server and handle all incoming SDO requests appropriately. This includes starting the application program if a write of 1 is detected to [1F51h,01h].

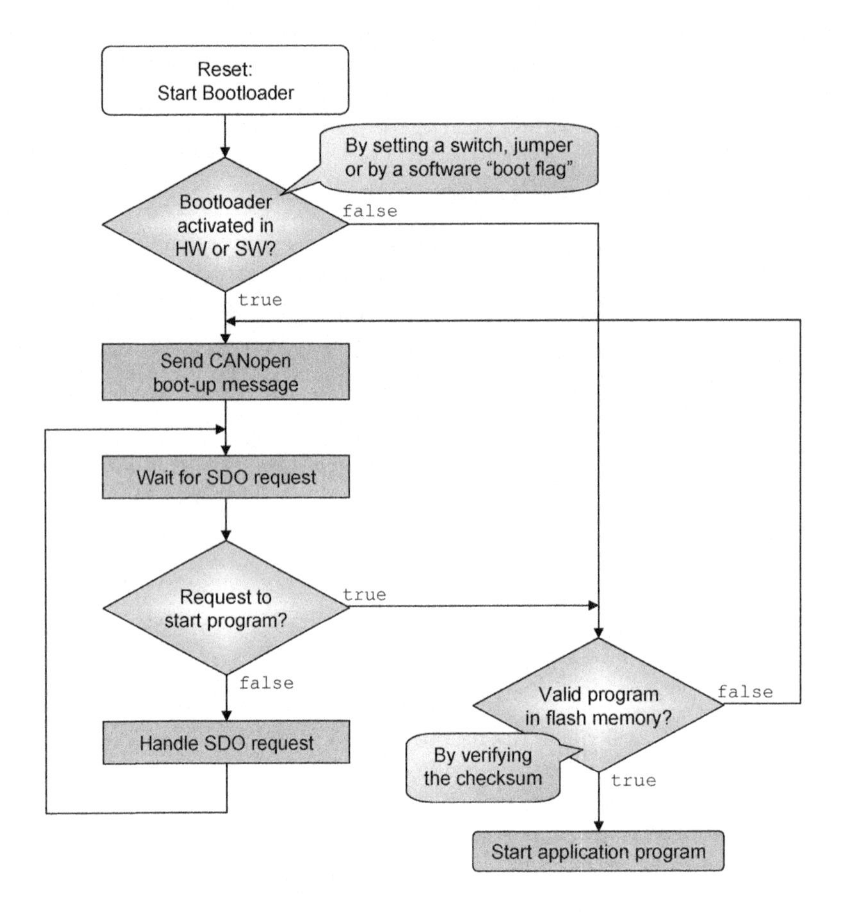

Flow Chart 6.6 Flow Diagram of a CANopen Bootloader

6.8.4 Handling the Bootloader

If a network is equipped with one or more nodes that have a bootloader as described in the previous sections, a standard CANopen configuration tool such as CANeds, proCANopen or PCANopenMagic can be used to handle the nodes. The following example assumes that only one program area is used and thus [1F50h,00h] and [1F51h,00h] are both 1. The following steps should be executed for an entire firmware/ software upgrade cycle.

1. The bootloader of any node becomes active by writing a 0 to [1F51h,01h].

2. To verify that a node is indeed in bootloader mode, the entry [1000h,00h] should be read (contents should be "boot").
3. If additional identification is required, the entries at [1018h,xxh] can be read.
4. If the flash memory needs to be erased first, a "clrf" can be written to to [1F50h,01h].
5. Now the hex file with the new code can be downloaded to [1F50h,01h]. The download is not considered successful if an SDO Abort or Emergency occurs during the download.
6. After completion, the error register should be read to verify that no error occurred.
7. Writing 1 to [1F51h,01h] starts the new code.

6.9 CANopen Implementation Example

Objective

This section includes text from the paper "Customizing CANopen for Use in an Automated Laboratory Instrument" by Michael B. Simmonds of Quantum Design, Alan Wilson of Quantum Design and Olaf Pfeiffer of Embedded Systems Academy. The paper was presented at the International CAN Conference 2002 in Las Vegas.

Due to its openness, there are many very different CANopen applications. Some engineers take specific advantage of the openness by customizing it towards a specific application, others need the standardization and remain well within specific device profiles to ensure inter-operation with third party products.

The following application example is that of a laboratory instrument. Because communication with third party products was not required, it would have been possible to customize the CANopen implementation to a point where it would not be CANopen compliant anymore. However, the engineers wanted to keep their application open for possible future enhancements involving other CANopen devices. As a result, the engineers came up with many customized features that still stayed within the CANopen specification.

This application uses optimization of the CAN physical layer as well as the CANopen application layer for use as an internal bus in a line of modularized laboratory instruments. Modifications and extensions are described for the pin assignments, Default

Connection Set, Emergency Object, and the Device Profile to better support the requirements of the hardware. In addition, a customized method for firmware updates via CANopen is implemented.

6.9.1 Background

The products of Quantum Design are relatively complex cryogenic instruments used by physicists and chemists to perform research in material science. These instruments contain several GPIB (IEEE-488) modules that are controlled by the operator from an application running on a PC. The GPIB was chosen primarily because it was widely used by the scientific/engineering community at that time, and enjoyed substantial hardware and software support. It also enabled users to integrate their own third party instruments into the measurement system.

As the engineers began looking toward more modular and modern architectures for their products, the shortcomings of the GPIB became more evident. The cost, complexity, and cable size for this 8-bit parallel bus becomes very unattractive when used with a larger number of modules. Even the size of the stacked 26-pin ribbon connectors became a major problem.

Furthermore, the protocols required for exchanging short packets with an array of modules is very time-consuming and negates all the advantages one would expect from a parallel bus; indeed, the effective bit-rate for actual data was only about 200kbps.

For these reasons, Quantum Design's engineers began searching for an alternative among the various serial buses that have become popular in the nearly two decades since the original design decision was made. They looked carefully at physical layers based on RS485, FireWire, Ethernet, USB, and CAN. CAN was chosen because of several perceived benefits including non-critical cables and connector impedance requirements, good hardware support at the chip level, excellent bus arbitration and error checking, and adequate bandwidth. While the engineers were initially impressed by the promise of very high bit-rates available with other buses, a closer evaluation showed that for this system they would be better off with the shorter frames and inherent collision-avoidance provided by CAN. Also, the high bit-rates of these other buses would limit cable lengths or turn impedance matching into a serious design concern.

Having chosen CAN for the lower-level protocols, the engineers needed to select (or invent) an "application layer" for the system. Several options were available, all based upon CAN: DeviceNet, CAN Kingdom, SDS, and CANopen. Here the decision became more a matter of taste since all of these approaches appeared to offer a reasonable set of features. The most important service requirement was a confirmed exchange of messages longer than 8 bytes; a Service Data Object in the terminology of CANopen. DeviceNet and CANopen appeared to be the most widely used and best supported of these options, with DeviceNet enjoying a much greater presence in the United States. But since it was not the intention to market fieldbus devices except as internal components in the laboratory instruments, this bias toward DeviceNet was not a particular concern. The higher bit rate and more efficient block transfers offered by CANopen were of greater importance.

6.9.2 Lab Instrumentation Requirements vs. CAN Physical Layer Specification

The modules comprising the instrument required several electrical services in addition to CAN communication, including 24VDC power, 50/60Hz line synchronization, hardware reset, and a low-jitter hardware sync signal. In addition, separate paths for returning unbalanced supply currents were needed for establishing system ground reference, and for dumping shield currents. Table 6.6 shows how CAN's 9-pin D-sub connector was adapted to fill all of these requirements. It should be noted that it is possible to connect a standard third party CANopen module into the network by using a cable with wires on only pins 2, 3, 7, and 9. In this case, the 24V supply would only provide power for the galvanically isolated CAN interface of the module.

The 50/60Hz sync line allows for very stable measurements in the presence of substantial line interference. Non-synchronous measurements are prone to exhibit low-frequency beats as their phase slowly slips with respect to the power lines.

Pin #	CAN Standard Pinout	QD-CAN Pinout
1	Reserved	-24VDC Supply
6	Optional Ground	System Ground
2	CAN-L Line	CAN-L Line
7	CAN-H Line	CAN-H Line
3	CAN Ground	24VDC Return
8	Reserved	SYNC-H/RS Line
4	Reserved	SYNC-L Line

Table 6.6 Comparison of Pin Assignments

Pin #	CAN Standard Pinout	QD-CAN Pinout
9	CAN_V+ Optional Supply	+24VDC Supply
5	CAN_SHLD Optional	Line-Sync (50/60 Hz)

Table 6.6 (Continued) Comparison of Pin Assignments

The SYNC-H/RS and SYNC-L lines allow a very accurate and stable timing signal to be distributed throughout the system. This differential signal can serve as a clock, sync, or trigger for various modules depending on their requirements. The sub-microsecond latency and jitter available through this SYNC mechanism is far better than could have been obtained through the CAN bus itself. Commands sent over the CAN interface can be used to configure or arm modules so they make use of this timing signal as desired.

CAN transceiver chips are used to control these SYNC lines, so in normal operation they will have the same electrical characteristics as the CAN bus. However, pulling Sync-H/RS to system ground level for a few microseconds will initiate a hardware reset of all modules connected to the bus.

6.9.3 Lab Instrumentation Requirements vs. CANopen Specification

As previously mentioned, the serial bus was selected for internal use in the instrument lines, therefore slavish adherence to an official specification was not required. Nevertheless, the engineers wished to avoid "reinventing the wheel" as much as possible. Earlier designs had suffered from incompletely engineered and under-documented interfaces between the components of the instruments. It was felt that such problems could be reduced by following an official standard that many people had already spent considerable time designing.

Typical CANopen Usage	Laboratory Instrumentation Bus
Every implementation quite different	Most instruments basically identical
Large number of simple modules	A few complex modules
Several interchangeable vendors	Vendor makes, uses own modules
Only a few generic module types	Unique, application-specific modules
Substantial module configuration req'd	Modules wake up knowing their role

Table 6.7 Different Networking Requirements

Typical CANopen Usage	Laboratory Instrumentation Bus
Modules exchange process data	User's computer collects process data
Minimal SDO traffic when operational	Commands continually sent via SDO
Computer used for config & diagnostics	User runs instrument through computer

Table 6.7 (Continued) Different Networking Requirements

It was also desirable to maintain the ability to run third party CANopen modules on the instrument's bus in the future. Therefore, any liberties taken with the DS-301 CANopen specification must be compatible with this requirement. The converse is not true, however; the engineers did not care that their own instruments would not function correctly in someone else's network or if the instruments failed to pass CiA conformance testing.

There is a substantial difference between the "flavor" of a typical CANopen fieldbus system and the bus required for the instruments. These differences are summarized in Table 6.8. As one can see, the developers of CANopen were attempting to solve a very different set of problems than the engineers at Quantum Design. Nevertheless, the CANopen application layer comes fairly close to providing Quantum Design with the necessary and sufficient services required.

	Std. CANopen	QD-CANopen
Maximum nodes in system	127	31
Default TPDOs / node	4	34
Default RPDOs / node	4	4
Default SDOs / node	1	1
Bit rates	10kbps - 1Mbps	500kbps - 1Mbps
Dynamic PDO Mapping	Optional	No
Variable COB IDs	Optional	No
Remote Response	Optional	No
29-Bit Identifiers	Optional	No
LMT Services	Optional	No

Table 6.8 Comparing Standard CANopen and QD-CANopen

	Std. CANopen	QD-CANopen
SDO Block Transfers	Optional	Mandatory
Error Control Protocol	Guarding or Heartbeat	Heartbeat
±24V System Power on Bus	No	Yes
Sync/Reset Signals on Bus	No	Yes
Line-sync Signal on Bus	No	Yes
Compatible with DS-301 Net	Yes	No
Compatible with QD-CANopen	Yes	Yes

Table 6.8 (Continued) Comparing Standard CANopen and QD-CANopen

The modules are quite application-specific and can be pre-configured to perform their assigned functions in the instruments. There is no need to have dynamic assignment of PDO data, nor is there even a need to have configurable COB IDs for the PDOs. In fact, it is desirable to have all these parameters "hard-wired" into the firmware so that the modules know everything about each other at power-on.

6.9.4 Optimizing the Default Connection Set

Using a fully static configuration of PDOs (the "default connection set" for the network), it needed to be ensured that it would provide maximum capability towards the specific application as far as the number of available communication channels was concerned. Modules for this application needed to be able to send as much process data as required. The CANopen specification only allows for four TPDOs and 4 RPDOs per node, a number that was felt to be insufficient for the system requirements. On the other hand, the number of nodes permitted by the CANopen specification was far in excess of what would be needed for the instruments.

QD TPDO	Default Connections	Assigned COB ID
1	TPDO 1 on N	180h + N
2	TPDO 1 on N + 32	1A0h + N
3	TPDO 1 on N + 64	1C0h + N
4	TPDO 1 on N + 96	1E0h + N

Table 6.9 QD-CANopen Connection Set on Node N (0 < N < 32)

QD TPDO	Default Connections	Assigned COB ID
5	RPDO 1 on N + 32	220h + N
6	RPDO 1 on N + 64	240h + N
7	RPDO 1 on N + 96	260h + N
8	TPDO 2 on N	280h + N
9	TPDO 2 on N + 32	2A0h + N
10	TPDO 2 on N + 64	2C0h + N
11	TPDO 2 on N + 96	2E0h + N
12	RPDO 2 on N + 32	320h + N
13	RPDO 2 on N + 64	340h + N
---	---	---
29	TSDO on N + 32	5A0h + N
30	TSDO on N + 64	5C0h + N
31	TSDO on N + 96	5E0h + N
32	RSDO on N + 32	620h + N
33	RSDO on N + 64	640h + N
34	RSDO on N + 96	660h + N

Table 6.9 (Continued) QD-CANopen Connection Set on Node N (0 < N < 32)

It was therefore decided to make a tradeoff: limit the nodes to 31 in order to expand the number of default TPDOs available on each node. Since the modules would serve primarily to control the instrument and report back process data, it was the TPDOs (as opposed to the RPDOs) that were in short supply. Therefore a strategy was devised for "stealing" COB IDs of the default PDOs excluded from the instrument network (32-127).

The technique allowed each node in the range 1-31 to have three additional images in the range of 32-127. Thus, node 1 also inherited the default PDOs for nodes 33, 65, and 97. The COB IDs for both RPDOs and TPDOs in this range were taken for use as TPDOs for the modules. In addition, there were the COB IDs of the default SDOs for these unused nodes. Thus a total of 34 separate Process Data Objects were made available on each module for reporting data back to the user's computer. Note that the four (4) RPDOs provided by the CANopen standard were retained as part of the modified default connection set. The order for assigning COB IDs to these 34 PDOs is shown in Table 6.9, and was chosen so that they would be used in order of decreasing priority.

Since the COB IDs were not allowed to be changed, the values listed in Table 6.9 could be relied upon at all times - the control computer and the other nodes automatically know a PDO's source node and number from its COB ID. And since dynamic data mapping is not allowed in the network, the type and meaning of the data payload is also immediately known throughout the network.

Although the COB IDs are not allowed to be changed, bit 31 in the dictionary entry for PDO communication parameter/COB ID can be set or cleared. According to DS-301, setting this bit disables the PDO and may prove useful in managing bus bandwidth with so many default TPDOs potentially defined.

Table 6.8 summarizes the differences described so far between the CANopen standard and Quantum Design's adaptation of it.

6.9.5 Enhancing the Role of the CANopen Emergency Object

Specification DS301 appears to leave quite a bit of flexibility in the use of the Emergency Object for device-specific purposes. There are several blocks of Error Codes that have been provided to facilitate this - F0xxh is for "Additional Functions," FFxxh covers "Device Specific" errors, 50xxh covers "Device Hardware" errors, and the entire "6xxxh" block is available for "Device Software" errors.

Quantum Design's engineers extended the definition of "emergency" to include any significant events or state changes that might occur in a module, but whose actual occurrence would not otherwise be known without performing continuous polling of the module. Having to do such polling is a substantial programming burden and adds unnecessarily to the loading on the CAN bus. Also, such polling cannot be done by another node on the network unless it has Client SDO capability, a service not supported by some commercial CANopen slave stacks.

The engineers proposed to use the block of codes from F000h to FFFFh to indicate when there had been a change-of-state in one of the modules subsystems. One bit (of the available 12) was assigned to each subsystem that could have externally significant state information. Whenever there was an event or state change in one of the module's subsystems, the corresponding bit-flag in the Error Code was set. An entry in the Object Dictionary was provided for the purpose of clearing the flag-bits of this Error Code, called the "Event Reset Register." Setting a bit of this object cleared the corresponding flag of the Error Code. According to the Emergency Object specification described in DS301, the EMCY (emergency) telegram is sent when (and only when) the Error Code changes. Thus clearing any bits in the Error Code will cause the EMCY telegram to be sent again. But rather than sending an Error Code of 0000h upon resetting one of these bits (as mentioned in the standard), the engineers proposed to send the new Fxxxh pattern. Clearing a bit in the Fxxxh group indicated that the module had been re-armed to send an EMCY telegram when another state change occurred on that subsystem. Otherwise no further state changes would be announced. A suitable EMCY 'inhibit time" was used in order to avoid consuming excessive bandwidth through this module-state reporting scheme.

The five bytes of the "Manufacturer Specific Error Field" provided a set of status flags and mode bit-fields. Up to 40 bits of state/mode information could be communicated with this scheme.

There has been considerable discussion about "borrowing" the official CANopen emergency protocol for the posting of state-change information. The alternative would have been to implement the above scheme using PDOs. Quantum Design elected to use the emergency protocol for several reasons - it gave these messages a higher priority than all normal PDOs, it allowed state information to be presented by a module even when that module was in the Pre-operational or Stopped mode, and it conserved COB IDs. In the case of the particular CANopen master API used, emergency messages had their own dedicated queue and callback function. This made them somewhat less likely to become lost.

6.9.6 Providing for Application Firmware Update via CANopen

Quantum Design needed the capability to update a device's firmware by loading new executable code directly through the device's own CAN interface. This requirement created an interesting challenge for the firmware architect since the CANopen stack is an integral part of the application firmware and must be compiled together with it. It was decided that the most reliable and robust method for implementing this capability was to have a "CANopen Loader" permanently available on the module. This minimal operating system only needed to provide a few services. It had to be able to implement an SDO-Server download, it needed to verify the checksum of the program it had downloaded, and it needed to transfer command to the downloaded program. Once the new downloaded program initialized and began execution, it completely replaced the loader and provided the code necessary to implement a CANopen interface for any further communications.

Two separate banks of flash memory were available on each module. One bank contained the CAN loader firmware in a write-protected area segment. The other bank was available for storing downloaded application code. When the device was first powered-up or after a hardware reset, program execution transferred to the loader program. The loader would then verify that the stored application had the correct checksum as part of its initialization process.

When the loader started, the node was in a special state not described within the CANopen specification. Entry into this mode was signaled by a Boot-up Message with a node number that was offset from the actual node by a value of 20 (720h+NodeID). This would not normally be a valid Boot-up Message within the restricted pre-defined connection set where only a range of Node IDs (1-31) is allowed, so it can be interpreted as an entry into the "System State." In this state, a node could receive data and report status via SDO, but it had no access to the application's Object Dictionary and could not process any PDOs. If the checksum of the current application firmware was determined to be correct by the system code, the node could be sent into its normal "Pre-operational" mode by sending the usual network command. Alternatively, new firmware could be downloaded by use of SDO writes. After the new firmware had been loaded, execution could be transferred over to it by bank-switching between the two memory blocks. After initialization, the "real" application sent a standard Boot-up telegram and entered into its Pre-operational mode. By using bank switching, having to re-map the interrupt vector table was avoided because a new table was automatically loaded in the operation.

6.9.7 Creating a Manufacturer's Device Profile

Quantum Design's modules were not intended for use on CANopen networks apart from their own internal instrument bus. Therefore the engineers were free to create their own device profile with a common set of dictionary entries in the range of 6000h – 9FFFh. According to the specification, non-standard device profiles should be indicated by a Device Profile Number of zero in the Device Type entry (1000h) of the devices Communication Profile. The 16 high bits of this entry are available to specify "Additional Information." A characteristic version number in this location is still used so that the system software can distinguish between different revisions of the device profile.

The device profile provides a device-independent structure for accessing common information such as module temperatures, module voltages, firmware checksums, error registers, and diagnostic test results. SDO writes to a standardized dictionary entry are used to command various levels of diagnostic tests.

6.9.8 Conclusions

Completely standardized CANopen would come remarkably close to filling the needs for the modularized instrumentation of Quantum Design. They used PDOs to report measurement results back to the master node in the PC. SDOs are used to set or read parameters as well as to issue confirmed commands to the nodes. The modification of the Default Connection Set, the expanded scope of the Emergency Object, the provision for CAN-based firmware updates, and the customization of the Device Profile go a long way toward making this high-level protocol a perfect fit for Quantum Design's requirements.

6.10 Example of an Entire Design Cycle

Objective

In this section we show you all steps involved in an entire design cycle of an embedded, completely pre-configured master-less system where individual nodes are part of the design and development.

For an example of a CANopen network configuration based on off-the-shelf CANopen components, see Chapter 4.

6.10.1 Defining Nodes and Process Variables

As discussed in Section 6.1, one should begin with the overall communication layout for any new CANopen development. In cases where the entire network is "embedded" and fairly fixed (not re-configured during operation) it has been proven useful to use the same Object Dictionary structure for all nodes. This means that one would define *all* process variables in the entire system as having unique Object Dictionary entries. The benefit of such an assignment is that the same Object Dictionary structure is used on every node. As an example, the "current speed" variable of a construction machine would have the same Object Dictionary entry number on both the producer and consumer of that variable.

> It should be noted that such a shared Object Dictionary limits the usage of generic I/O CANopen devices (DS401 compliant) in the same network. Generic I/O devices have their process variables at fixed locations that cannot be changed, so using two digital input devices would both produce inputs that are stored in their local Object Dictionaries at location 6000h.

6.10.2 Define Process Data Objects

Once the nodes and the process variables are defined, the next step is to define the PDOs used by each node. The usual approach is to start with all Transmit PDOs first, defining which nodes combine which process variables into which TPDO (TPDO mapping). The next step is defining the communication parameters, determining which CAN message IDs are used, as well as the transmission type (which determines when the TPDO is triggered). The list of TPDOs along with the selection of the transmission type also directly sets the bandwidth required to handle the PDO related communication.

Once the TPDOs are defined, the Receive PDOs are next. It needs to be decided which of the many PDOs used on the entire network need to be received and handled by each node.

> Defining the PDOs, both transmit and receive is a process that typically requires a few iterations of refinement. One of the challenges is to decide which process variables can be best combined into one PDO.
>
> A typical example would be command bytes sent by one device to several others. Should these bytes be sent one-by-one; one separate PDO to each receiving device? Or should they all be combined into one PDO and all sent at once?
>
> The latter approach has the benefit of optimizing the usage of the available network bandwidth. It means, however, that each receiver also receives data that it does not need (in this case the command bytes directed at the other devices). In general, the benefit of bandwidth optimization outweighs the disadvantage of handling some additional receive data. CANopen supports receiving such "unwanted" data by using dummy-mapping. A receiver may map unwanted data of a RPDO directly to so-called dummy entries – which basically means that the unwanted data is ignored.

6.10.3 Electronic Data Sheets, Device Configuration Files and Development Tools

Although tables and worksheets (discussed in Section 6.1) can be used for listing and defining nodes, Object Dictionary entries and PDOs, the final specification should always be made in the form of an Electronic Data Sheet (EDS) or Device Configuration File (DCF). They can be generated using an Electronic Data Sheet Editor such as Vector's CANeds. Because EDS and DCF are electronically readable, they can be used by standard CANopen tools during development and test of the network. They should simply be regarded as an electronically readable version of the Object Dictionary specification.

6.10.3.1 Configuration

Once the EDS and DCF files are in place, standard CANopen tools can use them. Configuration tools such as Vector's CANsetter or proCANopen use these files to provide the user with device access lists. For each node on the network, these configuration tools provide read and write access to each node's Object Dictionary as shown in Chapter 2, Figure 2.1.

6.10.3.2 Monitoring and Analyzing

In addition, the symbol information stored in the EDS and DCF files can be carried over to monitoring and analyzing tools such as Vector's CANalyzer. Instead of just displaying the "raw" CAN messages, the CANalyzer with the CANopen option can take the symbol information provided by the EDS and DCF and display it along with the data transmitted. This way a process data variable can be visualized directly in several windows along with its symbolic name. Windows using the symbol information include the data window, the trace window and the graphic window where variables can be tracked over time. Figure 5.20 in Chapter 5 shows a screen shot of the CANalyzer displaying the traffic on a CANopen network.

6.10.3.3 Simulation

In cases where simulation of a CANopen network is desirable, the CANoe simulation tool from Vector with its CANopen option can automatically simulate the network traffic simply by extracting the required communication information from the EDS and DCF files.

Even the simulation process of fairly complex networks can be setup within a few minutes. The first step is to define the network in proCANopen. Each node must be named, assigned a Node ID and an EDS and DCF file. Once all nodes are defined, a "make" process has to be started that produces all the files required for the simulation.

In CANoe, these files can be imported and used to simulate the entire network. The simulation includes all CANopen specific network traffic, including heartbeats, PDOs and even the entire SDO server of each simulated node. The configuration tools CANsetter and proCANopen can access the simulated nodes just as they would access the physical devices.

Naturally, the data within the PDOs is not simulated at this point, it is left at zero. The tools have no information about the specific data a node produces – however, they can simulate the communication behavior by producing the TPDOs according to the transmission type setting in the TPDO communication parameters.

It should be noted that the network traffic is not only simulated, it actually gets generated onto a CAN network if a CAN interface is connected to the system running the CANoe simulation model. Because single nodes within the simulation can be disabled one-by-one, it is possible to replace simulated nodes with the physical nodes in the network once they become available. This allows for a step-by-step migration from a simulated network to a physical network.

A more detailed simulation example with screen shots is shown in Chapter 4, Section 4.5.

Part Three: CANopen Reference

A Frequently Asked Questions

"The only way to discover the limits of the possible is to go beyond them into the impossible."
Arthur C. Clarke

This FAQ selection was adapted from the FAQ section of www.canbus.us and www.canopen.us, web pages dedicated to US users of CAN and CANopen. These web pages are maintained by the authors of this book.

A.1 General

A.1.1 What is the identifier of a node, message and/or variable?

In CAN and CANopen there are several identifiers used for different purposes. Beginners tend to mix these up, so pay close attention to the different meanings of the word "identifier":

1. On the CAN level (looking at CAN messages on the bus, generated by a CAN controller, no higher-layer protocol involved), the "identifier" is the CAN message identifier. Version CAN 2.0A allows for an 11-bit ID (theoretically up to 2048 dif-

ferent identifiers, some older CAN controllers might support less), version CAN 2.0B allows for a 29-bit ID.

2. Higher-layer protocols such as CANopen use node identifiers to address a specific node in the network. The Node ID is in the range of 1 to 127 in CANopen and 0 to 63 in DeviceNet. Sometimes, the Node ID is embedded into the CAN ID. The pre-defined connection set of CANopen places the Node ID into the lower 7 bits of the 11-bit CAN ID.

3. In CANopen, process variables have their own identifier. All process variables are located in the Object Dictionary, which is a look-up table using a 16-bit Index and a 8-bit Subindex. The Index and Subindex are used to identify one specific process variable in one specific node. A typical access (SDO access) to such a variable uses a CAN message that contains a Node ID within the CAN ID and the Index and Subindex (indexing a variable in the Object Dictionary) within the data field.

A.1.2 When and why would I need a higher-layer protocol such as CANopen instead of plain CAN?

CAN by itself only provides a method of exchanging up to 8 data bytes using message frames that have an identifier. Once you sit down and specify which identifier is used for which purpose - and what the contents of each message means (data types, byte order, variables) - you are already starting to specify your own higher-layer protocol.

As soon as a certain number of nodes, messages and process variables are involved, an in-house specification of that higher-layer protocol needs to be written and maintained.

With CANopen, all of that work has already been done. Instead of re-inventing existing technology, engineers can take advantage of CANopen and adopt existing technology.

Due to the "openness" of CANopen it is even possible to pick and choose the features required by an application and skip unwanted ones. CANopen literally reduces an in-house specification to a document that states which features of CANopen are used.

Most commercial CANopen source codes support the selection of features via "#define" statements in the source code.

If you are not yet certain if you want to use CANopen or implement an in-house higher-layer protocol, consider using www.MicroCANopen.com instead (see Section 6.3).

A.1.3 Do I need to have my node CANopen conformance tested?

If you are selling your node to 3rd parties as CANopen compliant, or if you purchase a CANopen node from a 3rd party, a CANopen conformance certificate gives both parties the extra insurance that the part actually behaves as specified. So if 3rd parties are involved (either selling or buying), CANopen nodes should be certified.

For in-house applications where all CANopen nodes come from the same manufacturer a conformance test would not really be required. However, if several engineers, teams or departments are involved, a conformance test can help, especially in the debug and test phase to confirm that a particular node behaves as expected.

In addition, the conformance test provides an aditional quailty check for all applications that require in-depth testing. These are often applications like medical devices and transport systems, but also include all applications or devices produced by an ISO9000-certified manufacturer.

An independent 3rd party such as the CiA (CAN in Automation) should do the conformance testing if 3rd parties are involved. This way an independent organization can be used as a mediator, in case the parties do not agree about the degree of conformance achieved. For details on CiA conformance testing see www.canopen.org/canopen/conformance.

For in-house applications, the conformance test can also be purchased and performed internally. The CANopen conformance test is sold by National Instruments.

A.1.4 Is 127 "really" the maximum number of nodes in a CANopen network?

Answer: Not really.

The original CANopen specification is limited to a maximum of 127 nodes, however it was kept "open" enough to provide room for extensions. In fact, there is so much room available that several very different solutions exist for this problem. There are application-specific, customized versions of CANopen networks installed that have more than 127 nodes. Contact your favourite CANopen consultant to learn how the support of more than 127 nodes can best be implemented in your application.

A.1.5 Can the Node IDs in a CANopen network be auto-assigned?

Although it is not part of the original standard, there are several, application-specific ID claiming implementations. Depending on the application requirements, several options are available. One software solution was introduced with device profile [CiADSP416].

Pure software solutions usually require that each node have a "unique number of bytes" either in the form of a serial number or a random number generator. Depending on bus speed and number of nodes, the claim-cycle may take several seconds to execute.

Other applications might require that the Node ID is related to the physical location in the network. So if the Node IDs should really be 1, 2, 3, etc, sorted by their physical location, then additional hardware is required.

There are many implementations that use an additional wire in the cabling for this purpose. And there are several options on how to use this wire - one is creating a daisy-chain (going in and out of each node, with each node having the ability to switch the signal for the next node in the chain). In this case the node "closest" to the master will be configured first. Once it is configured, it enables the next node in the chain.

Consultants can assist with any of the methods above.

A.2 Implementation Issues

A.2.1 How do I implement CANopen?

Depending on your expertise you might be tempted to simply buy the specification and start implementing it. Unless you already have a great deal of expertise with CAN and at least one other higher-layer protocol you should really evaluate this option carefully. If the project demands a limited CANopen implementation not requiring 100% CANopen compliance, then this might be a possible route.

However, as soon as more complex CANopen features or 100% CANopen conformance is required, the recommendation is to not start from scratch. The specification unfortunately does not contain all the details necessary, and many issues will only show up once the CANopen conformance test is started. Buying somebody else's implementation that has already passed the conformance test is a great shortcut, shaving several months off your development time.

A comparison of different implementation methods can be found in Chapter 6.

A.2.2 What are the memory requirements for a CANopen communication protocol stack?

The memory requirements differ a lot depending on the microcontroller architecture used and the CANopen features required by a particular node/application.

The nice thing about CANopen is that the set of mandatory functionality is very small and all the other functionality is optional. So a CANopen node can be built with exactly the required set of communication functions.

Although a minimal bootloader fits into 2kbytes of code, this does not really implement a true CANopen node, as there is no process data.

On an 8-bit microcontroller, take the following generalized rule-of-thumb:

Generic implementations require some 12-20kbytes of code space and about 500 to 1000 bytes of data memory. An implementation highly optimized towards a specific microcontroller can use 25% less code and data memory.

With MicroCANopen (as introduced in Section 6.3) code sizes stay in the 4-5kbytes area with about 200 bytes of RAM required.

A.2.3 Why do most CANopen applications use CAN 2.0A (base frames with 11-bit identifiers) and not CAN 2.0B (extended frames with 29-bit identifiers)?

CANopen was specified to support both protocol variants, but switching to the 29-bit identifiers has several consequences:

- Because only the address field is extended, but not the data field, the overall available data bandwidth decreases. More bits of overhead are added to each message.
- The overall reliability decreases, as the CRC checksum in each message now needs to cover 18 bits more.
- The worst-case delay for high priority messages becomes longer. Even a low priority message cannot be interrupted or aborted once it won arbitration to the bus. And the maximum length of messages on the bus is increased by 18 bits.

A.3 Performance

A.3.1 How do I calculate worst-case message delay times and data bandwidth?

The Embedded Systems Academy offers a free online worst-case calculator:

www.esacademy.com/faq/calc/can.htm

After entering the desired bit rate, the length of the shortest CAN message (enter number of data bytes in your shortest PDO) and the length of the longest CAN message (enter 8 - SDO message is always 8 bytes long) hit the "Calculate" button.

The form gets updated and shows you an approximation of the expected timing behavior. It is an approximation only - because CAN uses stuff bits the exact message length varies slightly with data contents.

A.3.2 How fast is a CAN/CANopen I/O cycle? (read INPUT, trasmit via CANopen, write OUTPUT)

Unfortunately there are MANY factors going into this formula. If you are looking at an entire I/O cycle, you have the following potential delays:

Input scan/recognizing loop until setting CAN transmit bit.
Depending on microcontroller performance and priorities this will be some 200us or more. With input filters, polarity changes or configurable "mapping" (which input goes to which CAN message) as provided by CANopen, this might more than double.

CAN message on bus delay.
All delays here are multiple bit times. At 1Mbps, a single bit time is 1us. At 250kbps, bit time is 4us. If there is currently a message on the bus, it cannot be aborted/interrupted. The maximum delay until any node gets a chance to try to arbitrate the bus depends on the longest possible message on the CANbus. With CANopen that is typically about 135 bit times.

CAN arbitration delay.
Assuming the message of our node has the highest priority, this delay will be zero. However, each message currently waiting to be transmitted anywhere on the bus having a higher priority will get the bus before our node. Each message delay is another 47 to 135 bit times.

Actual CAN transmit time.
See www.esacademy.com/faq/calc/can.htm: Some 47 to 135 bit times.

CAN receive interrupt delay in Output module.
Depending on microcontroller performance and interrupt priorities this will be some 100us or more. With output filters, polarity changes or configurable "mapping" (which CAN message contents goes to which output is configurable) as provided by CANopen, this might more than quadruple.

Many commercial CANopen stacks leave the CAN receive interrupt before applying the output "sometime" later in the background task, making the worst-case MUCH longer.

MINIMUM TOTAL (1Mbps example, highest priority):
200us + 135us + 0 + 60us + 100us = 495us

REALISTIC TOTAL (1Mbps example, medium priority):
300us + 135us + 135us + 60us + 300us = 930us

Conclusion: A complete I/O cycle can be completed within 1ms on a CANbus running at 1Mbps, if the priorities (interrupts on controllers and CAN message) are fairly high.

As soon as the bus' bitrate is slowed down or a lot of CANopen protocol functionality is added, the total I/O cycle time will be closer to 2-3ms.

A.3.3 How can the data bandwidth of a CAN/CANopen network be increased?

There are several options that can help to increase the bandwidth.

As the maximum possible bit rate depends on the maximum bus length, see if you can make your network shorter - and thus faster. Still, the maximum is about 1Mbps.

If you need the longer distance, see if a bridge/gateway can solve the problem. An existing 125kbps bus layout stretching to the maximum possible length can usually be doubled in speed if a bridge/gateway is introduced - separating the bus into two segments of 250kbps each.

Finally, there are several microcontrollers with multiple CAN interfaces. Consider using multiple CAN/CANopen networks to multiply the overall bandwidth.

A.4 Physical Layer

A.4.1 What is the difference between base frame format (CAN 2.0A) and extended frame format (CAN2.0B)?

Base frame format/CAN 2.0A uses an 11-bit ID in the CAN message identifier field allowing for 2048 different message IDs.

Extended frame format /CAN 2.0B uses a 29-bit ID in the CAN message identifier field allowing for more than 500 million different CAN IDs.

Extended frame format compatible devices can typically handle both the 11-bit and the 29-bit identifiers, even at the same time.

Both types of identifiers can be mixed on the same network. For the arbitration process the 11 most significant bits of the CAN 2.0B ID are arbitrated against the 11-bits of the CAN 2.0A.

Also see Section A.2.3.

A.4.2 What is the difference between "Basic CAN" and "Full CAN"?

Today these terms can be regarded as historic, as CAN interface implementations are continuously modified and updated by chip manufacturers adding more and more functionality that can no longer be accurately be described using the terms "Basic" or "Full".

Basic CAN:
The first "Basic CAN" implementation was made with Philips 82C200 CAN controller. Basic CAN controllers have primarily one message buffer each to transmit and receive messages. Unfortunately the microprocessor or microcontroller operating a Basic CAN interface needs to deal with many high-priority interrupts, since it gets an interrupt for every message received in the receive buffer. It then needs to decide in software if this message is of interest or not and, if it is, start the processing of the message.

The worst-case scenario for a Basic CAN controller is a CAN network running at 1Mbps with back-to back messages. In this scenario the shortest message on the bus is about 50 microseconds. That means that once a receive interrupt occurs, the receive buffer needs to be processed within 50 microseconds, otherwise the next message could potentially be received, overwriting the one previously received. For more details about Basic CAN, see Section 5.3.3.

"Full" CAN:
The "Full" CAN controller as first implemented by the Intel 82526 CAN controller eases the burden on the host microprocessor or microcontroller by offering extended hardware filtering capabilities. The "traditional" Full CAN controller has a total of 15 message buffers (called message objects) each of which can be configured to either transmit or receive. Thus each buffer can be configured to listen for exactly one specific CAN message identifier.

If the total number of CAN message IDs that a node needs to listen to can be kept below the number of message buffers available, the CAN controller will only issue a receive interrupt if a message is received that matches one of the specified CAN message IDs.

While this improves overall interrupt behavior (interrupts are not issued on messages that do not match any of the configured receive filters), the worst-case scenario for back-to-back messages does not change (compared to the Basic CAN controller). If back-to-back messages occur, an overrun can still occur within about 50 microseconds. For more details about Full CAN, see Section 5.3.3.

A.4.3 What is PeliCAN?

Philips came up with a solution to the back-to-back message problem as described in the previous paragraphs. The PeliCAN interface as implemented in the SJA1000 stand-alone CAN controller and the 8xC591 and LPC99x microcontrollers use a true FIFO buffer to receive messages.

This solves the back-to-back problem, as the worst-case timing for a potential buffer overrun is now about 500 microseconds instead of 50. This relieves the host processor, because the CAN receive interrupt does not need to be of the highest priority anymore. For more details about Philips' PeliCAN, see section Section 5.3.3.3.

A.4.4 How do I connect a CAN controller to the bus?

The most common form is to use a differential transceiver. One of the most popular transceivers is the Philips PCA82C251 high-speed, differential signal transceiver. See a datasheet at:

www.semiconductors.philips.com/pip/PCA82C251T

Notes:

The "Tx" pin of the CAN controller goes to the "Tx" pin of the transceiver.

The "Rx" pin of the CAN controller goes to the "Rx" pin of the transceiver.

The "Rs" (slope) control of the transceiver is set to GND (high speed mode) in most applications. If EMI is a problem in your application, consider other operating modes.

The "Vref" is an output of the transceiver that in many applications can be left unconnected.

A.4.5 How do I calculate the CAN bit timing of my CAN controller?

Either carefully read the data sheet of your CAN controller and go from there...

...or take a shortcut and use the program *CANtime* from Mike Schofield:

www.mjschofield.com/cantime.htm

CANtime is a shareware program that supports the following CAN controllers:

Intel 82527
Intel 87C196CA and 87C196CB (both variants)
Motorola 68HC08 (and all devices that use the msCAN08 module)
Motorola 68HC12 (and all devices that use the msCAN12 module)
Philips 82C200, 80C592, 80C598 and SJA1000 (all four devices)
Infineon C164CI and C167CR (both devices)
Infineon C505C and C515C (both devices)
ST Microelectronics ST10F168

B Physical Layer

Objective

This section gives you a quick reference to the CAN Physical Layer requirements for CANopen. For complete details please refer to the relelated CAN and CANopen specifications [CiADRP3031].

B.1 Recommended Bit Timings

Bit Rate	Maximum Bus Length	Bit Time	Time Quanta per bit	Length of 1 Time Quanta	Sample Point
1Mbps	25m	1µs	8	125ns	6 TQ (75%)
800kbps	50m	1.25µs	10	125ns	8 TQ (80%)
500kbps	100m	2µs	16	125ns	14 TQ (87.5%)
250kbps	250m	4µs	16	250ns	14 TQ (87.5%)

Table B.1 CANopen Recommended Bit Timings

Bit Rate	Maximum Bus Length	Bit Time	Time Quanta per bit	Length of 1 Time Quanta	Sample Point
125kbps	500m	8µs	16	500ns	14 TQ (87.5%)
50kbps	1km	20µs	16	1.25µs	14 TQ (87.5%)
20kbps	2.5km	50µs	16	3.125µs	14 TQ (87.5%)
10kbps	5km	100µs	16	6.25µs	14 TQ (87.5%)

Table B.1 (Continued) CANopen Recommended Bit Timings

C Data Types

Objective

The data types form the basis of the data storage in the Object Dictionary. This appendix provides a quick reference to the data types used in CANopen and examples of their usage. It can also be used to determine which types are right for manufacturer-specific entries.

C.1 Basic Data Types

Basic Data Types are the simplest types defined in CANopen. They can be used to construct Extended and Complex Data Types, and they may be stored in a single sub-entry of the Object Dictionary.

C.1.1 Boolean

Definition: A single bit value. The value zero indicates a false condition and the value one indicates a true condition.

Name Notation: BOOLEAN

Range: 0 to 1

Examples: 0
1

OD Location: 0001h

C.1.2 Void

Definition: A bit sequence of varying length. The value that a void type may have is undefined, and this type is commonly used as a place holder for reserved fields in complex data types or in the Object Dictionary.

Name Notation: VOID*n* represents a Void type with a bit sequence of *n* bits.

Range: Undefined

OD Location: Not defined in the Object Dictionary

C.1.3 Unsigned Integer

Definition: An non-negative integer value.

Name Notation: UNSIGNED*n* represents an unsigned integer value stored in *n* bits.

Range: 0 to $2^n - 1$ where *n* is the number of bits used to store the value.

Examples: 0
45
5611

OD Location: 0005h UNSIGNED8
0006h UNSIGNED16
0016h UNSIGNED24
0007h UNSIGNED32
0018h UNSIGNED40
0019h UNSIGNED48
001Ah UNSIGNED56
001Bh UNSIGNED64

C.1.4 Signed Integer

Definition:	An integer value.
Name Notation:	INTEGER*n* represents an integer value stored in *n* bits.
Range:	-2^{n-1} to $2^{n-1} - 1$
Examples:	-45 6234 -182
OD Location:	0002h INTEGER8 0003h INTEGER16 0010h INTEGER24 0004h INTEGER32 0012h INTEGER40 0013h INTEGER48 0014h INTEGER56 0015h INTEGER64

C.1.5 Floating Point (Real)

Definition:	A floating point/real value conforming to the IEEE 754-1985 standard.
Name Notation:	REAL32 represents a 32-bit value, usually called "single precision." REAL64 represents a 64-bit value, usually called "double precision."
Range:	$-(2-2^{-23})^{127}$ to $(2-2^{-23})^{127}$ single precision $-(2-2^{-52})^{1023}$ to $(2-2^{-52})^{1023}$ double precision
Examples:	-23.643732 0.117774 34562.545324

OD Location: 0008h REAL32
0011h REAL64

C.1.6 Visible Character

Definition: A non-negative integer value in the range 20h to 7Eh inclusive, corresponding to the non-control (printable) characters in the ASCII character set. Stored in eight bits (i.e. it is a limited value range UNSIGNED8).

Name Notation: VISIBLE_CHAR

Range: 20h to 7Eh

Examples: 45h (ASCII 'E')
7Ah (ASCII 'z')
21h (ASCII '!')

OD Location: Not defined in the Object Dictionary

C.2 Extended Data Types

Extended Data Types are types constructed from a collection of more than one Basic Data Type. They may be stored in a single subentry of the Object Dictionary.

C.2.1 Octet String

Description: A sequential collection (array) of UNSIGNED8 values of varying length. This allows sequences of 8-bit values to be used, for example storing binary data.

Note that although the word "string" appears in the name, the stored value may not be printable in the ASCII character set, which is limited to seven bits.

Name Notation: OCTET_STRING*n* where *n* is the number of UNSIGNED8 values in the array.

Examples: 9Ah,3Bh,11h
4, 251,45

OD Location: 000Ah

Note that although the Octet String type may have varying length, only one version of the type is stored in the Object Dictionary, omitting length indication. This is because an Octet String can be any length and it is simply not possible to store every possible length in the Object Dictionary.

C.2.2 Visible String

Description: A sequential collection (array) of VISIBLE_CHAR values of varying length. This allows sequences of printable characters from the ASCII character set to be used, for example storing names, descriptions, versions, etc.

Note that unlike C, a null terminator is not required on the end of the array.

Name Notation: VISIBLE_STRING*n* where *n* is the number of VISIBLE_CHAR values in the array.

Examples: Version 1.00
Embedded Systems Academy, Inc.
CANopen

OD Location: 0009h

Note that although the Visible String type may have varying length, only one version of the type is stored in the Object Dictionary, omitting length indication. This is because a Visible String can be any length and it is simply not possible to store every possible length in the Object Dictionary.

C.2.3 Unicode String

Description: A sequential collection (array) of UNSIGNED16 values of varying length. This allows sequences of 16-bit values to be used, for example storing text in languages that do not use the Roman alphabet such as Hebrew, Russian, Greek, etc.

Name Notation: UNICODE_STRING*n* where *n* is the number of UNSIGNED16 values in the array.

OD Location: 000Bh

Note that although the Unicode String type may have varying length, only one version of the type is stored in the Object Dictionary, omitting length indication. This is because a Unicode String can be any length and it is simply not possible to store every possible length in the Object Dictionary.

C.2.4 Time of Day

Description: The Time of Day type is a collection of basic types grouped together to store the date and time to the nearest millisecond since the epoch, which is midnight January 1, 1984.

The type is stored in 48 bits arranged as follows:

UNSIGNED28 ms
VOID4 reserved
UNSIGNED16 days

ms stores the current time in milliseconds since midnight of the day specified with the days value

days stores the number of whole days since the epoch

Note the void type forces the days value to be aligned on a 16-bit boundary and therefore for the Time of Day type to fit exactly into six bytes. Also note that the Time of Day type does not make any provision for time zones, therefore the epoch used is the local epoch.

Name Notation: TIME_OF_DAY

OD Location: 000Ch

Examples: ms = 20040000, days = 3 January 4th 1984, 5:34am
ms = 43320000, days = 8 Janurary 9th 1984, 12:02pm

C.2.5 Time Difference

Description: The Time Difference type is a collection of basic types grouped together to store a length of time to the nearest millisecond.

The type is stored in 48 bits arranged as follows:

UNSIGNED28 ms
VOID4 Reserved
UNSIGNED16 days

ms stores the current time in milliseconds since midnight of the day specified with the days value.

Days stores the number of whole days.

Note the void type forces the days value to be aligned on a 16-bit boundary and therefore for the Time Difference type to fit exactly into six bytes.

Name Notation: TIME_DIFFERENCE

OD Location: 000Dh

Examples: ms = 20040000, days = 3, 3 days, 5 hours, 34 minutes
ms = 43320000, days = 8, 8 days, 12 hours, 2 minutes

C.2.6 Domain

Description: A block of data of arbitrary length. The contents, size and format of the block of data are not defined in the CANopen specification. This type is especially useful for application specific data where the length of data may vary each time it is used, for example to store firmware.

Name Notation: DOMAIN

OD Location: 000Fh

Note that although the Domain type may have arbitary length, only one version of the type is stored in the Object Dictionary, omitting length indication. This is because a

Domain can be any length and it is simply not possible to store every possible length in the Object Dictionary.

C.3 Complex Data Types

Complex Data Types are types constructed from a collection of more than one Basic or Extended Data Type. They are stored in multiple entries of the Object Dictionary, with each Basic or Extended Data Type used occupying one Subentry in the Object Dictionary.

The first value in a Complex Data Type always indicates the number of types that follow for a specific value, i.e. the highest Subindex used to store the value of the type in the Object Dictionary.

C.3.1 PDO Communication Parameter Record

Description: This type contains a description of the communication characteristics for a PDO. It is constructed as follows:

UNSIGNED8 Number of Entries
UNSIGNED32 COB ID
UNSIGNED8 Transmission Type
UNSIGNED16 Inhibit Time
UNSIGNED8 Reserved
UNSIGNED16 Event Timer

COB ID stores the COB ID used for the PDO.

Transmission Type indicates how and when the PDO is transmitted.

Inhibit Time indicates if there is a limit on the maximum transmission frequency of the PDO.

Event Timer is used to determine a specific frequency of transmission of the PDO.

OD Location: 0020h

Name Notation: PDO_COMMUNICATION_PARAMETER

C.3.2 PDO Mapping Parameter Record

Description: The PDO Mapping Parameter type stores the mapping of process data into a specific PDO. It is constructed as follows:

UNSIGNED8 Number of Entries
UNSIGNED32 1st Object Mapped
UNSIGNED32 2nd Object Mapped
UNSIGNED32 3rd Object Mapped
UNSIGNED32 63rd Object Mapped
UNSIGNED32 64th Object Mapped

Up to 64 objects may be mapped into a PDO. There is one entry in the type for each mapped object, therefore the number of entries used for a value for this type depends on the number of objects mapped into a specific PDO.

OD Location: 0021h

Name Notation: PDO_MAPPING

C.3.3 SDO Parameter Record

Description: This type stores details of a specific SDO server implemented in a CANopen node. It is constructed as follows:

UNSIGNED8 Number of Entries
UNSIGNED32 COB ID Client to Server
UNSIGNED32 COB ID Server to Client
UNSIGNED8 Node ID of SDO Client/Server

The COB ID fields describe the COB IDs used for SDO communication in both directions to and from the node. The Node ID is the ID of the other node involved in the SDO communications.

OD Location: 0022h

Name Notation: SDO_PARAMETER

C.3.4 Identity Record

Description: The Identity Record type stores basic information about who manufactured the node, the product, revision and serial number. It is constructed as follows:

UNSIGNED8 Number of Entries
UNSIGNED32 Vendor ID
UNSIGNED32 Product Code
UNSIGNED32 Revision Number
UNSIGNED32 Serial Number

OD Location: 0023h

Name Notation: IDENTITY

C.3.5 Debugger Parameter Record

Description: This type defines a method of providing a command interface to a node, and allows the node to return responses to commands.

UNSIGNED8 Number of Entries
OCTET_STRING Command
UNSIGNED8 Status
OCTET_STRING Reply

OD Location: 0024h

Name Notation: DEBUGGER_PAR

C.3.6 Command Parameter Record

Description: This type defines a method of providing a command interface to a node, and allows the node to return responses to commands.

UNSIGNED8 Number of Entries
OCTET_STRING Command
UNSIGNED8 Status
OCTET_STRING Reply

OD Location: 0025h

Name Notation: COMMAND_PAR

C.4 Transfer Format

C.4.1 Basic Data Types

Data is transmitted in bytes. If the data is not a whole number of bytes, then it must be grouped with enough bits to construct a whole number of bytes. The other bits may be unused but must be present in the transmission.

Bits in a byte are transmitted with the most significant bit first.

Bytes are transmitted using the little-endian format with the least significant byte being transmitted first.

For a set of bits grouped into bytes the transmission format is as follows:

Byte	Contents	Order of Transmission
0	Bits 7 to 0	First
1	Bits 15 to 8	Second
2	Bits 23 to 16	Third
3	Bits 31 to 24	Fourth
4	Bits 39 to 32	Fifth
5	Bits 47 to 40	Sixth
6	Bits 55 to 48	Seventh
7	Bits 63 to 56	Eighth

Table C.1 Basic Data Type Transmission Order

For example, the UNSIGNED32 value 1A2B3C4Dh is transmitted as:

4Dh, 3Ch, 2Bh, 1Ah

The INTEGER16 value –266 (= FEF6h) is transmitted as:

F6h, FEh

The BOOLEAN value TRUE is transmitted as:

01h

C.4.2 Extended Data Types

The bit sequence for transmitting an Extended Data Type is formed by concatenating the bit sequences of each Basic Type used. The order of concatenation is from the first listed type to the last listed type in the definition.

For example, the Time of Day type is defined as:

UNSIGNED28 ms
VOID4 reserved
UNSIGNED16days

The bits used are shown in Table C.2.

Value	Bits Used
ms	Bits 27 to 0
reserved	Bits 31 to 28
days	Bits 47 to 32

Table C.2 Bit Allocation for Time of Day Type

The transmission order is shown in Table C.3.

Byte	Contents	Order of Transmission
0	Bits 7 to 0	First
1	Bits 15 to 8	Second
2	Bits 23 to 16	Third
3	Bits 31 to 24	Fourth

Table C.3 Transmission Order for Time of Day Type

Byte	Contents	Order of Transmission
4	Bits 39 to 32	Fifth
5	Bits 47 to 40	Sixth

Table C.3 (Continued) Transmission Order for Time of Day Type

Thus, if a Time of Day entry had the value:

ms = 1122334h
reserved = 0
days = 8899h

Then the value would be transmitted as:

34h, 23h, 12h, 01h, 99h, 88h

D The Object Dictionary

Objective

CANopen is based primarily around the Object Dictionary. Because of this the entries are numerous and varied. This appendix aims to provide a quick reference to many basic Object Dictionary entries and indicate how they are used.

D.1 Object Dictionary Organization

The Object Dictionary is divided into the sections shown in Table D.1.

Indexes Used	Description
0000h	Reserved
0001h – 025Fh	Data Type Definitions
0260h – 0FFFh	Reserved
1000h – 1FFFh	Communication Profile

Table D.1 Object Dictionary Sections

Indexes Used	Description
2000h – 5FFFh	Manufacturer Specific
6000h – 9FFFh	Standardized Device Profile
A000h – BFFFh	Standardized Interface Profile
C000h – FFFFh	Reserved

Table D.1 (Continued) Object Dictionary Sections

D.2 Data Type Definitions

D.2.1 Object Dictionary Sections

The Data Type Definitions area of the Object Dictionary is subdivided into the sections shown in Table D.2:

Indexes Used	Description
0001h – 001Fh	Basic and Extended Data Types
0020h – 003Fh	Complex Data Types
0040h – 005Fh	Manufacturer Specific Complex Data Types
0060h – 007Fh	Device Profile 0 Basic and Extended Data Types
0080h – 009Fh	Device Profile 0 Complex Data Types
00A0h – 00BFh	Device Profile 1 Basic and Extended Data Types
00C0h – 00DFh	Device Profile 1 Complex Data Types
00E0h – 00FFh	Device Profile 2 Basic and Extended Data Types
0100h – 011Fh	Device Profile 2 Complex Data Types
0120h – 013Fh	Device Profile 3 Basic and Extended Data Types
0140h – 015Fh	Device Profile 3 Complex Data Types
0160h – 017Fh	Device Profile 4 Basic and Extended Data Types
0180h – 019Fh	Device Profile 4 Complex Data Types
01A0h – 01BFh	Device Profile 5 Basic and Extended Data Types
01C0h – 01DFh	Device Profile 5 Complex Data Types
01E0h – 01FFh	Device Profile 6 Basic and Extended Data Types
0200h – 021Fh	Device Profile 6 Complex Data Types
0220h – 023Fh	Device Profile 7 Basic and Extended Data Types
0240h – 025Fh	Device Profile 7 Complex Data Types

Table D.2 Data Type Object Dictionary Sections

D.2.2 Object Dictionary Implementation

Any data type may optionally be implemented in the Object Dictionary of a node.

Basic or Extended Data Types are implemented as follows:

Index	Subindex	Type	Value	Access
See Table D.2	00h	UNSIGNED32	Bit size of type or zero for variable	Read Only

Table D.3 Basic and Extended Data Type Implementation

The following example shows how the Time Of Day type would be implemented in the Object Dictionary.

Index	Subindex	Type	Value	Access
000Ch	00h	UNSIGNED32	30h (48)	Read Only

Table D.4 Example Implementation of an Extended Data Type

Complex Data Types are implemented as follows:

Index	Subindex	Type	Value	Access
See Table D.2	00h	UNSIGNED8	Highest Subindex used by type (= *n*)	Read Only
	01h	UNSIGNED16	OD Index of Suben-try type	Read Only
	n	UNSIGNED16	OD Index of Suben-try type	Read Only

Table D.5 Complex Data Type Implementation

The following example shows how the PDO Communication Parameter Record would be implemented in the Object Dictionary.

Index	Subindex	Type	Value	Access
0020h	00h	UNSIGNED8	05h	Read Only
	01h	UNSIGNED16	0007h	Read Only
	02h	UNSIGNED16	0005h	Read Only
	03h	UNSIGNED16	0006h	Read Only
	04h	UNSIGNED16	0005h	Read Only
	05h	UNSIGNED16	0006h	Read Only

Table D.6 Example Implementation of a Complex Data Type

D.3 Communication Profile

D.3.1 Object Dictionary Entries

The following table gives an overview of all Object Dictionary entries in the Communication Profile section of the Object Dictionary.

Index	Name	Type	Access
1000h	Device Type	UNSIGNED32	Read Only
1001h	Error Register	UNSIGNED8	Read Only
1002h	Manufacturer Status Register	UNSIGNED32	Read Only
1003h	Pre-defined Error Field	UNSIGNED32	Read Only
1004h	Reserved	-	-
1005h	SYNC COB ID	UNSIGNED32	Read/ Write
1006h	Communication Cycle Period	UNSIGNED32	Read/ Write
1007h	Synchronous Window Length	UNSIGNED32	Read/ Write
1008h	Manufacturer Device Name	VISIBLE_STRING	Read Only
1009h	Manufacturer Hardware Version	VISIBLE_STRING	Read Only
100Ah	Manufacturer Software Version	VISIBLE_STRING	Read Only
100Bh	Reserved	-	-
100Ch	Guard Time	UNSIGNED16	Read/ Write

Table D.7 Communication Profile Object Dictionary Entries

Index	Name	Type	Access
100Dh	Life Time Factor	UNSIGNED8	Read/ Write
100Eh	Reserved	-	-
100Fh	Reserved	-	-
1010h	Store Parameters	UNSIGNED32	Read/ Write
1011h	Restore Default Parameters	UNSIGNED32	Read/ Write
1012h	TIME COB ID	UNSIGNED32	Read/ Write
1013h	High Resolution Time Stamp	UNSIGNED32	Read/ Write
1014h	Emergency COB ID	UNSIGNED32	Read/ Write
1015h	Emergency Inhibit Time	UNSIGNED16	Read/ Write
1016h	Consumer Heartbeat Time	UNSIGNED32	Read/ Write
1017h	Producer Heartbeat Time	UNSIGNED16	Read/ Write
1018h	Identity	IDENTITY (0023h)	Read Only
1019h	Reserved	-	-
1020h	Verify Configuration	UNSIGNED32	Read/ Write
1021h	Store EDS	DOMAIN	Read/ Write
1022h	Storage Format	UNSIGNED8	Read/ Write
1023h	OS Command	COMMAND_PAR (0025h)	Read/ Write

Table D.7 (Continued) Communication Profile Object Dictionary Entries

Index	Name	Type	Access
1024h	OS Command Mode	UNSIGNED8	Write Only
1025h	OS Debugger Interface	DEBUGGER_PAR (0024h)	Read/ Write
1026h	OS Prompt	UNSIGNED8	Read/ Write
1027h	Module List	UNSIGNED16	Read Only
1028h	Emergency Consumer	UNSIGNED32	Read/ Write
1029h	Error Behavior	UNSIGNED8	Read/ Write
102Ah to 11FFh	Reserved	-	-
1200h	1st SDO Server Parameters	SDO_PARAMETER (0022h)	Read Only
1201h to 127Fh	Additional SDO Server Parameters	SDO_PARAMETER (0022h)	Read/ Write
1280h	1st SDO Client Parameters	SDO_PARAMETER (0022h)	Read/ Write
1281h to 12FFh	Additional SDO Client Parameters	SDO_PARAMETER (0022h)	Read/ Write
1300h to 13FFh	Reserved	-	-
1400h	1st Receive PDO Parameter	PDO_COMMUNICATION_ PARAMETER (0020h)	Read/ Write
1401h	2nd Receive PDO Parameter	PDO_COMMUNICATION_ PARAMETER (0020h)	Read/ Write

Table D.7 (Continued) Communication Profile Object Dictionary Entries

Index	Name	Type	Access
1402h	3rd Receive PDO Parameter	PDO_COMMUNICATION_ PARAMETER (0020h)	Read/ Write
1403h	4th Receive PDO Parameter	PDO_COMMUNICATION_ PARAMETER (0020h)	Read/ Write
1404h to 15FFh	Additional Receive PDO Parameters	PDO_COMMUNICATION_ PARAMETER (0020h)	Read/ Write
1600h	1st Receive PDO Mapping	PDO_MAPPING (0021h)	Read/ Write
1601h	2nd Receive PDO Mapping	PDO_MAPPING (0021h)	Read/ Write
1602h	3rd Receive PDO Mapping	PDO_MAPPING (0021h)	Read/ Write
1603h	4th Receive PDO Mapping	PDO_MAPPING (0021h)	Read/ Write
1604h to 17FFh	Additional Receive PDO Mappings	PDO_MAPPING (0021h)	Read/ Write
1800h	1st Transmit PDO Parameter	PDO_MAPPING (0020h)	Read/ Write
1801h	2nd Transmit PDO Parameter	PDO_COMMUNICATION_ PARAMETER (0020h)	Read/ Write
1802h	3rd Transmit PDO Parameter	PDO_COMMUNICATION_ PARAMETER (0020h)	Read/ Write
1803h	4th Transmit PDO Parameter	PDO_COMMUNICATION_ PARAMETER (0020h)	Read/ Write
1804h to 19FFh	Additional Transmit PDO Parameters	PDO_COMMUNICATION_ PARAMETER (0020h)	Read/ Write
1A00h	1st Transmit PDO Mapping	PDO_MAPPING(0021h)	Read/ Write
1A01h	2nd Transmit PDO Mapping	PDO_MAPPING (0021h)	Read/ Write

Table D.7 (Continued) Communication Profile Object Dictionary Entries

Index	Name	Type	Access
1A02h	3rd Transmit PDO Mapping	PDO_MAPPING (0021h)	Read/ Write
1A03h	4th Transmit PDO Mapping	PDO_MAPPING (0021h)	Read/ Write
1A04h to 1BFFh	Additional Transmit PDO Mappings	PDO_MAPPING (0021h)	Read/ Write
1C00h to 1F9Fh	Reserved	-	-
1FA0h to 1FCFh	Object Scanner List	UNSIGNED32	Read/ Write
1FD0h to 1FFFh	Object Dispatching List	UNSIGNED64	Read/ Write

Table D.7 (Continued) Communication Profile Object Dictionary Entries

D.3.2 Device Type (1000h)

Index	1000h
Name	Device Type
Mandatory	Yes

Subindex	00h
Type	UNSIGNED32
Default Value	Determined by device profile used
Access	Read Only
Mandatory	Yes
Map to PDO	No

Description: This entry indicates the number of the device profile used and often provides some additional basic information about which features of the device profile are used in the node. The entry value is constructed as follows:

Bit	Description
0 – 15	Device Profile Number
16 - 31	Additional Information

Table D.8 Device Type Contents

The Additional Information that may be provided in this entry is defined in the Device Profile specification.

If the node does not use a device profile then the Device Profile Number is zero and the Additional Information value is undefined, but often set to zero as well.

If the node uses more than one device profile, then the Device Profile Number is the number of the first Device Profile used by the node, and Additional Information is FFFFh.

This entry is mandatory and must be implemented in all CANopen nodes. Because of this it is often used as a way of dynamically scanning for nodes connected to the network.

Example: 00030191h

Digital input/output module 191h = 401, which is the number of the digital input/output device profile. 0003h = the module implements both digital inputs and outputs.

D.3.3 Error Register (1001h)

Index	1001h
Name	Error Register
Mandatory	Yes

Subindex	00h
Name	Error Register
Type	UNSIGNED8
Default Value	Not defined
Access	Read Only
Mandatory	Yes
Map to PDO	Yes

Description: The error register value indicates if various types of errors have occurred. The following table indicates the bits used. Bit zero must be implemented. All other bits are optional.

Bit	Description	Mandatory
0	Generic Error	Yes
1	Current	No
2	Voltage	No
3	Temperature	No
4	Communication Error	No
5	Device Profile Defined Error	No
6	Reserved (always zero)	No
7	Manufacturer Specific Error	No

Table D.9 Error Register Contents

A set bit indicates the specified error has occurred.

The Generic Error bit is set when any type of error occurs.

The Error Register is included in byte two of the Emergency object, but may also be mapped into PDOs.

Example: 05h
Voltage error has occurred

D.3.4 Manufacturer Status Register (1002h)

Index	**1002h**
Name	Manufacturer Status Register
Mandatory	No

Subindex	**00h**
Name	Manufacturer Status Register
Type	UNSIGNED32
Default Value	Not defined
Access	Read Only
Mandatory	No
Map to PDO	Yes

Description: The Manufacturer Status register contents are undefined in the CANopen specification. Manufacturers may use this entry for any purpose desired.

D.3.5 Pre-Defined Error Field (1003h)

Index	1003h
Name	Pre-Defined Error Field
Mandatory	No

Subindex	00h
Name	Number of Errors
Type	UNSIGNED8
Default Value	0
Access	Read/Write
Mandatory	Yes if this entry is implemented
Map to PDO	No

Subindex	01h – FEh
Name	Standard Error Field
Type	UNSIGNED32
Default Value	Not defined
Access	Read Only
Mandatory	No
Map to PDO	No

Description: This entry contains up to 254 of the most recent errors that occurred in the node and resulted in the transmission of the Emergency Object.

Subentries 01h to FEh store information about the errors, with entry [1003h,01h] storing the most recent error and entry [1003h,FEh] storing the oldest error.

When a new error occurs it is stored in entry [1003h,01h] and any currently existing Subentries are shuffled down. For example, the error previously stored at [1003h,01h] will be moved to [1003h,02h], and the error previously stored at [1003h,02h] will be moved to [1003h,03h], etc.

Entry [1003h,00h] can be read to determine the number of errors currently stored. Writing zero to [1003h,00h] erases the error history.

Each entry is constructed as follows:

Bit	Description
0 – 15	Error Code as transmitted in the Emergency Object
16 - 31	Manufacturer Specific Additional Information

Table D.10 Pre-Defined Error Field Contents

Example: 00003000h
Voltage error occurred

D.3.6 SYNC COB ID (1005h)

Index	1005h
Name	SYNC COB ID
Mandatory	Yes if the node transmits or receives synchronous PDOs or if any PDO supports changing the transmission type to a synchronous type

Subindex	00h
Name	SYNC COB ID
Type	UNSIGNED32
Default Value	00000080h or 40000080h
Access	Read/Write
Mandatory	Yes if the node transmits or receives synchronous PDOs or if any PDO supports changing the transmission type to a synchronous type
Map to PDO	No

Description: Contains the COB ID used by the SYNC Object along with a flag to indicate if the node generates the SYNC Object or not.

If the PDOs supported by the node permit a changing of transmission type to one of the synchronous transmission types, then this entry must be implemented.

For an 11-bit COB ID, the value of the entry is constructed as follows:

Bit	Description
0 – 10	COB ID for SYNC Object
11 - 28	Set to 0
29	Set to 0 to select 11-bit COB ID
30	Set to 0 if the node does not generate the SYNC Object. Set to 1 if the node does generate the SYNC Object.
31	Not used. Recommendation: set to 0

Table D.11 SYNC COB ID Contents for 11-bit COB ID

For a 29-bit COB ID, the value of the entry is constructed as follows:

Bit	Description
0 – 28	COB ID for SYNC Object
29	Set to 1 to select 29-bit COB ID
30	Set to 0 if the node does not generate the SYNC Object. Set to 1 if the node does generate the SYNC Object.
31	Not used. Recommendation: set to 0

Table D.12 SYNC COB ID Contents for 29-bit COB ID

A node can optionally make bits 29 and 30 read only.

If a device cannot generate the SYNC Object, then attempting to set bit 30 can result in an abort message.

If a device cannot use 29-bit COB IDs, then it may either ignore an attempt to set bit 29 or generate an abort message in response.

In order to change the COB ID for the node that is currently generating the SYNC Object, bit 30 must first be set to 0.

Example: 000007FAh
11-bit SYNC COB ID of 7FAh

D.3.7 Communication Cycle Period (1006h)

Index	1006h
Name	Communication Cycle Period
Mandatory	Mandatory if the node generates the SYNC Object or allows bit 30 in entry 1005h to be set

Subindex	00h
Name	Communication Cycle Period
Type	UNSIGNED32
Default Value	0
Units	µs
Access	Read/Write
Mandatory	Mandatory if the node generates the SYNC Object or allows bit 30 in entry 1005h to be set
Map to PDO	No

Description: This entry defines the period between transmission of the SYNC Object by the node in µs, if the node is currently the SYNC Object producer.

A value of zero results in no transmission of the SYNC Object. Therefore writing the value zero to this entry of the SYNC Object producer will stop transmission of the SYNC Object.

Example: 00152622h = 1386018
Transmit every 1.386018 seconds

D.3.8 Synchronous Window Length (1007h)

Index	1007h
Name	Synchronous Window Length
Mandatory	No

Subindex	00h
Name	Synchronous Window Length
Type	UNSIGNED32
Default Value	0
Units	µs
Access	Read/Write
Mandatory	No
Map to PDO	No

Description: The Synchronous Window Length is the period of time in µs after a SYNC Object has been transmitted on the bus in which synchronous PDOs must be transmitted.

This period must be shorter than the Communication Cycle Period of the SYNC Object producer.

Each node using the same COB ID for the SYNC Object must use the same Synchronous Window Length. For example, if the network has two SYNC Objects and nodes 02h, 04h and 05h use COB ID 80h for the SYNC Object, and nodes 01h and 07h use COB ID 7Fh for the SYNC Object, then nodes 02h, 04h and 05h must use the same Synchronous Window Length and nodes 01 and 07h must use the same Synchronous Window Length. However, the two Synchronous Window Lengths used in the network may be different from each other.

If a node attempts to transmit a Synchronous PDO within the Synchronous Window Length but fails to do so (if, for example, higher priority messages were on the bus), then the node must not transmit the PDO. That is, synchronous PDOs must never be transmitted outside of the Synchronous Window Length.

Example: 00001432h = 5170
PDOs transmit within 5.17ms of SYNC occurance

D.3.9 Manufacturer Device Name (1008h)

Index	1008h
Name	Manufacturer Device Name
Mandatory	No

Subindex	00h
Name	Manufacturer Device Name
Type	VISIBLE_STRING
Default Value	Not defined
Access	Read only
Mandatory	No
Map to PDO	No

Description: Stores the name of the manufacturer of the node as an ASCII string. The length of the string is not limited by the CANopen specification, however minimal CANopen implementions that only support expedited SDO transfers limit the length to four characters.

Example: Embedded Systems Academy, Inc.

D.3.10 Manufacturer Hardware Version (1009h)

Index	**1009h**
Name	Manufacturer Hardware Version
Mandatory	No

Subindex	**00h**
Name	Manufacturer Hardware Version
Type	VISIBLE_STRING
Default Value	Not defined
Access	Read only
Mandatory	No
Map to PDO	No

Description: Stores the hardware version of the node as an ASCII string. The length of the string is not limited by the CANopen specification, however minimal CANopen implementions that only support expedited SDO transfers limit the length to four characters.

Example: Version 1.01

D.3.11 Manufacturer Software Version (100Ah)

Index	**100Ah**
Name	Manufacturer Software Version
Mandatory	No

Subindex	**00h**
Name	Manufacturer Software Version
Type	VISIBLE_STRING
Default Value	Not defined
Access	Read only

Subindex	00h
Mandatory	No
Map to PDO	No

Description: Stores the software version of the node as an ASCII string. The length of the string is not limited by the CANopen specification, however minimal CANopen implementions that only support expedited SDO transfers limit the length to four characters.

Example: Version 2.6.3 pre-release 5

D.3.12 Guard Time (100Ch)

Index	100Ch
Name	Guard Time
Mandatory	Yes if the node does not support heartbeats

Subindex	00h
Name	Guard Time
Type	UNSIGNED16
Default Value	0
Units	ms
Access	Read/Write if node guarding is supported Read Only if node guarding is not supported
Mandatory	Yes if the node does not support heartbeats
Map to PDO	No

Description: Specifies how long the period should be in milliseconds between node guarding requests sent to the node. If the NMT Master implements node guarding, then it should read this entry and send the node guarding requests to the node at the frequency indicated by the value of this entry.

If a response to a node guarding request is not transmitted within the node life time, then a node guarding event occurs, indicating that the node may have possibly stopped working. If a node guarding request from the NMT Master is not received within the node life time, then the node knows that the NMT Master may have possibly stopped working.

The node life time is the guard time multiplied by the life time factor (100Dh).

If the node does not support heartbeats then this entry must be implemented.

If this entry is implemented and the node does not support node guarding, then the access is read only and the value is zero to disable node guarding.

Note that a node must implement either heartbeats or node guarding or both heartbeats and node guarding.

Example: 1122h = 4386
Requests every 4.386 seconds

D.3.13 Life Time Factor (100Dh)

Index	100Dh
Name	Life Time Factor
Mandatory	Yes if the node does not support heartbeats

Subindex	00h
Name	Life Time Factor
Type	UNSIGNED8
Default Value	0
Access	Read/Write if node guarding is supported Read Only if node guarding is not supported

Subindex	**00h**
Mandatory	Yes if the node does not support heartbeats
Map to PDO	No

Description: Specifies the number of multiples of the guard time to wait for a response from the node to a node guarding request.

If a response to a node guarding request is not transmitted within the node life time, then a node guarding event occurs, indicating that the node may have possibly stopped working. If a node guarding request from the NMT Master is not received within the node life time, then the node knows that the NMT Master may have possibly stopped working.

The node life time is the guard time multiplied by the life time factor (100Ch).

If the node does not support heartbeats then this entry must be implemented.

If this entry is implemented and the node does not support node guarding, then the access is read only and the value is zero to disable node guarding.

Note that a node must implement either heartbeats, node guarding or both heartbeats and node guarding.

Example: 04h
Wait guard time x 4

D.3.14 Store Parameters (1010h)

Index	**1010h**
Name	Store Parameters
Mandatory	No

Subindex	**00h**
Name	Number of Entries
Type	UNSIGNED8
Default Value	Not defined
Access	Read Only
Mandatory	Yes if this entry is implemented
Map to PDO	No

Subindex	**01h**
Name	Save All Parameters
Type	UNSIGNED32
Default Value	Not defined
Access	Read/Write
Mandatory	Yes if this entry is implemented
Map to PDO	No

Subindex	**02h**
Name	Save Communication Parameters
Type	UNSIGNED32
Default Value	Not defined
Access	Read/Write
Mandatory	No
Map to PDO	No

Subindex	03h
Name	Save Application Parameters
Type	UNSIGNED32
Default Value	None specified
Access	Read/Write
Mandatory	No
Map to PDO	No

Subindex	04h – 7Fh
Name	Save Manufacturer Defined Parameters
Type	UNSIGNED32
Default Value	Not defined
Access	Read/Write
Mandatory	No
Map to PDO	No

Description: If a node contains non-volatile memory that can be used to store the settings, then this entry may optionally be implemented.

By writing to the Subentries, the node can be instructed to immediately store all or some of the settings in non-voltatile memory.

By reading the Subentries, non-volatile storage capabilities of the node may be determined.

By writing the value 65766173h (ASCII "save" transmitted with the "e" first) to a Subentry of 1010h, the related section of the Object Dictionary is stored in non-volatile memory, if the node supports this feature.

If a value other than "save" is written, or if the parameter storing fails for some reason, an SDO Abort message is transmitted.

The following table shows which sections of the Object Dictionary are saved when the different Subentries are written to.

Subindex	Object Dictionary Entries Saved
01h	All entries that can be saved
02h	All communication parameters that can be saved. Entries 1000h – 1FFFh and any manufacturer specific communication parameters
03h	All application specific parameters that can be saved. Entries 6000h – 9FFFh
04h – 7Fh	Writing to these entries saves manufacturer defined parameters.

Table D.13 Save Parameters Subentries

Reading a Subentry returns a value that indicates if the node can save parameters without manual intervention and if the node can save parameters manually. The returned value is structured as follows:

Bit	Description
0	Set to 1 if the device can save the parameters by writing to the Subentry Set to 0 if the device cannot save the parameters by writing to the Subentry
1	Set to 1 if the device can save the parameters autonomously Set to 0 if the device cannot save the parameters autonomously
2 - 31	Reserved. Set to 0

Table D.14 Store Parameters Read Value Contents

For example, if the value 00000003h is read from [1010h,02h], then the node can autonomously (and when requested) save application specific parameters to non-volatile memory. It is then known that writing the value "save" to [1010h,03h] will immediately cause the application specific parameters to be saved.

The CANopen specification does not define what will happen if an attempt is made to store parameters when that Subindex does not support writing "save." Therefore the entry with that Subindex should always be read first to ensure the operation is supported.

D.3.15 Restore Default Parameters (1011h)

Index	**1011h**
Name	Restore Default Parameters
Mandatory	No

Subindex	**00h**
Name	Number of Entries
Type	UNSIGNED8
Default Value	Not defined
Access	Read Only
Mandatory	Yes if this entry is implemented
Map to PDO	No

Subindex	**01h**
Name	Restore All Default Parameters
Type	UNSIGNED32
Default Value	Not defined
Access	Read/Write
Mandatory	Yes if this entry is implemented
Map to PDO	No

Subindex	**02h**
Name	Restore Communication Default Parameters
Type	UNSIGNED32

Subindex	02h
Default Value	Not defined
Access	Read/Write
Mandatory	No
Map to PDO	No

Subindex	03h
Name	Restore Application Default Parameters
Type	UNSIGNED32
Default Value	Not defined
Access	Read/Write
Mandatory	No
Map to PDO	No

Subindex	04h – 7Fh
Name	Restore Manufacturer Defined Default Parameters
Type	UNSIGNED32
Default Value	Not defined
Access	Read/Write
Mandatory	No
Map to PDO	No

Description: This entry provides a means to restore some or all of the default values for parameters in the Object Dictionary. By writing the value 64616F6Ch (ASCII "load" transmitted with the "d" first) to a Subindex, the parameters corresponding to the entry will be restored to their default values on the next reset of the node or the next power cycle, depending on which parameters were restored.

If a value other than "load" is written, or the restoring of the default parameters fails, then an SDO Abort message is transmitted.

The following table lists the parameters which will be restored when the value "load" is written, and which type of reset will cause the default values to be used by the node.

Subindex	Object Dictionary Entry Defaults Restored	Default values used after
01h	All entries that can be restored	Node reset
02h	All communication entries that can be restored. Entries 1000h – 1FFFh and any manufacturer specific communication entries.	Communication reset
03h	All application entries that can be restored. Entries 6000h – 9FFFh.	Node reset
04h – 7Fh	Manufacturer defined entries that can be restored.	Node reset

Table D.15 Restore Parameters Subentries

By reading the Subentries, information regarding the capabilities of the node to restore default values can be determined. The value read is constructed as shown in Table D.16:

Bit	Description
0	Set to 1 if node can restore default parameters. Set to 0 if node cannot restore default parameters.
1 - 31	Reserved. Set to 0.

Table D.16 Restore Parameters Read Value Contents

For example, if the value 00000001h is read from Subindex 01h, then the node is able to restore all default parameters that can be restored. If the value 00000000h is read from Subindex 03h, then the node is not able to restore application default parameters.

The CANopen specification does not define what will happen if an attempt is made to restore default parameters when that Subindex does not support writing "load", therefore the entry with that Subindex should always be read first to ensure the operation is supported.

D.3.16 TIME COB ID (1012h)

Index	1012h
Name	TIME COB ID
Mandatory	No

Subindex	00h
Name	TIME COB ID
Type	UNSIGNED32
Default Value	00000100h
Access	Read/Write
Mandatory	No
Map to PDO	No

Description: This entry specifies the COB ID for the Timestamp Object. It also specifies whether the node does or does not use the Timestamp Object and whether the node does or does not produce the Timestamp Object.

For an 11-bit COB ID the value of the entry is constructed as follows:

Bit	Description
0 – 10	COB ID for TIME Object
11 - 28	Set to 0
29	Set to 0 to select 11-bit COB ID
30	Set to 0 if the node does not generate the TIME Object. Set to 1 if the node does generate the TIME Object.
31	Set to 0 if the node does not use the TIME Object. Set to 1 if the node does use the TIME Object.

Table D.17 TIME COB ID Contents for 11-bit COB ID

For a 29-bit COB ID the value of the entry is constructed as follows:

Bit	Description
0 – 28	COB ID for TIME Object
29	Set to 1 to select 29-bit COB ID
30	Set to 0 if the node does not generate the TIME Object. Set to 1 if the node does generate the TIME Object.
31	Set to 0 if the node does not use the TIME Object. Set to 1 if the node does use the TIME Object.

Table D.18 TIME COB ID Contents for 29-bit COB ID

Optionally a node may not allow the values of bits 29 and 30 to be changed if it does not support dynamic configuration of how the node uses the TIME Object.

If the node does not support generating the TIME Object or does not support 29-bit COB IDs, then attempts to set bits 29 and 30 will result in SDO Abort messages being transmitted by the node.

Example: 400001A4h
Node generates TIME Object with 11-bit ID 1A4h. Node does not use the TIME Object.

D.3.17 High Resolution Timestamp (1013h)

Index	1013h
Name	High Resolution Timestamp
Mandatory	No

Subindex	00h
Name	High Resolution Timestamp
Type	UNSIGNED32
Default Value	0
Units	μs

Subindex	00h
Access	Read/Write
Mandatory	No
Map to PDO	Yes

Description: This entry contains a high resolution timestamp in μs, which may be mapped into a PDO. The high resolution timestamp allows for local clock synchronization with great precision. If only the SYNC signal is used for the synchronization of local clocks the deviation between the clocks could be several hundreds of microseconds, because even the high priority SYNC message can be delayed (for example because it has to wait until a message currently on the bus is transmitted).

When the SYNC Producer finishes transmitting the SYNC Object a local CAN message transmit complete interrupt is generated. In the interrupt service routine the high resolution timestamp is taken of that moment in time. This timestamp is then transmitted in a PDO after the SYNC Object.

When a SYNC Consumer receives the SYNC Object (recognized by a CAN message received interrupt) it also takes a high resolution timestamp of that moment in time. Shortly afterwards the SYNC Consumer will receive the PDO containing the high resolution timestamp from the SYNC Producer. The SYNC Consumer compares the two timestamps, which should be identical if the local clocks of the producer and the consumer are perfectly synchronized. If they are not identical, the SYNC consumer has to start a process of synchronizing itself with the clock of the SYNC producer.

Note however that there is still an error in the synchronization. It takes some time for the SYNC Producer to react to the SYNC Object transmission complete interrupt and generate the timestamp. Also it takes some time for the SYNC Consumer to react to the SYNC Object being received and generate a timestamp. These delays are typically different, however they can be calculated in advance based on the code executed and used to adjust the high resolution timestamps accordingly. This behavior, however, is application specific.

The timestamp allows for a maximum delay of 72 minutes before it is reset back to zero.

Example: 00152242h = 1385026
A time stamp of 1.385026 seconds

D.3.18 Emergency COB ID (1014h)

Index	1014h
Name	Emergency COB ID
Mandatory	Yes if the Emergency Object is supported by the node

Subindex	00h
Name	Emergency COB ID
Type	UNSIGNED32
Default Value	Node ID + 00000080h
Access	Read Only or Read/Write
Mandatory	Yes if the Emergency Object is supported by the node
Map to PDO	No

Description: The Emergency COB ID entry defines the Identifier used for the Emergency Object transmitted by the node. The node may allow the COB ID to be changed by writing to this entry, or it may be fixed and unchangeable.

The value stored in the entry also determines if the Emergency Object exists or not and whether an 11-bit or 29-bit Identifier is used.

For an 11-bit COB ID the value of the entry is constructed as follows:

Bit	Description
0 - 10	COB ID for Emergency Object
11 - 28	Set to 0
29	Set to 0 to select 11-bit COB ID

Table D.19 Emergency COB ID Contents for 11-bit COB ID

Bit	Description
30	Reserved. Set to 0
31	Set to 0 if the node does use the Emergency Object. Set to 1 if the node does not use the Emergency Object.

Table D.19 (Continued) Emergency COB ID Contents for 11-bit COB ID

For a 29-bit COB ID the value of the entry is constructed as follows:

Bit	Description
0 – 28	COB ID for Emergency Object
29	Set to 1 to select 29-bit COB ID
30	Reserved. Set to 0
31	Set to 0 if the node does use the Emergency Object. Set to 1 if the node does not use the Emergency Object.

Table D.20 Emergency COB ID Contents for 29-bit COB ID

If the node does not support 29-bit COB IDs and an attempt is made to write a 1 to bit 29, then the node will respond with an SDO Abort message.

In order to change the COB ID on nodes that support writing to this entry, the Emergency Object must first be disabled by writing a 1 to bit 31. Once the COB ID has been changed, bit 31 can be set back to 0.

Example: 80000082h
Emergency COB ID of 82h used by the node

D.3.19 Inhibit Time Emergency (1015h)

Index	1015h
Name	Inhibit Time Emergency
Mandatory	No

Subindex	00h
Name	Inhibit Time Emergency
Type	UNSIGNED16
Default Value	0
Units	100µs
Access	Read/Write
Mandatory	No
Map to PDO	No

Description: This entry specifies the inhibit time for the Emergency Object transmitted by the node in multiples of 100µs. If used, once the Emergency Object has been transmitted the next Emergency Object cannot be transmitted until the time specified by this entry has elapsed, even if another emergency occurs.

Implementation of this entry is optional. If this entry is not implemented then the Emergency Object does not have an inhibit time and can transmit messages as frequently as desired. If the entry is implemented then it must be writeable, allowing dynamic changing of the inhibit time.

Example: 023Ah = 570
The Emergency Object may be transmitted at most once every 57ms

D.3.20 Consumer Heartbeat Time (1016h)

Index	1016h
Name	Consumer Heartbeat Time
Mandatory	Yes if the node consumes at least one heartbeat

Subindex	00h
Name	Number of Entries
Type	UNSIGNED8
Default Value	Not defined
Access	Read Only
Mandatory	Yes if the node consumes at least one heartbeat
Map to PDO	No

Subindex	01h
Name	Consumer Heartbeat Time
Type	UNSIGNED32
Default Value	0
Units	ms
Access	Read/Write
Mandatory	Yes if the node consumes at least one heartbeat
Map to PDO	No

Subindex	02h – 7Fh
Name	Consumer Heartbeat Time
Type	UNSIGNED32
Default Value	Not defined
Units	ms
Access	Read/Write

Subindex	02h – 7Fh
Mandatory	No
Map to PDO	No

Description: The node may listen to the heartbeat messages generated by other nodes on the network. This entry specifies the maximum time to wait for a heartbeat from a specific node before generating an internal heartbeat event in milliseconds (called the Heartbeat Consumer Time). Measurement begins after reception of the first heartbeat message. It does not begin after reception of a bootup message.

Each Subindex specifies the Heartbeat Consumer Time for a specific node. The value of the entry is constructed as follows:

Bit	Description
0 – 15	Heartbeat Consumer Time
16 – 23	Node ID
24 - 31	Reserved. Set to 0

Table D.21 Heartbeat Consumer Time Value

The Heartbeat Consumer Time of a specific node must be greater than the Heartbeat Producer Time of the node. The Producer Time can be read from entry 1017h.

Specifying a Heartbeat Consumer Time of zero for a specific Node ID disables the heartbeat monitoring of that node.

For a specific Node ID there may be more than one Subindex specifying a Heartbeat Consumer Time of zero, however attempts to set more than one of those entries to a non-zero value will result in the node transmitting an SDO Abort message.

Example: 005A1122h (1122h = 4386)
Heartbeat Consumer Time of 4.386 seconds for node 5Ah

D.3.21 Producer Heartbeat Time (1017h)

Index	1017h
Name	Producer Heartbeat Time
Mandatory	Yes if node guarding is not supported

Subindex	00h
Name	Producer Heartbeat Time
Type	UNSIGNED16
Default Value	0
Units	ms
Access	Read/Write
Mandatory	Yes if node guarding is not supported
Map to PDO	No

Description: A node must support either node guarding or heartbeat generation. If the node generates heartbeats then this entry must be implemented. The value of the entry specifies in milliseconds the time between transmission of heartbeat messages. A value of zero disables transmission of heartbeat messages by the node.

Because the entry is writeable, the value of the entry may change at any time.

Note that a node must implement either heartbeats, node guarding or both heartbeats and node guarding.

Example: 4455h = 17493
The node will transmit a heartbeat message every 17.493 seconds

D.3.22 Identity (1018h)

Index	1018h
Name	Identity
Mandatory	Yes

Subindex	00h
Name	Number of Entries
Type	UNSIGNED8
Default Value	Not defined
Access	Read Only
Mandatory	Yes
Map to PDO	No

Subindex	01h
Name	Vendor ID
Type	UNSIGNED32
Default Value	Not defined
Access	Read Only
Mandatory	Yes
Map to PDO	No

Subindex	02h
Name	Product Code
Type	UNSIGNED32
Default Value	Not defined
Access	Read Only
Mandatory	No
Map to PDO	No

Subindex	03h
Name	Revision Number
Type	UNSIGNED32
Default Value	Not defined
Access	Read Only
Mandatory	No
Map to PDO	No

Subindex	04h
Name	Serial Number
Type	UNSIGNED32
Default Value	Not defined
Access	Read Only
Mandatory	No
Map to PDO	No

Description: The Identity entry provides some basic information about the node in order to provide a standard way of differentiating between different versions of a node.

All nodes must implement Subindexes 00h and 01h. The remaining Subindexes are optional.

The Vendor ID is a unique ID assigned to each CANopen vendor by CAN in Automation. This allows the source of the node to be identified.

The product code and serial number formats are manufacturer specific, however the revision number has the following format:

Bits	Description
0 – 15	Minor Revision Number Identifies different versions of the node where the CANopen behavior has not changed.
16 – 31	Major Revision Number Identifies different versions of the node where the CANopen behavior has changed.

Table D.22 Revision Number Format

For example, if a new version of the node is produced with any difference in the CANopen messages, transmission types, Object Dictionary entries, etc., then the major revision number must be increased, otherwise the minor revision number must be increased.

Example: 00050001h
Revision number 5.1

D.3.23 Verify Configuration (1020h)

Index	1020h
Name	Verify Configuration
Mandatory	No, but recommended when 1010h and 1011h are implemented

Subindex	00h
Name	Number of Entries
Type	UNSIGNED8
Default Value	02h
Access	Read Only
Mandatory	Yes if this entry is implemented
Map to PDO	No

Subindex	01h
Name	Configuration Date
Type	UNSIGNED32
Default Value	Not defined
Units	days
Access	Read/Write
Mandatory	Yes if this entry is implemented
Map to PDO	No

Subindex	02h
Name	Configuration Time
Type	UNSIGNED32
Default Value	Not defined
Units	ms
Access	Read/Write
Mandatory	Yes if this entry is implemented
Map to PDO	No

Description: This entry allows an NMT Master to determine if the configuration of the device matches a known Device Configuration File.

When storing a new configuration for the node by writing to entry 1010h, the NMT Master can first write the current date and time to this entry along with storing the current date and time in a local copy of the Device Configuration File. The values in this entry will be saved along with the other parameters when entry 1010h is written.

Whenever any new values are written to the Object Dictionary of the node, it must set the current date and time stored in this entry to zero to indicate that the configuration has changed.

The next time the node is started, the NMT Master can read this entry and compare the date and time with the date and time stored in the

Device Configuration File. If the times match, the NMT Master knows the current configuration of the node without having to read any further Object Dictionary entries.

The date value is the number of whole days since January 1, 1984. The time value is the number of milliseconds since midnight.

Example: 23 in Subentry 01h, 6212000 in Subentry 02h
January 24, 1984, 1:43:32 am

D.3.24 Store EDS (1021h)

Index	**1021h**
Name	Store EDS
Mandatory	No

Subindex	**00h**
Name	Store EDS
Type	DOMAIN
Default Value	Not defined
Access	Read/Write
Mandatory	No
Map to PDO	No

Description: The Store EDS entry allows the Electronic Datasheet for the node to be stored in the node. This removes the requirement that the Electronic Datasheet files are supplied separately and removes any possible confusion as to which version of an Electronic Datasheet is to be used for the node.

The Electronic Data Sheet is read and written according to the format specified in entry 1022h.

D.3.25 Storage Format (1022h)

Index	1022h
Name	Storage Format
Mandatory	Yes if entry 1021h Store EDS is implemented

Subindex	00h
Name	Storage Format
Type	UNSIGNED16
Default Value	Not defined
Access	Read/Write
Mandatory	Yes if entry 1021h Store EDS is implemented
Map to PDO	No

Description: This entry defines the format that the Electronic Datasheet is read from and written to entry 1021h Store EDS. The following table lists the supported formats and values.

Value	Description
0000h	Uncompressed ASCII
0001h – FFFFh	Reserved

Table D.23 EDS Read and Write Formats

D.3.26 OS Command (1023h)

Index	1023h
Name	OS Command
Mandatory	No

Subindex	00h
Name	Number of Entries
Type	UNSIGNED8
Default Value	03h
Access	Read Only
Mandatory	Yes if this entry is implemented
Map to PDO	No

Subindex	01h
Name	Command
Type	OCTET_STRING
Default Value	Not defined
Access	Read/Write
Mandatory	Yes if this entry is implemented
Map to PDO	No

Subindex	02h
Name	Status
Type	UNSIGNED8
Default Value	Not defined
Access	Read Only
Mandatory	Yes if this entry is implemented
Map to PDO	No

Subindex	03h
Name	Reply
Type	OCTET_STRING
Default Value	Not defined
Access	Read Only
Mandatory	Yes if this entry is implemented
Map to PDO	No

Description: This entry allows a node to provide a command based interface. The command is written to the Command Subentry. The Status Subentry is then polled to determine the status of the command. Once the command has been processed the Status will indicate if there is a reply. The reply can be read from the Reply entry.

The format of the commands and replies are manfacturer specific and may be ASCII or binary.

Table D.24 lists the possible values for the Status Subentry.

Value	Description
00h	Last command completed. No error occurred. No reply.
01h	Last command completed. No error occurred. The reply can now be read.
02h	Last command completed. Error occured. No reply.
03h	Last command completed. Error occurred. The reply can now be read.
04h – FEh	Reserved
FFh	Command is executing

Table D.24 OS Command Status Values

D.3.27 OS Command Mode (1024h)

Index	1024h
Name	OS Command Mode
Mandatory	No

Subindex	00h
Name	OS Command Mode
Type	UNSIGNED8
Default Value	Not defined
Access	Write Only
Mandatory	No
Map to PDO	No

Description: This entry controls whether the node buffers the OS Commands written to entry 1023h or not, and provides some degree of control over the buffer. Table D.25 describes the possible values that may be written to this entry, and their effect.

Value	Description
00h	Execute the next command immediately (no buffering of commands)
01h	Buffer the next command
02h	Execute the commands in the buffer
03h	Abort the current command and flush the buffer
04h – FFh	Manufacturer specific

Table D.25 OS Command Modes

D.3.28 OS Debugger Interface (1025h)

Index	**1025h**
Name	OS Debugger Interface
Mandatory	No

Subindex	**00h**
Name	Number of Entries
Type	UNSIGNED8
Default Value	03h
Access	Read Only
Mandatory	Yes if this entry is implemented
Map to PDO	No

Subindex	**01h**
Name	Command
Type	OCTET_STRING
Default Value	Not defined
Access	Read/Write
Mandatory	Yes if this entry is implemented
Map to PDO	No

Subindex	**02h**
Name	Status
Type	UNSIGNED8
Default Value	Not defined
Access	Read Only
Mandatory	Yes if this entry is implemented
Map to PDO	No

Subindex	03h
Name	Reply
Type	OCTET_STRING
Default Value	Not defined
Access	Read Only
Mandatory	Yes if this entry is implemented
Map to PDO	No

Description: This entry allows a node to provide a debugger interface. The command is written to the Command Subentry. The Status Subentry is then polled to determine the status of the command. The reply can be read from the Reply entry.

The format of the commands and replies are manfacturer specific and may be ASCII or binary. Table D.26 lists the possible values for the Status Subentry.

Value	Description
00h	Last command completed. No error occurred.
01h	Last command completed. Error occurred.
02h – FEh	Reserved
FFh	Command is executing

Table D.26 OS Debugger Interface Status Values

D.3.29 OS Prompt (1026h)

Index	1026h
Name	OS Prompt
Mandatory	No

Subindex	00h
Name	Number of Entries
Type	UNSIGNED8
Default Value	Not defined
Access	Read Only
Mandatory	Yes if this entry is implemented
Map to PDO	No

Subindex	01h
Name	StdIn
Type	UNSIGNED8
Default Value	Not defined
Access	Write Only
Mandatory	Yes if this entry is implemented
Map to PDO	Yes

Subindex	02h
Name	StdOut
Type	UNSIGNED8
Default Value	Not defined
Access	Read Only
Mandatory	Yes if this entry is implemented
Map to PDO	Yes

Subindex	03h
Name	StdErr
Type	UNSIGNED8
Default Value	Not defined
Access	Read Only
Mandatory	No
Map to PDO	Yes

Description: This entry provides a command prompt type interface, where characters are sent and received one at a time. Characters are written to StdIn either by SDO or PDO, and the response is read from StdOut either by SDO or PDO. The error output appears on StdErr.

D.3.30 Module List (1027h)

Index	1027h
Name	Module List
Mandatory	Yes if modular devices are supported

Subindex	00h
Name	Number of Connected Modules
Type	UNSIGNED8
Default Value	Not defined
Access	Read Only
Mandatory	Yes if modular devices are supported
Map to PDO	No

Subindex	01h
Name	Module 2
Type	UNSIGNED16
Default Value	Not defined

Subindex	01h
Access	Read/Write
Mandatory	Yes if modular devices are supported
Map to PDO	No

Subindex	02h – FEh
Name	Module 3 – 255
Type	UNSIGNED16
Default Value	Not defined
Access	Read/Write
Mandatory	No
Map to PDO	No

Description: This entry allows modules to be dynamically added to a node. A module can consist of the device profile sections of the Object Dictionary, or manfacturer specific entries. Each module type must have a unique number, although a specific module may be added to a node multiple times, effectively allowing scaling of the node's capabilities. To add a module, the modules' unique identifying number must be written to a Subentry.

D.3.31 Emergency Consumer (1028h)

Index	1028h
Name	Emergency Consumer
Mandatory	No

Subindex	00h
Name	Number of Consumed Emergency Objects
Type	UNSIGNED8
Default Value	Not defined

Subindex	00h
Access	Read Only
Mandatory	Yes if this entry is implemented
Map to PDO	No

Subindex	01h
Name	Emergency Consumer 1
Type	UNSIGNED32
Default Value	Not defined
Access	Read/Write
Mandatory	Yes if this entry is implemented
Map to PDO	No

Subindex	02h – 7Fh
Name	Emergency Consumer 2 – 127
Type	UNSIGNED32
Default Value	Not defined
Access	Read/Write
Mandatory	No
Map to PDO	No

Description: Specifies which Emergency Objects the node consumes. For each Emergency Object consumed by the node a Subentry is implemented specifying the COB ID of the Emergency Object. The Subindex of the Subentry specifies which node generates the Emergency Object. For example, Subentry 4Ah contains the COB ID of the Emergency Object generated by node 4Ah.

For an 11-bit COB ID the value of the Subentries are constructed as follows:

Bit	Description
0 - 10	COB ID
11 - 28	Set to 0
29	Set to 0 to select 11-bit COB ID
30	Reserved. Set to 0
31	Set to 0 if the node does consume the Emergency Object. Set to 1 if the node does not consume the Emergency Object.

Table D.27 Emergency Consumer COB ID Contents for 11-bit COB ID

For a 29-bit COB ID the value of the Subentries are constructed as follows:

Bit	Description
0 – 28	COB ID
29	Set to 1 to select 29-bit COB ID
30	Reserved. Set to 0
31	Set to 0 if the node does consume the Emergency Object. Set to 1 if the node does not consume the Emergency Object.

Table D.28 Emergency Consumer COB ID Contents for 29-bit COB ID

To change a COB ID, bit 31 must first be set to 1 to disable consuming of the Emergency Object. Once the COB ID has been changed the entry can be reenabled by clearing bit 31.

Attempting to set bit 29 on a node that does not support 29-bit COB IDs, will result in the node transmitting an SDO Abort message.

Example: 0000073Ch
Node consumes the Emergency Object 73Ch

D.3.32 Error Behavior (1029h)

Index	1029h
Name	Error Behavior
Mandatory	No

Subindex	00h
Name	Number of Error Classes
Type	UNSIGNED8
Default Value	Not defined
Access	Read Only
Mandatory	Yes if this entry is implemented
Map to PDO	No

Subindex	01h
Name	Communication Error
Type	UNSIGNED8
Default Value	00h
Access	Read/Write
Mandatory	Yes if this entry is implemented
Map to PDO	No

Subindex	02h – FEh
Name	Device Profile or Manufacturer Specific Error
Type	UNSIGNED8
Default Value	00h
Access	Read/Write

Subindex	02h – FEh
Mandatory	No
Map to PDO	No

Description: When a node encounters a serious internal error while in the Operational state, it must switch to the Pre-operational state. By implementing this entry the node may be configured to enter the Stopped state instead of Pre-operational, or may not change states at all.

Subentry 01h defines the behavior of the node when a communication error is encountered. Communication errors include Bus Off, node guarding events and heartbeat events.

Subentries 02H to FFh define the behavior for other severe errors and the exact errors are manufacturer-specific.

The following table shows the allowed values that may be stored in the Subentries to configure the node behavior.

Value	Node Behavior When an Error is Encountered
00h	Switches to Pre-operational
01h	Does not change states
02h	Switches to Stopped
03h - FFh	Reserved

Table D.29 Error Behavior Values

D.3.33 Server SDO Parameters (1200h)

Index	1200h
Name	Server SDO Parameter
Mandatory	No

Subindex	00h
Name	Number of Entries
Type	UNSIGNED8
Default Value	Not defined
Access	Read Only
Mandatory	Yes if this entry is implemented
Map to PDO	No

Subindex	01h
Name	COB ID Client to Server (Receive SDO)
Type	UNSIGNED32
Default Value	Node ID + 00000600h
Access	Read Only
Mandatory	Yes if this entry is implemented
Map to PDO	No

Subindex	02h
Name	COB ID Server to Client (Transmit SDO)
Type	UNSIGNED32
Default Value	Node ID + 00000580h
Access	Read Only
Mandatory	Yes if this entry is implemented
Map to PDO	No

Description: This entry describes the mandatory default SDO communication channel for the node. Because all entries are read only, it is optional for a node to implement this entry. Often it is omitted.

Subentry 01h holds the COB ID of the SDO used to access the Object Dictionary of the node.

Subentry 02h holds the COB ID of the SDO used by the node to respond to Object Dictionary requests.

Example: 00000611h in Subentry 01h
Receive SDO is 611h for Node 11h

D.3.34 Server SDO Parameters (1201h – 127Fh)

Index	**1201h – 127Fh**
Name	Server SDO Parameter
Mandatory	Yes for each additional Server SDO Channel supported by the node

Subindex	**00h**
Name	Number of Entries
Type	UNSIGNED8
Default Value	Not defined
Access	Read Only
Mandatory	Yes for each additional Server SDO Channel supported by the node
Map to PDO	No

Subindex	**01h**
Name	COB ID Client to Server (Receive SDO)
Type	UNSIGNED32
Default Value	00000000h
Access	Read/Write

Subindex	01h
Mandatory	Yes for each additional Server SDO Channel supported by the node
Map to PDO	No

Subindex	02h
Name	COB ID Server to Client (Transmit SDO)
Type	UNSIGNED32
Default Value	00000000h
Access	Read/Write
Mandatory	Yes for each additional Server SDO Channel supported by the node
Map to PDO	No

Subindex	03h
Name	Node ID of the SDO Client
Type	UNSIGNED8
Default Value	Not defined
Access	Read/Write
Mandatory	No
Map to PDO	No

Description: If a node implements more than one SDO channel for access to the Object Dictionary, then for each additional channel one Server SDO Parameter entry must be implemented. For example, if a node implements two SDO channels, then entry 1201h must be implemented. If a node implements three SDO channels, then entries 1201h and 1202h must be implemented.

The entry defines the COB IDs used for the SDO channel along with (optionally) the Node ID of the client which will use the channel.

Subentry 01h holds the COB ID of the SDO used to access the Object Dictionary of the node.

Subentry 02h holds the COB ID of the SDO used by the node to respond to Object Dictionary requests.

For an 11-bit COB ID the value of Subentries 01h and 02h are constructed as follows:

Bit	Description
0 - 10	COB ID for SDO
11 - 28	Set to 0
29	Set to 0 to select 11-bit COB ID
30	Reserved. Set to 0
31	Set to 0 if the node does use the SDO. Set to 1 if the node does not use the SDO.

Table D.30 SDO Parameters COB ID Contents for 11-bit COB ID

For a 29-bit COB ID the value of Subentries 01h and 02h are constructed as follows:

Bit	Description
0 – 28	COB ID for SDO
29	Set to 1 to select 29-bit COB ID
30	Reserved. Set to 0
31	Set to 0 if the node does use the SDO. Set to 1 if the node does not use the SDO.

Table D.31 SDO Parameters COB ID Contents for 29-bit COB ID

To change a COB ID, bit 31 must first be set to 1 to disable the SDO. Once the COB ID has been changed the SDO can be reenabled by clearing bit 31.

Each SDO is only valid and usable if bit 31 in Subentry 01h and Subentry 02h are cleared. For example, the receive SDO cannot be used if bit 31 of the transmit SDO is set to 1.

Attempting to set bit 29 on a node that does not support 29-bit COB IDs, will result in the node transmitting an SDO Abort message.

Subentry 03h stores the Node ID of the client, which is the node that will send the Transmit SDO and process the Receive SDO. This Subentry is optional.

Note that this entry is usually written to by a CANopen Manager during the allocation of dynamic SDO channels.

Example: 0000051Eh in Subentry 01h
Node uses Receive SDO 51Eh

D.3.35 Client SDO Parameters (1280h – 12FFh)

Index	**1280h – 12FFh**
Name	Client SDO Parameter
Mandatory	Yes for each supported SDO Client Channel

Subindex	**00h**
Name	Number of Entries
Type	UNSIGNED8
Default Value	03h
Access	Read Only
Mandatory	Yes for each supported SDO Client Channel
Map to PDO	No

Subindex	**01h**
Name	COB ID Client to Server (Transmit SDO)
Type	UNSIGNED32
Default Value	00000000h

Subindex	01h
Access	Read/Write
Mandatory	Yes for each supported SDO Client Channel
Map to PDO	No

Subindex	02h
Name	COB ID Server to Client (Receive SDO)
Type	UNSIGNED32
Default Value	00000000h
Access	Read/Write
Mandatory	Yes for each supported SDO Client Channel
Map to PDO	No

Subindex	03h
Name	Node ID of the SDO Server
Type	UNSIGNED8
Default Value	Not defined
Access	Read/Write
Mandatory	Yes for each supported SDO Client Channel
Map to PDO	No

Description: If the node accesses the Object Dictionary of another node then it is an SDO Client. For each SDO Client Channel the node supports (transmit and receive SDO pair) a Client SDO Parameter entry must be implemented. The Subentries define the COB IDs used for the transmit and receive SDOs along with the Node ID of the node containing the Object Dictionary.

Subentry 01h stores the COB ID of the SDO used to access the Object Dictionary of the node specified in Subentry 03h.

Subentry 02h stores the COB ID of the SDO returned from the node specified in Subentry 03h.

For an 11-bit COB ID the value of Subentries 01h and 02h are constructed as follows:

Bit	Description
0 - 10	COB ID for SDO
11 - 28	Set to 0
29	Set to 0 to select 11-bit COB ID
30	Reserved. Set to 0
31	Set to 0 if the node does use the SDO. Set to 1 if the node does not use the SDO.

Table D.32 SDO Parameters COB ID Contents for 11-bit COB ID

For a 29-bit COB ID the value of Subentries 01h and 02h are constructed as follows:

Bit	Description
0 – 28	COB ID for SDO
29	Set to 1 to select 29-bit COB ID
30	Reserved. Set to 0
31	Set to 0 if the node does use the SDO. Set to 1 if the node does not use the SDO.

Table D.33 SDO Parameters COB ID Contents for 29-bit COB ID

To change a COB ID, bit 31 must first be set to 1 to disable the SDO. Once the COB ID has been changed the SDO can be reenabled by clearing bit 31.

Attempting to set bit 29 on a node that does not support 29-bit COB IDs, will result in the node transmitting an SDO Abort message.

Note that this entry is usually written to by a CANopen Manager during the allocation of dynamic SDO channels.

Example: 000004EDh in SubentrySubentry 01h
Node uses Transmit SDO 4EDh

D.3.36 Receive PDO Parameters (1400h – 15FFh)

Index	1400h - 15FFh
Name	Receive PDO Parameter
Mandatory	Yes for each supported Receive PDO

Subindex	00h
Name	Highest Subindex Supported
Type	UNSIGNED8
Default Value	Not defined
Access	Read Only
Mandatory	Yes for each supported Receive PDO
Map to PDO	No

Subindex	01h
Name	COB ID used by PDO
Type	UNSIGNED32
Default Value	See description
Access	Read Only or Read/Write
Mandatory	Yes for each supported Receive PDO
Map to PDO	No

Subindex	02h
Name	Transmission Type
Type	UNSIGNED8
Default Value	Determined by the device profile used
Access	Read Only or Read/Write
Mandatory	Yes for each supported Receive PDO
Map to PDO	No

Subindex	03h
Name	Inhibit Time
Type	UNSIGNED16
Default Value	Determined by the device profile used
Units	100µs
Access	Read/Write
Mandatory	No
Map to PDO	No

Subindex	04h
Name	Compatibility Entry
Type	UNSIGNED8
Default Value	Not defined
Access	Read/Write
Mandatory	No
Map to PDO	No

Subindex	05h
Name	Event Timer
Type	UNSIGNED16
Default Value	Determined by the device profile used
Units	ms
Access	Read/Write
Mandatory	No
Map to PDO	No

Description: This entry must be implemented for each Receive PDO supported by the node. The entry describes the communication configuration of the PDO.

Subentry 01h defines the COB ID of the PDO. The default value depends on the Index of the entry, as shown in the following table.

Index	Default Value
1400h	Node ID + 00000200h
1401h	Node ID + 00000300h
1402h	Node ID + 00000400h
1403h	Node ID + 00000500h
1404h – 15FFh	80000000h

Table D.34 Receive PDO Default COB IDs

The COB ID entry also indicates if the PDO is used or not, the size of the identifier and whether remote transmit requests are allowed for the PDO.

For an 11-bit COB ID the value of the Subentry is constructed as follows:

Bit	Description
0 - 10	COB ID for PDO
11 - 28	Set to 0
29	Set to 0 to select 11-bit COB ID
30	Set to 1
31	Set to 0 if the node does use the PDO Set to 1 if the node does not use the PDO

Table D.35 PDO Parameters COB ID Contents for 11-bit COB ID

For a 29-bit COB ID the value of the Subentry is constructed as follows:

Bit	Description
0 – 28	COB ID for PDO
29	Set to 1 to select 29-bit COB ID
30	Set to 1
31	Set to 0 if the node does use the PDO Set to 1 if the node does not use the PDO

Table D.36 PDO Parameters COB ID Contents for 29-bit COB ID

To change a COB ID, bit 31 must first be set to 1 to disable the SDO. Once the COB ID has been changed the SDO can be reenabled by clearing bit 31.

Attempting to set bit 29 on a node that does not support 29-bit COB IDs, or attempting to clear bit 30 on a node that does not support remote transmission requests, will result in the node transmitting an SDO Abort message.

Subentry 02h specifies the transmission type of the Receive PDO. The following table lists the available transmission types for a Receive PDO.

Transmission Type	Description
0 – 240	The Receive PDO is synchronous. The data in the PDO is processed on reception of the next SYNC Object. The actual value of the transmission type is not relevant.
241 – 253	Not used for Receive PDOs
254	The transmission type of the Receive PDO is manufacturer specific.
255	The Recieve PDO is asynchronous. As soon as the PDO arrives the data is processed by the node.

Table D.37 Receive PDO Transmission Types

If the PDO is a Destination Addressing Mode Multiplexed PDO then it must have transmission type 254. If the PDO is a Source Addressing Mode Multiplexed PDO then it must have transmission type 254 or 255.

Subentries 03h and 04h are not used and any attempt to read or write to these entries will return an SDO Abort message from the node.

Subentry 05h may optionally be implemented. It is an event timer which configures an event to occur after the specified number of milliseconds. A value of zero disables the event timer. The functionality of the event timer with regard to Receive PDOs is not described in the CANopen specification, however it may be used for several purposes, including generating an error if the PDO has not been received within a specific time.

Example: 00000201h in Subindex 01h, 0h in Subindex 02h
Node uses Receive PDO 201h with the data applied to the outputs upon reception of a SYNC message.

D.3.37 Receive PDO Mapping (1600h – 17FFh)

Index	**1600h – 17FFh**
Name	Receive PDO Mapping
Mandatory	Yes for each supported Receive PDO

Subindex	**00h**
Name	Number of Entries (Number of objects mapped into the PDO)
Type	UNSIGNED8
Default Value	Defined in the device profile
Access	Read Only if dynamic mapping is not supported. Read/Write if dynamic mapping is supported.
Mandatory	Yes for each supported Receive PDO
Map to PDO	No

Subindex	**01h – 40h**
Name	PDO Mapping for an application object
Type	UNSIGNED32
Default Value	Defined in the device profile

Subindex	01h – 40h
Access	Read/Write
Mandatory	No
Map to PDO	No

Description: This entry defines which process data is stored in a single PDO, along with the position of the process data in the eight data bytes of the PDO.

Each Receive PDO supported by the node must have a corresponding Receive PDO Mapping parameter entry implemented. The entry at 1600h is for the first Receive PDO whose communication parameters are defined at 1400h. The entry at 1601h is for the second Receive PDO whose communication parameters are defined at 1401h, etc.

A PDO may have 1 to 64 process data variables mapped to it, with each variable having any length from 1 to 64 bits, however the total size of all the process data mapped to a single PDO may not exceed 64 bits (eight bytes). Each Subentry defines a process data variable, therefore Subentry 00h holds the total number of process data variables mapped to the PDO.

The value of each Subentry defines the process data variable to be mapped and the size of the process data variable in bits. The process data variable is defined by specifying the Object Dictionary location where the data is stored. The value is constructed as follows:

Bit	Description
0 – 7	Data length in bits
8 – 15	Subindex data can be read at in Object Dictionary
16 - 31	Index data can be read at in Object Dictionary

Table D.38 PDO Mapping Entry Value

For example, if a 16-bit process data variable was stored in the Object Dictionary at Index 6001h, Subindex 04h, then it can be mapped into a PDO using the value 60010410h.

The Subentry number indicates the process data variable position in the eight bytes of the PDO. The process data variable at Subentry 01h is located in the first bits of the PDO. Table D.39 shows a mapping example and the location of the data in the PDO.

Subentry	Contents	Description	Location in PDO
01h	20010008h	8 bits of entry [2001h,00h]	Bits 0 – 7
02h	2002000Ch	12 bits of entry [2002h,00h]	Bits 8 – 19
03h	20030008h	8 bits of entry [2003h,00h]	Bits 20 - 27
04h	20040004h	4 bits of entry [2004h,00h]	Bits 28 - 31

Table D.39 PDO Mapping Example

It is possible to create gaps in the mapping by using dummy entries. A dummy entry is created by mapping one of the data types located at indexes 0001h – 0007h into the PDO. For example, to create a gap of 16 bits in the PDO, the UNSIGNED16 data type must be defined in a Subentry. This is achieved using the UNSIGNED16 Object Dictionary location of Index 0006h Subindex 00h, giving a value for the Subentry of 00060010h.

If Subentry 00h contains the value zero (i.e. no process data variables), then the PDO is disabled. In order to change the current mapping of a PDO, the PDO must first be disabled by writing zero to Subentry 00h. Once the new values for the Subentries have been written, Subentry 00h can be written with the number of process data variables mapped to the PDO. Attempting to write a non-zero value to Subentry 00h will cause the node to check and ensure that the entire mapping is valid. For example, the total number of bits mapped to the PDO does not exceed 64, each mapped process data variable exists in the Object Dictionary and can be mapped to a PDO. If the mapping is not valid, then the node will return an SDO Abort message in response to any attempt to set Subentry 00h to a non-zero value.

Each time a mapping entry is written, the node will check and ensure that the process data exists and can be mapped. If it does not exist or cannot be mapped then an SDO Abort message will be returned.

If Subentry 00h contains the value FEh then the PDO is a Source Addressing Mode Multiplexed PDO (SAM-MPDO). Subentries 01h – 40h are not used for SAM-MPDOs and any values stored there are ignored.

If Subentry 00h contains the value FFh then the PDO is a Destination Addressing Mode Multiplexed PDO (DAM-MPDO). For DAM-MPDOs only Subentry 01h is used and must be implemented. It defines the Object Dictionary entry that the DAM-MPDO data will be written to upon reception.

Example: 62000108h
8-bit digital output entry located at Index 6200h, Subindex 01h is mapped to the PDO

D.3.38 Transmit PDO Parameters (1800h – 19FFh)

Index	**1800h – 19FFh**
Name	Transmit PDO Parameter
Mandatory	Yes for each supported Transmit PDO

Subindex	**00h**
Name	Highest Subindex Supported
Type	UNSIGNED8
Default Value	Not defined
Access	Read Only
Mandatory	Yes for each supported Transmit PDO
Map to PDO	No

Subindex	**01h**
Name	COB ID used by PDO
Type	UNSIGNED32
Default Value	See description
Access	Read Only or Read/Write

Subindex	01h
Mandatory	Yes for each supported Transmit PDO
Map to PDO	No

Subindex	02h
Name	Transmission Type
Type	UNSIGNED8
Default Value	Determined by the device profile used
Access	Read Only or Read/Write
Mandatory	Yes for each supported Transmit PDO
Map to PDO	No

Subindex	03h
Name	Inhibit Time
Type	UNSIGNED16
Default Value	Determined by the device profile used
Units	100µs
Access	Read/Write
Mandatory	No
Map to PDO	No

Subindex	04h
Name	Compatibility Entry
Type	UNSIGNED8
Default Value	Not defined
Access	Read/Write
Mandatory	No
Map to PDO	No

Subindex	05h
Name	Event Timer
Type	UNSIGNED16
Default Value	Determined by the device profile used
Units	ms
Access	Read/Write
Mandatory	No
Map to PDO	No

Description: This entry must be implemented for each Transmit PDO supported by the node. The entry describes the communication configuration of the PDO.

Subentry 01h defines the COB ID of the PDO. The default value depends on the Index of the entry, as shown in the following table.

Index	Default Value
1400h	Node ID + 00000180h
1401h	Node ID + 00000280h
1402h	Node ID + 00000380h
1403h	Node ID + 00000480h
1404h – 15FFh	80000000h

Table D.40 Transmit PDO Default COB IDs

The COB ID entry also indicates if the PDO is used or not, size of the identifier and whether remote transmit requests are allowed for the PDO.

For an 11-bit COB ID the value of the Subentry is constructed as follows:

Bit	Description
0 - 10	COB ID for PDO
11 - 28	Set to 0
29	Set to 0 to select 11-bit COB ID
30	Set to 0 if remote transmit requests are allowed for the PDO. Set to 1 if remote transmit requests are not allowed for the PDO.
31	Set to 0 if the node does use the PDO. Set to 1 if the node does not use the PDO.

Table D.41 PDO Parameters COB ID Contents for 11-bit COB ID

For a 29-bit COB ID the value of the Subentry is constructed as follows:

Bit	Description
0 – 28	COB ID for PDO
29	Set to 1 to select 29-bit COB ID
30	Set to 0 if remote transmit requests are allowed for the PDO. Set to 1 if remote transmit requests are not allowed for the PDO.
31	Set to 0 if the node does use the PDO. Set to 1 if the node does not use the PDO.

Table D.42 PDO Parameters COB ID Contents for 29-bit COB ID

To change a COB ID, bit 31 must first be set to 1 to disable the SDO. Once the COB ID has been changed the SDO can be reenabled by clearing bit 31.

Attempting to set bit 29 on a node that does not support 29-bit COB IDs, or attempting to clear bit 30 on a node that does not support remote transmission requests, will result in the node transmitting an SDO Abort message.

Subentry 02h specifies the transmission type of the Transmit PDO. Table D.43 lists the available transmission types for a Transmit PDO.

Transmission Type	Description
0	The Transmit PDO is synchronous. Which specific SYNC Object occurrence triggers the transmission is given in the device profile. Additional details of the PDO transmission are given in the device profile.
1 – 240	The Transmit PDO is synchronous. It is transmitted after every *n*th SYNC Object within the Synchronous Window Length, where *n* is the transmission type. For example, when using transmission type 34, the PDO is transmitted after every 34th SYNC Object.
241 – 251	Not used for Transmit PDOs
252	The data for the PDO is updated on reception of a SYNC Object, but the PDO is not transmitted. The PDO is only transmitted on reception of a Remote Transmission Request.
253	The data for the PDO is updated and the PDO is transmitted on reception of a Remote Transmission Request.
254	The conditions that cause the Transmit PDO to be transmitted are manufacturer specific.
255	The Transmit PDO is asynchronous. Details of when the PDO is transmitted is given in the device profile.

Table D.43 Transmit PDO Transmission Types

If the PDO is a Destination Addressing Mode Multiplexed PDO then it must have transmission type 254. If the PDO is a Source Addressing Mode Multiplexed PDO then it must have transmission type 254 or 255.

Subindex 03h is optional and defines the inhibit time for the PDO. The inhibit time specifies the minimum time between transmissions of the PDO. Once the PDO is transmitted, any additional transmissions of the PDO will not take place during the inhibit time.

The inhibit time is a multiple of 100μs. For example, a value of 173Ah would give an inihibit time of 594.6ms. A value of zero disables the inhibit time functionality.

Note that the inhibit time is measured from the time when the node first attempts to send the PDO. If the PDO is blocked from being sent because of higher priority messages on the bus, then the delay before the PDO is actually transmitted is included in the inhibit time. Therefore the inhibit time must be greater than the worst case transmission time of the PDO.

The inhibit time may not be changed while the PDO is being used by the node. To change the inhibit time the PDO must first be disabled by setting bit 31 of Subentry 01h.

Subentry 04h is not used. It may optionally be implemented, but if it is not implemented then any attempt to read or write the entry will return an SDO Abort message from the node. If the Subentry is not implemented, Subentry 05h may still be implemented if desired.

Subentry 05h defines the optional event time for a Transmit PDO. A value of zero disables the event timer.

If the event timer is used, then the PDO is periodically transmitted. The value of the event timer entry is the number of milliseconds between transmissions. Each time the PDO is transmitted as a result of the event timer expiring, the event timer is reset.

Example: 00000181h in Subentry 01h, 100 in Subentry 02h, 1000 in Subentry 03h, 3000 in Subentry 04h
The PDO is transmitted with a COB ID of 181h every 100 SYNC messages, sampling the data to be transmitted in the PDO at the SYNC. The PDO will be transmitted at most every 1 second and transmitted every 3 seconds if 100 SYNCS have not occurred.

D.3.39 Transmit PDO Mapping (1A00h – 1BFFh)

Index	1A00h – 1BFFh
Name	Transmit PDO Mapping
Mandatory	Yes for each supported Transmit PDO

Subindex	00h
Name	Number of Entries (Number of objects mapped into the PDO)
Type	UNSIGNED8
Default Value	Defined in the device profile
Access	Read Only if dynamic mapping is not supported. Read/Write if dynamic mapping is supported.
Mandatory	Yes for each supported Transmit PDO
Map to PDO	No

Subindex	01h – 40h
Name	PDO Mapping for a process data variable
Type	UNSIGNED32
Default Value	Defined in the device profile
Access	Read/Write
Mandatory	No
Map to PDO	No

Description: This entry defines which process data is stored in a single PDO, along with the position of the process data in the eight data bytes of the PDO.

Each Transmit PDO supported by the node must have a corresponding Transmit PDO Mapping parameter entry implemented. The entry at 1A00h is for the first Transmit PDO whose communication parameters are defined at 1800h. The entry at 1A01h is for the second Transmit PDO whose communication parameters are defined at 1801h, etc.

A PDO may have 1 to 64 process data variables mapped to it, with each variable having any length from 1 to 64 bits, however the total size of all the process data mapped to a single PDO may not exceed 64 bits (eight bytes). Each Subentry defines a process data variable, therefore Subentry 00h holds the total number of variables mapped to the PDO.

The value of each Subentry defines the process data variable to be mapped and the size of the variable in bits. The process data variable is defined by specifying the Object Dictionary location where the data is stored. The value is constructed as follows:

Bit	Description
0 – 7	Data length in bits
8 – 15	Subindex data can be read at in Object Dictionary
16 - 31	Index data can be read at in Object Dictionary

Table D.44 PDO Mapping Entry Value

For example, if a 16-bit process data variable was stored in the Object Dictionary at Index 6001h, Subindex 04h, then it can be mapped into a PDO using the value 60010410h.

The Subentry number indicates the process data variable position in the eight bytes of the PDO. The process data variable at Subentry 01h is located in the first bits of the PDO. For example, the following table shows an example mapping and the location of the data in the PDO.

Subentry	Contents	Description	Location in PDO
01h	20010008h	8 bits of entry [2001h,00h]	Bits 0 – 7
02h	2002000Ch	12 bits of entry [2002h,00h]	Bits 8 – 19
03h	20030008h	8 bits of entry [2003h,00h]	Bits 20 - 27
04h	20040004h	4 bits of entry [2004h,00h]	Bits 28 - 31

Table D.45 PDO Mapping Example

Mapping dummy entries to a Transmit PDO is not permitted. No gaps may appear in the mapped data in a Transmit PDO.

If Subentry 00h contains the value zero (i.e. no process data variables), then the PDO is disabled. In order to change the current mapping of a PDO, the PDO must first be disabled by writing zero to Subentry 00h. Once the new values for the Subentries have been written, Subentry 00h can be written with the number of process data variables mapped to the PDO. Attempting to write a non-zero value to Subentry 00h will cause the node to check and ensure that the entire mapping is valid. For example, the total number of bits mapped to the PDO does not exceed 64, each mapped process data variable exists in the Object Dictionary and can be mapped to a PDO. If the mapping is not valid, then the node will return an SDO Abort message in response to any attempt to set Subentry 00h to a non-zero value.

Each time a mapping entry is written, the node will check and ensure that the process data exists and can be mapped. If it does not exist or cannot be mapped then an SDO Abort message will be returned.

If Subentry 00h contains the value FEh then the PDO is a Source Addressing Mode Multiplexed PDO (SAM-MPDO). Subentries 01h – 40h are not used for SAM-MPDOs and any values stored there are ignored.

If Subentry 00h contains the value FFh then the PDO is a Destination Addressing Mode Multiplexed PDO (DAM-MPDO). For DAM-MPDOs only Subentry 01h is used and must be implemented. It defines the Object Dictionary entry that the DAM-MPDO data will contain when transmitted. DAM-MPDOs also contain the Node ID of the DAM-MPDO consumer and the Index and Subindex of the Consumer's Object Dictionary entry where the data will be stored. How these values are specified is manfuacturer-specific and not covered by the CANopen specification.

Example: 60000108h
Transmit the 8-bit digital input located at Index 6000h, Subindex 01h in the PDO.

D.3.40 Object Scanner List (1FA0h – 1FCFh)

Index	**1FA0h – 1FCFh**
Name	Object Scanner List
Mandatory	Yes if Source Address Mode Multiplexed PDOs are transmitted by the node

Subindex	**00h**
Name	Number of Entries
Type	UNSIGNED8
Default Value	Not defined
Access	Read Only
Mandatory	Yes if Source Address Mode Multiplexed PDOs are transmitted by the node
Map to PDO	No

Subindex	**01h**
Name	Scan 1
Type	UNSIGNED32
Default Value	Not defined
Access	Read/Write
Mandatory	Yes if Source Address Mode Multiplexed PDOs are transmitted by the node
Map to PDO	No

Subindex	**02h – FEh**
Name	Scan 2 – 254
Type	UNSIGNED32
Default Value	Not defined
Access	Read/Write

Subindex	02h – FEh
Mandatory	No
Map to PDO	No

Description: If a node transmits Source Address Mode Multiplexed PDOs (SAM-MPDOs) then it must implement the Object Scanner List to specify which process data is transmitted in the SAM-MPDO.

The scanner list specifies the Object Dictionary Index and Subindex of the process data, and may optionally specify ranges of Subindexes. This allows for a large range of data to be specified in the Object Scanner List.

An entry in the Object Scanner List is constructed as shown in the following table:

Bit	Description
0 – 7	Subindex
8 – 23	Index
24 - 31	Number of Subindexes

Table D.46 Object Scanner List Entry

A SAM-MPDO producer scans the Object Scanner List and decides on a specific Object Dictionary entry to transmit in a SAM-MPDO. The transmission type of a SAM-MPDO is manufacturer-specific, therefore the method of determining which process data to transmit is not covered by the CANopen specification. When the SAM-MPDO is transmitted, the SAM-MPDO contains the process data along with the Index and Subindex of where in the node's Object Dictionary the process data is located.

Example: 03600102h
Entry 6001h, Subentries 02h to 04h

D.3.41 Object Dispatching List (1FD0h – 1FFFh)

Index	1FD0h – 1FFFh
Name	Object Dispatching List
Mandatory	Yes if Source Address Mode Multiplexed PDOs are received by the node

Subindex	00h
Name	Number of Entries
Type	UNSIGNED8
Default Value	Not defined
Access	Read Only
Mandatory	Yes if Source Address Mode Multiplexed PDOs are received by the node
Map to PDO	No

Subindex	01h
Name	Dispatch 1
Type	UNSIGNED64
Default Value	Not defined
Access	Read/Write
Mandatory	Yes if Source Address Mode Multiplexed PDOs are received by the node
Map to PDO	No

Subindex	02h – FEh
Name	Dispatch 2 – 254
Type	UNSIGNED64
Default Value	Not defined
Access	Read/Write

Subindex	02h – FEh
Mandatory	No
Map to PDO	No

Description: If a node receives Source Address Mode Multiplexed PDOs (SAM-MPDOs) then it must implement the Object Dispatching List to specify which process data is received in the SAM-MPDO.

Each entry in the Object Dispatching List specifies where in the Object Dictionary the received process data should be stored. It is cross referenced with the Object Dictionary entry of the node that produces the SAM-MPDO. The following table shows the contents of an entry in the Object Dispatching List.

Bit	Description
0 – 7	Sender Node ID
8 – 15	Sender Subindex
16 – 31	Sender Index
32 – 39	Local Subindex
40 – 55	Local Index
56 - 63	Number of Subindexes

Table D.47 Object Dispatching List Entry

When a SAM-MPDO is received, the Index, Subindex and sender Node ID in the SAM-MPDO are looked up in the Object Dispatching List. When an entry is found, the node can then determine in which local Object Dictionary entry to store the data.

It is possible for ranges of entries to be specified allowing more complex mapping of the sender (SAM-MPDO producer) to the local (SAM-MPDO consumer) Object Dictionary.

For example, assume the following value is used for an Object Dispatching List entry:

Sender Node ID	0Ah
Sender Subindex	01h

Sender Index	6001h
Local Subindex	2101h
Local Index	06h
Number of Subindexes	03h

Then the sender's Object Dictionary entry 6001h Subindexes 01h to 03h are mapped to the local Object Dictionary entry 2101h Subindexes 06h to 08h.

D.4 CANopen Managers and Programmable CANopen Devices

D.4.1 Object Dictionary Entries

The following table gives an overview of all Object Dictionary entries in the Programmable CANopen Devices section of the Object Dictionary.

Index	Name	Type	Access
1F00h	Request SDO	UNSIGNED32	Write Only
1F01h	Release SDO	UNSIGNED32	Write Only
1F02h	SDO Manager COB IDs	UNSIGNED32	Read/Write
1F03h	SDO Connections Part 1	UNSIGNED32	Read Only
1F04h	SDO Connections Part 2	UNSIGNED32	Read Only
1F05h	SDO Connections Part 3	UNSIGNED32	Read Only
1F06h	SDO Connections Part 4	UNSIGNED32	Read Only
1F10h	Dynamic SDO Connection State	UNSIGNED32	Read/Write
1F11h	Slave Failed	UNSIGNED16	Read Only
1F20h	Store DCF	DOMAIN	Read/Write
1F21h	Storage Format	UNSIGNED8	Read/Write
1F22h	Concise DCF	DOMAIN	Read/Write
1F23h	Store Slave EDS	DOMAIN	Read/Write
1F24h	Slave EDS Storage Format	UNSIGNED8	Read/Write
1F25h	Configure Slave	UNSIGNED32	Read/Write
1F26h	Expected Configuration Date	UNSIGNED32	Read/Write
1F27h	Expected Configuraton Time	UNSIGNED32	Read/Write
1F50h	Download Program Data	DOMAIN	Read/Write
1F51h	Program Control	UNSIGNED8	Read/Write
1F52h	Verify Application Software	UNSIGNED32	Read/Write

Table D.48 Programmable CANopen Devices Object Dictionary Entries

Index	Name	Type	Access
1F53h	Expected Application SW Date	UNSIGNED32	Read/Write
1F54h	Expected Application SW Time	UNSIGNED32	Read/Write
1F70h	Process Picture	RECORD	Read/Write
1F80h	NMT Startup	UNSIGNED32	Read/Write
1F81h	Slave Assignment	UNSIGNED32	Read/Write
1F82h	Request NMT	UNSIGNED8	Read/Write
1F83h	Request Guarding	UNSIGNED8	Read/Write
1F84h	Device Type Identification	UNSIGNED32	Read/Write
1F85h	Vendor Identification	UNSIGNED32	Read/Write
1F86h	Product Code	UNSIGNED32	Read/Write
1F87h	Revision Number	UNSIGNED32	Read/Write
1F88h	Serial Number	UNSIGNED32	Read/Write
1F89h	Boot Time	UNSIGNED32	Read/Write
1F90h	Flying Master Timing Parameters	UNSIGNED16	Read/Write
1F91h	Startup-capable Device Timing	UNSIGNED16	Read/Write

Table D.48 (Continued) Programmable CANopen Devices Object Dictionary

D.4.2 Request SDO (1F00h)

Index	1F00h
Name	Request SDO
Mandatory	Yes for SDO Managers

Subindex	00h
Name	Request SDO
Type	UNSIGNED32
Default Value	Not defined
Access	Write Only

Subindex	00h
Mandatory	Yes for SDO Managers
Map to PDO	No

Description: If a node referred to in this process as the SDO Requesting Device (SRD) wishes to request an SDO channel to another node (called the slave), then it must write to this entry on the SDO Manager. Section D.5.4 shows the sequence of Object Dictionary accesses involved.

The value of the entry is constructed as follows:

Bit	Description
0 – 7	Slave Node ID
8 – 15	SRD Node ID
16 – 31	Index of a free Client SDO Entry in the SRD's Object Dictionary (1280h – 12FFh)

Table D.49 Request SDO Entry

Example: 12800612h
The SRD with Node ID 06h wishes to request an SDO channel to Node ID 12h. Client SDO entry 1280h is free to be used.

D.4.3 Release SDO (1F01h)

Index	1F01h
Name	Release SDO
Mandatory	Yes for SDO Managers

Subindex	00h
Name	Release SDO
Type	UNSIGNED32
Default Value	Not defined
Access	Write Only

Subindex	00h
Mandatory	Yes for SDO Managers
Map to PDO	No

Description: If a node referred to in this process as the SDO Requesting Device (called the SRD) wishes to release the SDO channels it is using to connect to another node (called the slave) or release all SDO channels, then it writes to this entry on the SDO Manager. Section D.5.4 shows the sequence of Object Dictionary accesses involved.

The value of the entry is constructed as follows:

Bit	Description
0 – 7	Slave Node ID or zero to release all connections and un-register as an SRD
8 – 15	SRD Node ID
16 – 31	Index of a Client SDO Entry in the SRD's Object Dictionary (1280h – 12FFh) being used to connect to the Slave or zero to release all SDO channels to the Slave

Table D.50 Request SDO Entry

Example: 12800612h
The SRD with Node ID 06h wishes to release the SDO channel it used to communicate with Node 12. The Client SDO at 1280h is being used.

D.4.4 SDO Manager COB IDs (1F02h)

Index	1F02h
Name	SDO Manager COB IDs
Mandatory	Yes for SDO Managers

Subindex	00h
Name	Number of Entries
Type	UNSIGNED8
Default Value	Not defined
Access	Read/Write
Mandatory	Yes for SDO Managers
Map to PDO	No

Subindex	01h - FEh
Name	COB ID 1 - 254
Type	UNSIGNED32
Default Value	Not defined
Access	Read/Write
Mandatory	No
Map to PDO	No

Description: This entry allows a Configuration Tool to specify to the SDO Manager which COB IDs are available to be used for SDO channels, and also for reading which COB IDs are currently in use for SDO channels.

For an 11-bit COB ID the value of the Subentries are constructed as follows:

Bit	Description
0 - 10	COB ID
11 - 28	Set to 0
29	Set to 0 to select 11-bit COB ID
30	Set to 0 if the COB ID is free to be used for an SDO channel Set to 1 if the COB ID is currently in use for an SDO channel
31	Set to 0 if the COB ID is valid. i.e. this Subentry is being used. Set to 1 if the COB ID is not valid. i.e. this Subentry is not used.

Table D.51 SDO Manager COB ID Contents for 11-bit COB ID

For a 29-bit COB ID the value of the Subentries are constructed as follows:

Bit	Description
0 – 28	COB ID
29	Set to 1 to select 29-bit COB ID
30	Set to 0 if the COB ID is free to be used for an SDO channel Set to 1 if the COB ID is currently in use for an SDO channel
31	Set to 0 if the COB ID is valid. i.e. this Subentry is being used. Set to 1 if the COB ID is not valid. i.e. this Subentry is not used.

Table D.52 SDO Manager COB ID Contents for 29-bit COB ID

To change a COB ID, bit 31 must first be set to 1 to disable the Subentry. Once the COB ID has been changed the entry can be reenabled by clearing bit 31.

Attempting to set bit 29 on a node that does not support 29-bit COB IDs, will result in the node transmitting an SDO Abort message.

Note that when writing to this entry, it must be ensured that there are no dynamic SDO connections being used at the time.

Example: 00000412h
COB ID 412h is available and free to be used

D.4.5 SDO Connections Part 1 (1F03h)

Index	**1F03h**
Name	SDO Connections Part 1
Mandatory	Yes for SDO Managers

Subindex	**00h**
Name	Number of Entries
Type	UNSIGNED8
Default Value	Not defined
Access	Read Only
Mandatory	Yes for SDO Managers
Map to PDO	No

Subindex	**01h - FEh**
Name	SDO Connection 1 - 254
Type	UNSIGNED32
Default Value	Not defined
Access	Read Only
Mandatory	No
Map to PDO	No

Description: This entry describes the first 254 dynamic SDO connections between nodes in the CANopen network. This entry is implemented on the SDO Manager.

The Subentries are constructed as follows:

Bit	Description
0 – 7	Server Node ID
8 – 15	Server Offset
16 – 23	Client Node ID
24 - 31	Client Offset

Table D.53 SDO Connection Entry

Each entry specifies the Server and Client Node IDs.

The Server Offset is added to 1200h to obtain the Index of the Server SDO in the Server's Object Dictionary.

The Client Offset is added to 1280h to obtain the Index of the Client SDO in the Client's Object Dictionary.

If either the Client or Server Node ID is zero, then the connection described is not valid and not in use.

Section D.5.4 shows the sequence of Object Dictionary accesses used for dynamic SDO channel assignment.

Example: 03060112h
Node 06h using client SDO 1283h has established an SDO connection with node 12h using server SDO 1201h.

D.4.6 SDO Connections Part 2 (1F04h)

Index	**1F04h**
Name	SDO Connections Part 2
Mandatory	No

Subindex	**00h**
Name	Number of Entries
Type	UNSIGNED8
Default Value	Not defined
Access	Read Only
Mandatory	Yes if this entry is implemented
Map to PDO	No

Subindex	**01h - FEh**
Name	SDO Connection 255 - 508
Type	UNSIGNED32
Default Value	Not defined
Access	Read Only
Mandatory	No
Map to PDO	No

Description: This entry describes up to 254 dynamic SDO connections between nodes in the CANopen network. This entry is implemented on the SDO Manager.

The Subentries are constructed as follows:

Bit	Description
0 – 7	Server Node ID
8 – 15	Server Offset
16 – 23	Client Node ID
24 - 31	Client Offset

Table D.54 SDO Connection Entry

Each entry specifies the Server and Client Node IDs.

The Server Offset is added to 1200h to obtain the Index of the Server SDO in the Server's Object Dictionary.

The Client Offset is added to 1280h to obtain the Index of the Client SDO in the Client's Object Dictionary.

If either the Client or Server Node ID is zero, then the connection described is not valid and not in use.

Section D.5.4 shows the sequence of Object Dictionary accesses used for dynamic SDO channel assignment.

Example: 03060112h
Node 06h using client SDO 1283h has established an SDO connection with node 12h using server SDO 1201h.

D.4.7 SDO Connections Part 3 (1F05h)

Index	1F05h
Name	SDO Connections Part 3
Mandatory	No

Subindex	00h
Name	Number of Entries
Type	UNSIGNED8
Default Value	Not defined
Access	Read Only
Mandatory	Yes if this entry is implemented
Map to PDO	No

Subindex	01h - FEh
Name	SDO Connection 509 - 762
Type	UNSIGNED32
Default Value	Not defined
Access	Read Only
Mandatory	No
Map to PDO	No

Description: This entry describes up to 254 dynamic SDO connections between nodes in the CANopen network. This entry is implemented on the SDO Manager.

The Subentries are constructed as follows:

Bit	Description
0 – 7	Server Node ID
8 – 15	Server Offset
16 – 23	Client Node ID
24 - 31	Client Offset

Table D.55 SDO Connection Entry

Each entry specifies the Server and Client Node IDs.

The Server Offset is added to 1200h to obtain the Index of the Server SDO in the Server's Object Dictionary.

The Client Offset is added to 1280h to obtain the Index of the Client SDO in the Client's Object Dictionary.

If either the Client or Server Node ID is zero, then the connection described is not valid and not in use.

Section D.5.4 shows the sequence of Object Dictionary accesses used for dynamic SDO channel assignment.

Example: 03060112h
Node 06h using client SDO 1283h has established an SDO connection with node 12h using server SDO 1201h.

D.4.8 SDO Connections Part 4 (1F06h)

Index	**1F06h**
Name	SDO Connections Part 4
Mandatory	No

Subindex	**00h**
Name	Number of Entries
Type	UNSIGNED8
Default Value	Not defined
Access	Read Only
Mandatory	Yes if this entry is implemented
Map to PDO	No

Subindex	**01h - FEh**
Name	SDO Connection 763 - 1016
Type	UNSIGNED32
Default Value	Not defined
Access	Read Only
Mandatory	No
Map to PDO	No

Description: This entry describes up to 254 dynamic SDO connections between nodes in the CANopen network. This entry is implemented on the SDO Manager.

The Subentries are constructed as follows:

Bit	Description
0 – 7	Server Node ID
8 – 15	Server Offset
16 – 23	Client Node ID
24 - 31	Client Offset

Table D.56 SDO Connection Entry

Each entry specifies the Server and Client Node IDs.

The Server Offset is added to 1200h to obtain the Index of the Server SDO in the Server's Object Dictionary.

The Client Offset is added to 1280h to obtain the Index of the Client SDO in the Client's Object Dictionary.

If either the Client or Server Node ID is zero, then the connection described is not valid and not in use.

Section D.5.4 shows the sequence of Object Dictionary accesses used for dynamic SDO channel assignment.

Example: 03060112h
Node 06h using client SDO 1283h has established an SDO connection with node 12h using server SDO 1201h.

D.4.9 Dynamic SDO Connection State (1F10h)

Index	1F10h
Name	Dynamic SDO Connection State
Mandatory	Yes for nodes using dynamic SDO channels

Subindex	00h
Name	Dynamic SDO Connection State
Type	UNSIGNED32
Default Value	Not defined
Access	Read/Write
Mandatory	Yes for nodes using dynamic SDO channels
Map to PDO	No

Description: This entry is implemented in a node (SDO Requesting Device - SRD) that wishes to obtain an SDO channel to another node (slave). It allows the SRD to provide information to the SDO Manager as well as receive configuration data back from the SDO Manager. Therefore, during the process of obtaining dynamic SDO channels, the SRD must make specific values available for reading at this entry, and operate on values written to this entry by the SDO Manager.

This entry is constructed as follows:

Bit	Description
0	Rq Indication
1 - 2	Cnxn State
3	Req. EC
4 - 7	Reserved. Always zero
8 - 15	Error code
16 - 31	Index

Table D.57 Dynamic SDO Connection State Entry

The SRD sets the Rq Indication flag when it wishes to be registered with the SDO Manager as an SRD. It sets this bit before sending the Dynamic SDO Request message. The Request Message has the COB ID 6E0h and contains no data. When the SDO Manager successfully recognizes the node as an SRD by scanning the Rq Indication Flag of all nodes on the network, it writes a zero to this bit. Note that the SDO Manager only scans until it finds the first node with the Rq Indication Flag set. Therefore if there is more than one node wishing to set up a dynamic SDO channel, each node may have to send the Request Message repeatedly until the SDO Manager recognizes the node as an SRD.

If the SRD wishes to establish a connection with a slave using a single SDO channel, then the Cnxn State value is set to 0h before the Dynamic SDO Request message is transmitted.

If the SRD wishes to obtain all default SDO channels that are currently unused, then it sets the Cnxn State to 1h instead.

The SDO Manager writes various values to the Cnxn State to indicate the result of operations.

A value of 0h indicates that the SDO Manager failed to establish an SDO connection between the SRD and SDO Manager or failed to establish an SDO connection between the SRD and slave. The reason for the failure is given in the Error code field.

A value of 1h indicates that the SDO Manager has successfully established an SDO connection between the SRD and SDO Manager.

A value of 2h indicates that the SDO Manager is allowing the SRD to obtain all default SDO Channels that are currently unused.

A value of 3h indicates that the SDO connection between the SRD and slave has been established.

The Req EC value must be set to 1h if the SRD wishes the SDO Manager to perform error control on the slave. This will result in the SDO Manager using either heartbeat or node guarding to determine that the slave node is present and operational while the SRD has an SDO channel to the slave.

If the SDO Manager does not support error control on slaves then it will write 0h to this bit, otherwise it will write 1h.

The error code field is only used if the SDO Manager fails to establish an SDO connection between itself and the SRD or between the SRD and slave. It has one of the following values:

Error Code	Description
00h	Unspecified error
01h	There was no free SDO channel to create a connection between the SDO Manager and SRD
02h	There were no more free SDO channels in the CANopen network
03h	The Slave does not have any free Server SDOs
04h	The Slave node is not available
05h – FFh	Reserved

Table D.58 Dynamic SDO Connection Error Codes

The SRD specifies the Index (1280h – 12FFh) of the Client SDO to use to communicate with the SDO Manager in the Index field. If all unused default SDOs are being requested then the Index field is ignored by the SDO Manager.

The SDO Manager writes the Client SDO Index of the Client SDO that will actually be used to communicate with the SDO Manager.

Section D.5.4 lists the sequence of Object Dictionary accesses used for dynamic SDO channel assignment.

D.4.10 Slave Failed (1F11h)

Index	1F11h
Name	Slave Failed
Mandatory	No

Subindex	00h
Name	Slave Failed
Type	UNSIGNED16

Subindex	00h
Default Value	Not defined
Access	Write Only
Mandatory	No
Map to PDO	No

Description: If a node (SDO Requesting Device - SRD) has established an SDO Connection with another node (slave) and specified that the SDO Manager should use error control, then this entry which is implemented on the SRD will be written to by the SDO Manager should a heartbeat or node guarding event occur.

If this entry is written to, then the SRD must assume that the SDO connection it has with the slave is no longer valid. The SDO Manager will automatically take steps to release the SDO Connection between the SRD and the slave.

The value written is the Node ID of the slave.

D.4.11 Store DCF (1F20h)

Index	1F20h
Name	Store DCF
Mandatory	No

Subindex	00h
Name	Number of Entries
Type	UNSIGNED8
Default Value	7Fh
Access	Read Only
Mandatory	Yes if this entry is implemented
Map to PDO	No

Subindex	01h – 7Fh
Name	Store DCF Node 1 – 127
Type	DOMAIN
Default Value	Not defined
Access	Read/Write
Mandatory	Yes if this entry is implemented
Map to PDO	No

Description: This entry is implemented on a Configuration Manager and allows the Device Configuration Files (DCF) to be written to and read from the manager. Each Subentry corresponds to a node on the network, with the Subindex specifying the Node ID. For example, to store the DCF for node 3Ah in the Configuration Manager, the DCF is written to Subentry 3Ah.

The format that the DCF is read and written is specified by OD entry 1F21h.

D.4.12 Storage Format (1F21h)

Index	1F21h
Name	Storage Format
Mandatory	No

Subindex	00h
Name	Number of Entries
Type	UNSIGNED8
Default Value	7Fh
Access	Read Only
Mandatory	Yes if this entry is implemented
Map to PDO	No

Subindex	01h – 7Fh
Name	Storage Format Node 1 - 127
Type	UNSIGNED8
Default Value	Not defined
Access	Read/Write
Mandatory	Yes if this entry is implemented
Map to PDO	No

Description: This entry specifies the storage format of the DCF when read to and written from OD entry 1F20h. Currently the following values are implemented. All other values are reserved.

Value	Format
00h	Non-compressed ASCII

Table D.59 DCF Storage Formats

Note that the internal storage format of the DCFs is manufacturer-specific and may be compressed if desired.

D.4.13 Concise DCF (1F22h)

Index	1F22h
Name	Concise DCF
Mandatory	No

Subindex	00h
Name	Number of Entries
Type	UNSIGNED8
Default Value	7Fh
Access	Read Only

Subindex	00h
Mandatory	Yes if this entry is implemented
Map to PDO	No

Subindex	01h – 7Fh
Name	Concise DCF Node 1 - 127
Type	DOMAIN
Default Value	Not defined
Access	Read/Write
Mandatory	Yes if this entry is implemented
Map to PDO	No

Description: If a Configuration Manager does not have enough disk or non-volatile memory space to store the full Device Configuration Files (DCFs), then it may optionally implement this entry to store concise versions of the DCFs.

There is one Subentry for each possible node on the network, allowing a DCF to be read or written for each node. For example, to store the DCF for node 3Ah on the Configuration Manager, the Concise DCF is written to Subentry 3Ah.

The concise version of the DCF is stored as a stream of data containing information about where the Object Dictionary data is stored and how large it is.

The data stream is structured as follows:

Number of supported entries *n* (UNSIGNED32)
Entry 1
Entry 2
...
Entry *n*

Table D.60 Concise DCF Data Stream

Where each entry has the following data format:

Name	Type
Index	UNSIGNED16
Subindex	UNSIGNED8
Data size	UNSIGNED32
Data	DOMAIN

Table D.61 Concise DCF Entry

The first item of data indicates the number of entries that are contained in the data stream and has the type UNSIGNED32. Each entry then follows.

In order to simplify the operation of the Configuration Manager when using the concise DCF to configure nodes, the Configuration Manager writes to the Object Dictionary of the node by processing each entry in the data stream one at a time. For example, if entry 1 in the data stream specifies the value 12h for entry [2001h,04h] and entry 2 specifies the value 6Ah for entry [25AFh,00h] then the Configuration Manager will first write 12h to [2001h,04h] then write 6Ah to [25AFh,00h]. This means that when writing to entries that require a flag to be set or cleared before the contents can be changed (for example COB ID entries), multiple entries in the data stream must be used. For example to change the COB ID of a PDO there must be two entries specified in the data stream. The first only sets bit 31 to 1 to disable the PDO. The second sets the new COB ID and clears bit 31 to enable the PDO.

An empty stream can be written by specifying zero for the number of entries in the stream. Reading an unused entry will result in a stream with zero for the number of entries.

D.4.14 Store Slave EDS (1F23h)

Index	**1F23h**
Name	Store Slave EDS
Mandatory	No

Subindex	**00h**
Name	Number of Entries
Type	UNSIGNED8
Default Value	7Fh
Access	Read Only
Mandatory	Yes if this entry is implemented
Map to PDO	No

Subindex	**01h – 7Fh**
Name	Store Slave EDS Node 1 - 127
Type	DOMAIN
Default Value	Not defined
Access	Read/Write
Mandatory	Yes if this entry is implemented
Map to PDO	No

Description: This entry is implemented on a Configuration Manager, and allows the Electronic Datasheets (EDSs) to be written to and read from the manager, usually by a configuration tool. Each Subentry corresponds to a node on the network, with the Subindex specifying the Node ID. For example, to store the EDS for node 3Ah in the Configuration Manager, the EDS is written to Subentry 3Ah.

The format in which the EDS is read and written is specified by OD entry 1F24h.

D.4.15 Slave EDS Storage Format (1F24h)

Index	1F24h
Name	Slave EDS Storage Format
Mandatory	No

Subindex	00h
Name	Number of Entries
Type	UNSIGNED8
Default Value	7Fh
Access	Read Only
Mandatory	Yes if this entry is implemented
Map to PDO	No

Subindex	01h – 7Fh
Name	Slave EDS Storage Format Node 1 - 127
Type	UNSIGNED8
Default Value	Not defined
Access	Read/Write
Mandatory	Yes if this entry is implemented
Map to PDO	No

Description: This entry specifies the storage format of the Electronic Datasheet (EDS) when read to and written from OD entry 1F23h. Currently the following values are implemented. All other values are reserved.

Value	Format
00h	Non-compressed ASCII

Table D.62 EDS Storage Formats

Note that the internal storage format of the EDSs is manufacturer-specific and may be compressed if desired.

D.4.16 Configure Slave (1F25h)

Index	**1F25h**
Name	Configure Slave
Mandatory	No

Subindex	**00h**
Name	Number of Entries
Type	UNSIGNED8
Default Value	80h
Access	Read Only
Mandatory	Yes if this entry is implemented
Map to PDO	No

Subindex	**01h – 7Fh**
Name	Configure Slave 1 - 127
Type	UNSIGNED32
Default Value	Not Defined
Access	Write Only
Mandatory	Yes if this entry is implemented
Map to PDO	No

Subindex	**80h**
Name	Configure All Slaves
Type	UNSIGNED32
Default Value	Not Defined
Access	Write Only

Subindex	80h
Mandatory	Yes if this entry is implemented
Map to PDO	No

Description: Nodes may write to this entry, implemented on the Configuration Manager, to request that the manager configure a specific node or all nodes. By writing the value "conf" (666E6F63h) to Subentries 01h – 7Fh, the corresponding node whose Node ID matches the Subindex will be subsequently configured by the Configuration Manager. By writing "conf" to Subentry 80h, all nodes will be configured.

D.4.17 Expected Configuration Date (1F26h)

Index	1F26h
Name	Expected Configuration Date
Mandatory	No

Subindex	00h
Name	Number of Entries
Type	UNSIGNED8
Default Value	7Fh
Access	Read Only
Mandatory	Yes if this entry is implemented
Map to PDO	No

Subindex	01h – 7Fh
Name	Expected Configuration Date Node 1 - 127
Type	UNSIGNED32
Default Value	Not defined
Units	Days
Access	Read/Write

Subindex	01h – 7Fh
Mandatory	Yes if this entry is implemented
Map to PDO	No

Description: This entry is implemented on the Configuration Manager and stores the expected configuration date of each node. Each Subentry corresponds to a node on the network, with the Subindex indicating the Node ID.

When the Configuration Manager wishes to check if the node is using the currently known configuration it may read this entry to determine if an expected configuration date exists. This is indicated by a non-zero value. If an expected configuration date exists then it is compared with the date stored in the corresponding node in OD entry 1020h. If the two dates match (along with the expected configuration time) then the configuration of the node is known to match with the Device Configuration File stored in the Configuration Manager. If the dates (or times) do not match or entry 1020h could not be read, or if there is no expected configuration date (or time) stored, then the Configuration Manager can proceed to download the configuration to the node using the DCF.

The date is stored as the number of whole days since January, 1984.

Example: 23
January 24, 1984

D.4.18 Expected Configuration Time (1F27h)

Index	1F27h
Name	Expected Configuration Time
Mandatory	No

Subindex	00h
Name	Number of Entries
Type	UNSIGNED8

Subindex	**00h**
Default Value	7Fh
Access	Read Only
Mandatory	Yes if this entry is implemented
Map to PDO	No

Subindex	**01h – 7Fh**
Name	Expected Configuration Time Node 1 - 127
Type	UNSIGNED32
Default Value	Not defined
Units	ms
Access	Read/Write
Mandatory	Yes if this entry is implemented
Map to PDO	No

Description: This entry is implemented on the Configuration Manager and stores the expected configuration time of each node. Each Subentry corresponds to a node on the network, with the Subindex indicating the Node ID.

When the Configuration Manager wishes to check if the node is using the currently known configuration it may read this entry to determine if an expected configuration time exists. This is indicated by a non-zero value. If an expected configuration time exists then it is compared with the time stored in the corresponding node in OD entry 1020h. If the two times match (along with the expected configuration date) then the configuration of the node is known to match with the Device Configuration File stored in the Configuration Manager. If the times (or dates) do not match or entry 1020h could not be read, or if there is no expected configuration time (or date) stored, then the Configuration Manager can proceed to download the configuration to the node using the DCF.

The date is stored as the number of milliseconds since midnight.

Example: 6212000
1:43:32 am

D.4.19 Download Program Data (1F50h)

Index	**1F50h**
Name	Download Program Data
Mandatory	No

Subindex	**00h**
Name	Number of Entries
Type	UNSIGNED8
Default Value	Not defined
Access	Read Only
Mandatory	Yes if this entry is implemented
Map to PDO	No

Subindex	**01h – FEh**
Name	Program 1 – 254
Type	DOMAIN
Default Value	Not defined
Access	Read/Write
Mandatory	No
Map to PDO	No

Description: This entry allows firmware to be programmed into a node. Each node may support up to 254 programs. This can be used, for example, to re-program individual tasks. The firmware is written to the appropriate Subentry and the data format used is not defined by the CANopen specification. For example, raw binary, Intel HEX File, etc. could be used.

Program execution is controlled by Object Dictionary entry 1F51h.

D.4.20 Program Control (1F51h)

Index	**1F51h**
Name	Program Control
Mandatory	No

Subindex	**00h**
Name	Number of Entries
Type	UNSIGNED8
Default Value	Not defined
Access	Read Only
Mandatory	Yes if this entry is implemented
Map to PDO	No

Subindex	**01h – FEh**
Name	Control Program 1 – 254
Type	UNSIGNED8
Default Value	Not defined
Access	Read/Write
Mandatory	No
Map to PDO	No

Description: By writing values to this entry, the corresponding program is controlled. Programs are written using Object Dictionary entry 1F50h. Once a program has been written, it may be stopped, started or reset by writing to the program control Subentry for that program. The following table lists the values that may be written.

Value	Description
00h	Stop program
01h	Start program
02h	Reset Program

Table D.63 Program Control Values

When the Subentries are read, information is given on the current state of the corresponding program. The following table lists the meanings of the values that can be read.

Value	Description
00h	Program is stopped
01h	Program is running
02h	Program is stopped

Table D.64 Program Control States

D.4.21 Verify Application Software (1F52h)

Index	1F52h
Name	Verify Application Software
Mandatory	No

Subindex	00h
Name	Number of Entries
Type	UNSIGNED8
Default Value	02h
Access	Read Only

Subindex	00h
Mandatory	Yes if this entry is implemented
Map to PDO	No

Subindex	01h
Name	Application Software Date
Type	UNSIGNED32
Default Value	Not defined
Units	Days
Access	Read/Write
Mandatory	No
Map to PDO	No

Subindex	02h
Name	Application Software Time
Type	UNSIGNED32
Default Value	Not defined
Units	ms
Access	Read/Write
Mandatory	No
Map to PDO	No

Description: This entry allows the date and time of program 1 to be stored. Program 1 is loaded into the node by writing to Object Dictionary entry [1F50h,01h]. It allows another node to determine the current version of a programmable portion of the firmware for the node.

The date is the number of days since January 1, 1984. The time is the number of milliseconds since midnight.

Example: 23 in Subentry 01h, 6212000 in Subentry 02h
January 24, 1984, 1:43:32 am

D.4.22 Expected Application SW Date (1F53h)

Index	**1F53h**
Name	Expected Application SW Date
Mandatory	No

Subindex	**00h**
Name	Number of Entries
Type	UNSIGNED8
Default Value	7Fh
Access	Read Only
Mandatory	Yes if this entry is implemented
Map to PDO	No

Subindex	**01h – 7Fh**
Name	Expected Application SW Date Node 1 - 127
Type	UNSIGNED32
Default Value	Not defined
Units	Days
Access	Read/Write
Mandatory	Yes if this entry is implemented
Map to PDO	No

Description: This entry is implemented on the CANopen Manager and stores the expected application software date of each node. Each Subentry corresponds to a node on the network, with the Subindex indicating the Node ID. When the CANopen Manager wishes to check if the node is using the currently known software version, it is compared with the date stored in the corresponding node in OD entry 1F52h.

The date is stored as the number of whole days since January 1, 1984.

Example: 23
January 24, 1984

D.4.23 Expected Application SW Time (1F54h)

Index	**1F54h**
Name	Expected Application SW Time
Mandatory	No

Subindex	**00h**
Name	Number of Entries
Type	UNSIGNED8
Default Value	7Fh
Access	Read Only
Mandatory	Yes if this entry is implemented
Map to PDO	No

Subindex	**01h – 7Fh**
Name	Expected Application SW Time Node 1 - 127
Type	UNSIGNED32
Default Value	Not defined
Units	ms
Access	Read/Write
Mandatory	Yes if this entry is implemented
Map to PDO	No

Description: This entry is implemented on the CANopen Manager and stores the expected application software time of each node. Each Subentry corresponds to a node on the network, with the Subindex indicating the Node ID. When the CANopen Manager wishes to check if the node is using the currently known software version, it is compared with the time stored in the corresponding node in OD entry 1F52h.

The time is stored as the number of milliseconds since midnight.

Example: 6212000
1:43:32 am

D.4.24 Process Picture / Process Image (1F70h)

Index	1F70h
Name	Process Picture / Process Image
Mandatory	No

Subindex	00h
Name	Number of Entries
Type	UNSIGNED8
Default Value	02h
Access	Read Only
Mandatory	Yes if this entry is implemented
Map to PDO	No

Subindex	01h
Name	Selected Range
Type	UNSIGNED32
Default Value	0h
Access	Read/Write
Mandatory	Yes if this entry is implemented
Map to PDO	No

Subindex	02h
Name	Process Picture Domain / Process Image Domain
Type	DOMAIN

Subindex	02h
Default Value	Not defined
Access	Read/Write
Mandatory	Yes if this entry is implemented
Map to PDO	No

Description: This entry allows Object Dictionary entries to be treated like variables. By writing to this entry the Object Dictionary containing these variables can be configured.

Sections of the Object Dictionary are grouped together into segments. The segment to read or write is first specified by writing to the Selected Range Subentry. Once the segment has been specified, the segment data can be read and written using the Process Image Domain Subentry.

The following table shows the structure of the value written to the Selected Range Subentry.

Bit	Description
0 – 15	Object Segment
16 – 31	Data Length

Table D.65 Process Image Selected Range Value

The Object Segment is the Index of the segment to read or write. The Data Length is the number of bytes to read or write. If a Data Length of zero is used then the complete segment may be accessed.

The segment is written as a stream of bytes.

D.4.25 NMT Startup (1F80h)

Index	**1F80h**
Name	NMT Startup
Mandatory	No

Subindex	**00h**
Name	NMT Startup
Type	UNSIGNED32
Default Value	Not defined
Access	Read/Write
Mandatory	No
Map to PDO	No

Description: This entry configures the startup of a device that is able to operate as an NMT Master. Each bit is writable unless the node does not support that particular feature, in which case that bit is read only. The following table describes the meaning of each bit.

Bit	**Description**
0	If 0 the device is not the NMT Master. If 1 the device is the NMT Master.
1	If 0 then start only explicitly assigned nodes. If 1 then start all nodes. If bit 3 is 1 then this bit is ignored.
2	If 0 then automatically enter the Operational state on bootup. If 1 then do not automatically enter the Operational state on bootup.
3	If 0 then the NMT Master may automatically start nodes. The behavior is configured using bit 1. If 1 then the NMT Master not not automatically start nodes. Bit 1 is ignored.

Table D.66 NMT Master Startup

Bit	Description
4	If 0 and a node fails to respond to node guarding or heartbeat, only handle that node. If 1 and a node fails to respond to node guarding or heartbeat, reset all nodes. If bit 6 is 1 then this bit is ignored.
5	If 0 then the NMT Master will not participate in the Flying Master process. If 1 then the NMT Master will participate in the Flying Master process.
6	If 0 then use the configuaration specified by bit 4. If 1 then ignore bit 4 and if a node fails to respond to node guarding or heartbeat, stop all nodes.
7 – 31	Reserved. Always zero.

Table D.66 (Continued) NMT Master Startup

Bit 0 indicates if the node is an NMT Master or not. If the node also participates in the Flying Master process (bit 5 is set to 1), but loses out in the process it should not clear bit 0.

Bit 2 determines if the node should automatically enter the operational state on bootup, or whether it should wait until it is told to enter the operational state. This feature is useful for networks without NMT Masters as it allows the node to startup autonomously.

Bit 3 controls whether or not the node may start nodes automatically. If it may, then the behavior of this functionality is configured using bit 1.

Bits 4 and 6 configure how the node should operate in the event of a node guarding or heartbeat event, whether it should only handle the node that failed to transmit a heartbeat or respond to a node guarding request, or whether all nodes on the network should be reset or stopped.

Bit 5 indicates if the node should participate in the Flying Master process and attempt to become the NMT Master for the network.

Example: 00000017h
NMT Master, starts all nodes, resets all nodes if node guarding or heartbeat event occurs, not a flying master.

D.4.26 Slave Assignment (1F81h)

Index	**1F81h**
Name	Slave Assignment
Mandatory	No

Subindex	**00h**
Name	Number of Entries
Type	UNSIGNED8
Default Value	7Fh
Access	Read Only
Mandatory	Yes if this entry is implemented
Map to PDO	No

Subindex	**01h – 7Fh**
Name	Slave Assigment Node 1 – 127
Type	UNSIGNED32
Default Value	Not defined
Access	Read/Write
Mandatory	No
Map to PDO	No

Description: This entry defines which slaves are assigned to the NMT Master and how the NMT Master controls the slave.

Each entry corresponds to one node on the network, with the Subindex indicating the Node ID. For example, Subentry 1Ah contains information relating to Node 1Ah.

The values stored in this entry are constructed as shown in the following table.

Bit	Description
0	Set to 0 if the node is not a slave for this NMT Master. Set to 1 if the node is a slave for this NMT Master.
1	Reserved.
2	Set to 0 if the node should not be automatically configured and started when a bootup message is detected being transmitted from the node. Set to 1 if the node should be automatically configured and started when a bootup message is detected being transmitted from the node.
3	Set to 0 if the node is an optional slave. The network may be started if this node cannot be contacted. Set to 1 if the node is a mandatory slave. Do not start the network if this node cannot be contacted.
4	Set to 0 if the node may be reset regardless of the current state of the node. Set to 1 if the node may only be reset if the node is currently not operational.
5	Set to 0 if application software version verification is not required for the node. Set to 1 if application software version verification is required for the node.
6	Set to 0 if automatic software update of the node is not allowed. Set to 1 if automatic software update of the node is allowed.
7	Reserved.
8 – 15	Retry Factor.
16 – 31	Guard Time.

Table D.67 Slave Assignment Entry

If the NMT Master does not support specific features, then those bits are read only.

If the NMT Master transmits a node guarding request to a node and does not receive a reply it will keep retrying until it has sent the request the number of times specified by the Retry Factor. The interval between transmission of node guarding request is specified by the Guard Time value for the node. If either Retry Factor or Guard Time are zero for a specific node, then the NMT Master will not perform node guarding on that node.

Example: 03E80101h
Node is an optional slave, not automatically configured and started, may be reset regardless of state, no software verification, no automatic update, 1 second guard time, 1 retry.

D.4.27 Request NMT (1F82h)

Index	1F82h
Name	Request NMT
Mandatory	No

Subindex	00h
Name	Number of Entries
Type	UNSIGNED8
Default Value	80h
Access	Read Only
Mandatory	Yes if this entry is implemented
Map to PDO	No

Subindex	01h – 7Fh
Name	Request NMT for Node 1 - 127
Type	UNSIGNED8
Default Value	Not defined
Access	Read/Write
Mandatory	Yes if this entry is implemented
Map to PDO	No

Subindex	80h
Name	Request NMT for All Nodes
Type	UNSIGNED8

Subindex	80h
Default Value	Node defined
Access	Write Only
Mandatory	Yes if this entry is implemented
Map to PDO	No

Description: A CANopen network only allows one NMT Master at any one time. This ensures that only one node transmits messages with ID 000h. If another node wishes to perform NMT operations then it must write to this entry on the NMT Master requesting an NMT operation take place. The NMT Master will then transmit the NMT command.

Writing to Subentries 01h – 7Fh will result in the NMT Master sending the NMT command to the node whose ID matches the Subindex written to. For example, writing to Subentry 5Ah will result in the NMT Master sending an NMT command to node 5Ah.

Writing to Subentry 80h sends an NMT command to all nodes.

The following table lists the values that may be written to this entry.

Value	NMT Command
04h	Stop
05h	Enter Operational
06h	Reset
07h	Reset Communication
7Fh	Enter Pre-operational

Table D.68 Request NMT Commands

Subentries 01h – 7Fh may be read to find out the current state of a node. The following table lists the values that may be read from these Subentries.

Value	NMT State
00h	State not known
01h	Node missing
04h	Stopped
05h	Operational
7Fh	Pre-operational

Table D.69 Request NMT Read Values

D.4.28 Request Guarding (1F83h)

Index	1F83h
Name	Request Guarding
Mandatory	No

Subindex	00h
Name	Number of Entries
Type	UNSIGNED8
Default Value	80h
Access	Read Only
Mandatory	Yes if this entry is implemented
Map to PDO	No

Subindex	01h – 7Fh
Name	Request Guarding for Node 1 - 127
Type	UNSIGNED8
Default Value	Not defined
Access	Read/Write

Subindex	01h – 7Fh
Mandatory	Yes if this entry is implemented
Map to PDO	No

Subindex	80h
Name	Request Guarding for All Nodes
Type	UNSIGNED8
Default Value	Node defined
Access	Write Only
Mandatory	Yes if this entry is implemented
Map to PDO	No

Description: A CANopen network only allows one NMT Master at any one time. This ensures that only one node transmits the node guarding messages. If another node wishes to perform node guarding then it must write to this entry on the NMT Master requesting that node guarding take place. The NMT Master will then perform the node guarding.

Writing to Subentries 01h – 7Fh will result in the NMT Master sending the node guarding requests to the node whose ID matches the Subindex written to. For example, writing to Subentry 5Ah will result in the NMT Master sending node guarding requests to node 5Ah.

Writing to Subentry 80h sends node guarding requests to all nodes.

The following table lists the values that may be written to this entry.

Value	NMT Command
00h	Stop node guarding
01h	Start node guarding

Table D.70 Request Guarding Commands

Subentries 01h – 7Fh may be read to find out whether a node is being guarded or not. The following table lists the values that may be read from these Subentries.

Value	NMT State
00h	Node is not being guarded
01h	Node is being guarded

Table D.71 Request Guarding Read Values

D.4.29 Device Type Identification (1F84h)

Index	1F84h
Name	Device Type Identification
Mandatory	No

Subindex	00h
Name	Number of Entries
Type	UNSIGNED8
Default Value	7Fh
Access	Read Only
Mandatory	Yes if this entry is implemented
Map to PDO	No

Subindex	01h – 7Fh
Name	Device Type Identification for Node 1 - 127
Type	UNSIGNED32
Default Value	Not defined
Access	Read/Write
Mandatory	Yes if this entry is implemented
Map to PDO	No

Description: This entry lists the expected Device Type values for slave nodes. Each Subentry correponds to the node with an ID the same as the Subindex. For example, Subindex 31h holds the expected Device Type for Node 31h.

If the value stored is zero, then the Device Type of the node is marked as "don't care." If the value stored is not zero, then the Device Type read from the node must match the expected value stored in this entry. If the values do not match then the node bootup is not completed.

The Subentry that corresponds to the NMT Master is ignored.

D.4.30 Vendor Identification (1F85h)

Index	**1F85h**
Name	Vendor Identification
Mandatory	No

Subindex	**00h**
Name	Number of Entries
Type	UNSIGNED8
Default Value	7Fh
Access	Read Only
Mandatory	Yes if this entry is implemented
Map to PDO	No

Subindex	**01h – 7Fh**
Name	Vendor Identification for Node 1 - 127
Type	UNSIGNED32
Default Value	Not defined
Access	Read/Write

Subindex	01h – 7Fh
Mandatory	Yes if this entry is implemented
Map to PDO	No

Description: This entry lists the expected Vendor ID values for slave nodes. Each Subentry correponds to the node with an ID the same as the Subindex. For example, Subindex 31h holds the expected Vendor ID for Node 31h.

If the value stored is zero, then the Vendor ID of the node is marked as "don't care." If the value stored is not zero, then the Vendor ID read from the node must match the expected value stored in this entry. If the values do not match then the node bootup is not completed.

The Subindex that corresponds to the NMT Master is ignored.

D.4.31 Product Code (1F86h)

Index	1F86h
Name	Product Code
Mandatory	No

Subindex	00h
Name	Number of Entries
Type	UNSIGNED8
Default Value	7Fh
Access	Read Only
Mandatory	Yes if this entry is implemented
Map to PDO	No

Subindex	01h – 7Fh
Name	Product Code for Node 1 - 127
Type	UNSIGNED32

Subindex	01h – 7Fh
Default Value	Not defined
Access	Read/Write
Mandatory	Yes if this entry is implemented
Map to PDO	No

Description: This entry lists the expected Product Code values for slave nodes. Each Subentry correponds to the node with an ID the same as the Subindex. For example, Subindex 31h holds the expected Product Code for Node 31h.

If the value stored is zero, then the Product Code of the node is marked as "don't care." If the value stored is not zero, then the Product Code read from the node must match the expected value stored in this entry. If the values do not match then the node bootup is not completed.

The Subindex that corresponds to the NMT Master is ignored.

D.4.32 Revision Number (1F87h)

Index	1F87h
Name	Revision Number
Mandatory	No

Subindex	00h
Name	Number of Entries
Type	UNSIGNED8
Default Value	7Fh
Access	Read Only
Mandatory	Yes if this entry is implemented
Map to PDO	No

Subindex	01h – 7Fh
Name	Revision Number for Node 1 – 127
Type	UNSIGNED32
Default Value	Not defined
Access	Read/Write
Mandatory	Yes if this entry is implemented
Map to PDO	No

Description: This entry lists the expected Revision Number values for slave nodes. Each Subentry correponds to the node with an ID the same as the Subindex. For example, Subindex 31h holds the expected Revision Number for Node 31h.

If the value stored is zero, then the Revision Number of the node is marked as "don't care." If the value stored is not zero, then the Revision Number read from the node must match the expected value stored in this entry. If the values do not match then the node bootup is not completed.

The Subindex that corresponds to the NMT Master is ignored.

D.4.33 Serial Number (1F88h)

Index	1F88h
Name	Serial Number
Mandatory	No

Subindex	00h
Name	Number of Entries
Type	UNSIGNED8
Default Value	7Fh
Access	Read Only

Subindex	**00h**
Mandatory	Yes if this entry is implemented
Map to PDO	No

Subindex	**01h – 7Fh**
Name	Serial Number for Node 1 - 127
Type	UNSIGNED32
Default Value	Not defined
Access	Read/Write
Mandatory	Yes if this entry is implemented
Map to PDO	No

Description: This entry lists the expected Serial Number values for slave nodes. Each Subentry corresponds to the node with an ID the same as the Subindex. For example, Subindex 31h holds the expected Serial Number for Node 31h.

If the value stored is zero, then the Serial Number of the node is marked as "don't care." If the value stored is not zero, then the Serial Number read from the node must match the expected value stored in this entry. If the values do not match then the node bootup is not completed.

The Subindex that corresponds to the NMT Master is ignored.

D.4.34 Boot Time (1F89h)

Index	1F89h
Name	Boot Time
Mandatory	No

Subindex	00h
Name	Boot Time
Type	UNSIGNED32
Default Value	0h
Units	ms
Access	Read/Write
Mandatory	No
Map to PDO	No

Description: This entry defines the maximum time the NMT Master will wait when trying to read the Device Type of a mandatory slave. The time is in milliseconds. A value of zero indicates that the NMT Master should wait forever. The timing starts from the first attempt to read the Device Type of the slave. If the time elapses without a successful read of the Device Type then the NMT Master will give up on attempting to start the network, enter an error state and inform the application.

Example: 3000
NMT Master will wait 3 seconds

D.4.35 Flying Master Timing Parameters (1F90h)

Index	1F90h
Name	Flying Master Timing Parameters
Mandatory	Yes if the NMT Master supports the Flying Master Mechanism

Subindex	00h
Name	Number of Entries
Type	UNSIGNED8
Default Value	06h
Access	Read Only
Mandatory	Yes if the NMT Master supports the Flying Master Mechanism
Map to PDO	No

Subindex	01h
Name	Timeout for Detection of an Active NMT Master
Type	UNSIGNED16
Default Value	100
Units	ms
Access	Read/Write
Mandatory	Yes if the NMT Master supports the Flying Master Mechanism
Map to PDO	No

Subindex	02h
Name	NMT Master Negotiation Time Delay
Type	UNSIGNED16
Default Value	500
Units	ms
Access	Read/Write

Subindex	**02h**
Mandatory	Yes if the NMT Master supports the Flying Master Mechanism
Map to PDO	No

Subindex	**03h**
Name	Master Priority Level
Type	UNSIGNED8
Default Value	Not defined
Access	Read/Write
Mandatory	Yes if the NMT Master supports the Flying Master Mechanism
Map to PDO	No

Subindex	**04h**
Name	Priority Time Slot
Type	UNSIGNED16
Default Value	1500
Units	ms
Access	Read/Write
Mandatory	Yes if the NMT Master supports the Flying Master Mechanism
Map to PDO	No

Subindex	**05h**
Name	Node Time Slot
Type	UNSIGNED16
Default Value	10
Units	ms
Access	Read/Write
Mandatory	Yes if the NMT Master supports the Flying Master Mechanism
Map to PDO	No

Subindex	06h
Name	Multiple Master Detect Cycle Time
Type	UNSIGNED16
Default Value	(4000 + Node ID) x 10
Units	ms
Access	Read/Write
Mandatory	Yes if the NMT Master supports the Flying Master Mechanism
Map to PDO	No

Description: This entry must be implemented if the NMT Master supports the Flying Master functionality. It specifies the timing parameters and priority to be used in the Flying Master protocol.

All values except for the Master Priority Level are times given in milliseconds.

The Timeout for Detection of an Active NMT Master is the timeout period in which any currently active NMT Master must respond to the request for the NMT Master Priority Level.

The NMT Master Negotiation Time Delay is the time which NMT Master capable devices must wait after a cold or warm boot. This is used to ensure that other devices have completed resets and initialization before an NMT Master capable device proceeds with the negotiation.

The Master Priority Level, Priority Time Slot and Node Time Slot are combined with the Node ID of the NMT capable device to calculate a wait time. After receiving the Trigger Timeslot message, each NMT Master capable device transmits an identification message after the wait time has elapsed. This ensures that the NMT Master capable device with the lowest wait time will transmit an identification message first. The wait time is calculated as follows:

Wait time = (Master Priority Level x Priority Time Slot) + (Node ID X Node Time Slot)

The Priority Time Slot must be greater than the the Node Time Slot multiplied by 127.

Priority time slot > 127 x Node Time Slot

The Master Priority level may have the value 0 to 2, with 0 being the highest priority level and 2 being the lowest priority level.

To allow an NMT Master to detect the presence of other NMT Masters, they each must periodically transmit the Forcing New NMT Master Negotiation Protocol message. The period between transmission of these messages is the Multiple Master Detect Cycle Time.

D.4.36 Startup-capable Device Timing (1F91h)

Index	**1F91h**
Name	Startup-capable Device Timing
Mandatory	Yes if the node is Startup Capable

Subindex	**00h**
Name	Number of Entries
Type	UNSIGNED8
Default Value	03h
Access	Read Only
Mandatory	Yes if the node is Startup Capable
Map to PDO	No

Subindex	**01h**
Name	Timeout for selection of an NMT Master Capable Device
Type	UNSIGNED16
Default Value	100
Units	ms
Access	Read/Write

Subindex	01h
Mandatory	Yes if the node is Startup Capable
Map to PDO	No

Subindex	02h
Name	Delay time for an NMT Master Capable Device Request
Type	UNSIGNED16
Default Value	500
Units	ms
Access	Read/Write
Mandatory	Yes if the node is Startup Capable
Map to PDO	No

Subindex	03h
Name	Node Time Slot
Type	UNSIGNED16
Default Value	15
Units	ms
Access	Read/Write
Mandatory	Yes if the node is Startup Capable
Map to PDO	No

Description: This entry specifies timing for a node that is capable of starting up without an NMT Master. Nodes that are Startup Capable automatically enter the operational state and optionally start a group of nodes.

All values are times given in milliseconds.

After initialization of a startup capable node, it must wait before starting the protocol to determine if there is a NMT Master capable node on the bus. This delay is configurable and stored in Subentry 02h – Delay time for an NMT Master Capable Device Request.

Once the node has transmitted the request for any NMT Master capable devices to identify themselves, it waits for a timeout period specified in Subentry 01h - Timeout for selection of an NMT Master Capable Device.

If no NMT Master capable devices have been found on the bus then the node should wait for a delay period before transmitting the NMT message to start all nodes. In order to avoid multiple nodes transmitting the message at the same time, the delay is configurable and dependent on the Node ID. It is calculated as follows.

Delay = Node ID x Node Time Slot

The Node Time Slot is specified in Subentry 03h.

D.5 Object Dictionary Access Sequences

D.5.1 PDO Communication Parameters

When changing the PDO Communication Parameters (Transmit or Receive) the following procedure is used.

1. Write to Subentry 01h setting bit 31 to disable the PDO.
2. Perform writes to the Subentries that should be changed (02h, 03h and 05h).
3. Write to Subentry 01h specifying the new COB ID to be used, ensuring bit 31 is cleared to enable the PDO.

D.5.2 PDO Mapping Parameters

To change the PDO Mapping Parameters (Transmit or Receive) the following procedure is used.

1. Write to Subentry 01h of the PDO Communication Parameters setting bit 31 to disable the PDO.
2. Write 00h to Subentry 00h of the Mapping Parameters to disable the mapping.
3. Write the new mapping to the Mapping Parameters.
4. Write the highest Subindex used in the mapping to Subentry 00h of the Mapping Parameters to enable the mapping.
5. Write to Subentry 01h of the PDO Communication Parameters clearing bit 31 to enable the PDO and specifying the desired COB ID to use for the PDO.

D.5.3 Dynamic SDO Channel Request – All Channels

The following procedure is used for a node (referred to as the SDO Requesting Device – SRD) to obtain all available SDO channels from the SDO Manager.

1. SRD sends Dynamic SDO Request (COB ID 6E0h, no data).
2. SDO Manager reads [1F10h,00h] of each node in turn until the SRD is found.
3. Upon read from [1F10h,00h] SRD response is 00000003h.
4. SDO Manager writes 00000004h to [1F10h,00h] of SRD to confirm.

5. SRD can use all SDO channels.
6. When done, SRD writes 0000xx00h (xx being the SRD Node ID) to [1F01h,00h] of SDO Manager .

D.5.4 Dynamic SDO Channel Request – Single Channel

The following procedure is used for a node (referred to as the SDO Requesting Device – SRD) to obtain SDO channels from the SDO Manager to a slave node.

Register as an SRD

1. SRD sends Dynamic SDO Request (COB ID 6E0h, no data).
2. SDO Manager reads [1F10h,00h] of each node in turn until the SRD is found.
3. Upon read from [1F10h,00h] SRD response is aaaa0001h, with aaaa being the Index of an SDO client communication parameter record in the SRD.
4. SDO Manager configures an SDO client channel in SRD at aaaa with the SDO Manager itself being the server to that client.
5. SDO Manager writes 00000002h to [1F10h,00h] of SRD to confirm that SDO channel to SDO Manager is established.

Request a Channel

6. Using the new SDO channel, SRD writes xxxxyyzzh to [1F00h,00h] of SDO Manager with xxxx being another SDO client communication parameter record (must be different than previous one), yy the Node ID of the SRD and zz the Node ID of the slave node.
7. SDO Manager now sets up an additional SDO server in zz. If the slave node only has one SDO channel, then the SDO Manager sets up the additional SDO server on itself and acts as a relay to the slave.
8. SDO Manager now sets up the SDO client xxxx in SRD.
9. The SDO Manager writes 00000006h to [1F10h,00h] of SRD to confirm that the SDO channel to the slave is established. The channel is not valid until this value is written. If there was an error in configuring the SDO channel, then the SDO Manager writes 0000xx00h to [1F10h,00h] to notify the SRD of the error.

To Release the Channel

10. The SRD still owns the SDO client channel aaaa to SDO Manager and can use it to send a release request to [1F01h,00h] using the same value as used in 6.
11. SDO Manager now releases the SDO client xxxx in SRD.
12. SDO Manager now releases the SDO server in zz

To De-register as an SRD

13. The SRD still owns the SDO client channel aaaa to SDO Manager and can use it to send a release request to [1F01h,00h] using 0000yy00h.
14. SDO Manager accesses the SRD and removes the SDO client aaaa.

E Minimal Object Dictionaries

Objective

One of the first questions that comes up when developing a CANopen node is *Which Object Dictionary entries should I implement?* This appendix gives you general guidelines about particular I/O functionality and which entries are recommended for implementation.

Important: All numbers in this section are hexadecimal. For a more detailed description of some of the individual entries, see the Object Dictionary entry reference section.

E.1 Standard Object Dictionary Entries

It is recommended that the Object Dictionary entries listed below be available in all CANopen nodes.

Index	Description
[1000h,00h]	Device Type Information
[1001h,00h]	Error Register
[1008h,00h]	Device Name Although not mandatory, this ASCII-string entry is read by many configuration tools and managers and displayed to their users.
[1016h,xxh]	Heartbeat: Consumer Time This entry is only needed if the local node needs to be able to monitor the heartbeats of other nodes.
[1017h,00h]	Heartbeat: Producer Time Node guarding or heartbeat or both must be supported by all CANopen compliant nodes.
[1018h,xxh]	Identity Object Only the Subentry with the Vendor ID must be implemented
[1F80h,00h]	NMT Startup Nodes that autostart (go to operational without waiting for the NMT startup message) should report 0 in this entry.

E.2 Digital Input Entries

The Object Dictionary entries listed here are those that need to be implemented for DS401 compliant generic digital input.

Index	Description
[1800h,xxh]	1st TPDO Communication Parameters
[180xh,xxh]	Additional TPDOs as required by the application
[1A00h,xxh]	1st TPDO Mapping Parameters
[1A0xh,xxh]	Additional TPDOs as required by the application
[6000h,xxh]	Read Digital Input (8bit)
[6002h,xxh]	Polarity Digital Input Only if needed by the application.

E.3 Digital Output Entries

The Object Dictionary entries listed here are those that need to be implemented for DS401 compliant generic digital output.

Index	Description
[1400h,xxh]	1st RPDO Communication Parameters
[140xh,xxh]	Additional RPDOs as required by the application
[1600h,xxh]	1st RPDO Mapping Parameters
[160xh,xxh]	Additional RPDOs as required by the application
[6200h,xxh]	Write Digital Output (8 bit)
[6202h,xxh]	Polarity Digital Output Only if needed by the application.
[6206h,xxh]	Error Mode Output Although not mandatory, this is a useful entry to enable default output values to be used when an error occurs.
[6207h,xxh]	Error Value Output Although not mandatory, this is a useful entry to set the default output values to be used when an error occurs.

E.4 Analog Input Entries

The Object Dictionary entries listed here are those that need to be implemented for DS401 compliant generic analog input.

Index	Description
[180xh,xxh]	TPDO Communication Parameters Per default TPDO 2, 3 and 4 are used for analog data. However, if a node does not have digital inputs TPDO 1 may be used for analog inputs, too.
[1A0xh,xxh]	TPDOs Mapping Parameters
[6401h,xxh]	Read Analog Input (16 bit)

E.5 Analog Output Entries

The Object Dictionary entries listed here are those that need to be implemented for DS401 compliant generic analog output.

Index	Description
[140xh,xxh]	RPDO Communication Parameters Per default RPDO 2, 3 and 4 are used for analog data. However, if a node does not have digital outputs, RPDO 1 may be used for analog outputs, too.
[160xh,xxh]	RPDOs Mapping Parameters
[6411h,xxh]	Write Analog Output (16 bit)
[6443h,xxh]	Error Mode Output Although not mandatory, this is a useful entry to enable default output values to be used when an error occurs.
[6444h,xxh]	Error Value Output Although not mandatory, this is a useful entry to set the default output values to be used when an error occurs.

E.6 Encoder Input Entries

The Object Dictionary entries listed here are those that need to be implemented for DS406 compliant encoders.

Index	Description
[1801h,xxh]	1st TPDO Communication Parameters
[1A01h,xxh]	1st TPDO Mapping Parameters
[6000h,00h]	Operating Parameters
[6003h,00h]	Preset Value This entry is not mandatory but very useful to many applications, as it allows to set the mechanical zero point of the encoder.
[6004h,00h]	Position Value

Index	Description
[6500h,00h]	Operating Status
[6501h,00h]	Resolution
[6502h,00h]	Number of Revolutions Only needed for rotary encoders

E.7 Support of Code Download

The Object Dictionary entries listed here are those that need to be implemented if a node is to receive code updates through the CANopen interface.

Index	Description
[1F50h,xxh]	Download Program Data In general it is sufficient to implement Subentry 00h and 01h.
[1F51h,xxh]	Program Control In general it is sufficient to implement Subentry 00h and 01h.

F Communication Object Identifiers (COB IDs)

Objective

Each message transmitted must have a Communications Object Identifier (COB ID). This appendix lists both the default COB IDs used in CANopen, and the reserved COB IDs, providing at a glance what may need to be configured and what cannot be configured.

F.1 Pre-defined Connection Set

The following table lists the default COB IDs used for the various CANopen communication objects. This collection of defaults is referred to as the Pre-defined Connection Set.

COB ID	Used For	Constructed Using
000h	NMT (Network Management)	-
001h	Global Failsafe Command	-

Table F.1 Pre-defined Connection Set

COB ID	Used For	Constructed Using
071h – 076h	Flying Master Protocol	-
080h	SYNC	-
081h – 0FFh	Emergency	80h + Node ID
100h	Time Stamp	-
101h – 180h	Safety Relevent Data Objects	100h + Node ID
181h – 1FFh	Transmit PDO 1	180h + Node ID
201h – 27Fh	Receive PDO 1	200h + Node ID
281h – 2FFh	Transmit PDO 2	280h + Node ID
301h – 37Fh	Receive PDO 2	300h + Node ID
381h – 3FFh	Transmit PDO 3	380h + Node ID
401h – 47Fh	Receive PDO 3	400h + Node ID
481h – 4FFh	Transmit PDO 4	480h + Node ID
501h – 57Fh	Receive PDO 4	500h + Node ID
581h – 5FFh	Transmit SDO	580h + Node ID
601h – 67Fh	Receive SDO	600h + Node ID
6E0h	Dynamic SDO Request	-
701h – 77Fh	NMT Error Control (Heartbeat and Node Guarding)	700h + Node ID

Table F.1 Pre-defined Connection Set

F.2 Reserved COB IDs

The following table lists the COB IDs that may not be used by objects which allow the COB ID to be configured.

COB ID	Used For
000h	NMT
001h	Reserved
101h – 180h	Reserved
581h – 5FFh	Transmit SDOs
601h – 67Fh	Receive SDOs
6E0h	Reserved
701h – 77Fh	NMT Error Control
780h – 7FFh	Reserved

Table F.2 Reserved COB IDs

G Emergency Objects

Objective

It is often useful when working with CANopen to be able to interpret values in the CAN messages. This appendix lists the codes that may be transmitted in Emergency Objects, along with their meanings

G.1 Emergency Object Error Codes

The following table lists the standard error codes that may be transmitted in Emergency Objects. Several Device Profiles define additional error codes.

Error Code	Description
0000h – 00FFh	No error (or Error reset)
1000h – 10FFh	Generic
2000h – 20FFh	Current

Table G.1 Emergency Message Error Codes

Error Code	Description
2100h – 21FFh	Current – device inputs
2200h – 22FFh	Current – inside the device
2300h – 23FFh	Current – device outputs
3000h – 30FFh	Voltage
3100h – 31FFh	Voltage – mains voltage
3200h – 32FFh	Voltage – inside the device
3300h – 33FFh	Voltage – output
4000h – 40FFh	Temperature
4100h – 41FFh	Temperature – Ambient
4200h – 42FFh	Temperature – Device
5000h – 50FFh	Device Hardware
6000h – 60FFh	Device Software
6100h – 61FFh	Device Software – internal
6200h – 62FFh	Device Software – user
6300h – 63FFh	Device Software – data set
7000h – 70FFh	Additional Modules
8000h – 80FFh	Monitoring
8100h – 81FFh	Monitoring - communication
8110h	Monitoring – CAN Overrun (objects lost)
8120h	Monitoring – CAN in error passive mode
8130h	Monitoring – Node Guarding or Heartbeat Error
8140h	Monitoring – recovering from bus off
8150h	Monitoring – COB ID
8200h – 82FFh	Protocol
8210h	Protocol – PDO not processed due to length error
8220h	Protocol – PDO length exceeded
9000h – 90FFh	External

Table G.1 (Continued) Emergency Message Error Codes

Error Code	Description
F000h – F0FFh	Additional functions
FF00h – FFFFh	Device specific

Table G.1 (Continued) Emergency Message Error Codes

H SDO Abort Messages

Objective

This appendix provides a quick reference to the SDO Abort codes and their meanings.

H.1 SDO Abort Codes

The following table lists the abort codes that may be transmitted by Clients and Servers when implementing the SDO Protocol.

Abort Code	Description
05030000h	Toggle bit not alternated
05040000h	SDO Protocol timed out
05040001h	Client/Server command specifier not valid or unknown
05040002h	Invalid block size (block mode)

Table H.1 SDO Abort Codes

Abort Code	Description
05040003h	Invalid sequence number (block mode)
05040004h	CRC error (block mode)
05040005h	Out of memory
06010000h	Unsupported access to an object
06010001h	Attempt to read a write-only object
06010002h	Attempt to write a read-only object
06020000h	Object does not exist in the Object Dictionary
06040041h	Object cannot be mapped to the PDO
06040042h	The number and length of the objects to be mapped would exceed PDO length
06040043h	General parameter incompatibility
06040047h	General internal incompatibility in the device
06060000h	Access failed due to a hardware error
06070010h	Data type does not match. Length of service parameter does not match.
06070012h	Data type does not match. Length of service parameter is too high.
06070013h	Data type does not match. Length of service parameter is too low.
06090011h	Subindex does not exist
06090030h	Value range of parameter exceeded (write access only)
06090031h	Value of parameter written is too high
06090032h	Value of parameter written is too low
06090036h	Maximum value is less than the minimum value
08000000h	General error
08000020h	Data cannot be transferred or stored to the application
08000021h	Data cannot be transferred or stored to the application because of local control

Table H.1 (Continued) SDO Abort Codes

Abort Code	Description
08000022h	Data cannot be transferred or stored to the application because of the present device state
08000023h	Object Dictionary dynamic generation failed or no Object Dictionary is present

Table H.1 (Continued) SDO Abort Codes

I Node States

Objective

Only certain objects may be transmitted while in specific node states. The aim of this appendix is to provide a description of what may be transmitted for each state.

I.1 Node State Functionality

The following table shows which communication objects a node may transmit and process when in the different states. A "Yes" indicates that the node may use that communication object.

	Initializing	Pre-operational	Operational	Stopped
PDOs	No	No	Yes	No
SDOs	No	Yes	Yes	No
SYNC	No	Yes	Yes	No
Time Stamp	No	Yes	Yes	No
Emergency	No	Yes	Yes	No
Bootup	Yes	No	No	No
Node Guarding and Heart-beat	No	Yes	Yes	Yes

Table I.1 Communication Objects Used in Different States

J References

[Barr99]
Barr, Michael. *Programming Embedded Systems in C and C++*. 1999, O'Reilly & Associates.

[Bentham02]
Bentham, Jeremy, and Michael Barr. *TCP/IP Lean: Web Servers for Embedded Systems*. 2002, CMP Books.

[Berger01]
Berger, Arnold S. *Embedded Systems Design: An Introduction to Processes, Tools and Techniques*. 2001, CMP Books.

[Charzinski]
J. Charzinski "Performance of the the Error Detection Mechanisms in CAN"

[CiADRP3031]
"CANopen Cabling and Connector Pin Assignment", CiA Draft Recommendation Proposal 303-1, Version 1.1.1

[CiADRP3032]
"CANopen representation of SI units and prefixes", CiA Draft Recommendation Proposal 303-2, Version 1.1

[CiADRP3033]
"CANopen indicator specification", CiA Draft Recommendation Proposal 303-3, Version 1.0

[CiADS301]
"CANopen application layer and communication profile", CiA Draft Standard 301, Version 4.02

[CiADS401]
"CANopen device profile for generic I/O modules", CiA Draft Standard 401, Version 2.1

[CiADS404]
"CANopen device profile measuring devices and closed loop controllers", CiA Draft Standard 404, Version 1.2

[CiADS405]
"CANopen interface and device profile for IEC 61131-3 programmable devices", CiA Draft Standard 405, Version 2.0

[CiADS406]
"CANopen device profile for encoders", CiA Draft Standard 406, Version 3.0

[CiADSP302]
"CANopen framework for CANopen managers and programmable CANopen devices", CiA Draft Standard Proposal 302, Version 3.2

[CiADSP304]
"CANopen framework for safety-relevant communication", CiA Draft Standard Proposal 304, Version 1.0

[CiADSP305]
"CANopen layer setting services and protocols (LSS)", CiA Draft Standard Proposal 305, Version 1.1.1

[CiADSP306]
"CANopen electronic data sheet (EDS) specification for CANopen", CiA Draft Standard Proposal 306, Version 1.1

[CiADSP307]
"CANopen framework for maritime electronics", CiA Draft Standard Proposal 307, Version 1.01

[CiADSP402]
"CANopen device profile for drives and motion control", CiA Draft Standard Proposal 402, Version 2.0

[CiADSP406]
"CANopen device profile for encoders", CiA Draft Standard Proposal 406, Version 2.0

[CiADSP407]
"CANopen application profile for passenger information", CiA Draft Standard Proposal 407, Version 1.0

[CiADSP410]
"CANopen device profile for inclinometer", CiA Draft Standard Proposal 410, Version 1.0

[CiADSP413]
"CANopen device profiles for truck gateways", CiA Draft Standard Proposal 413, Version 1.0

[CiADSP414]
"CANopen device profiles for weaving machines", CiA Draft Standard Proposal 414, Version 1.0

[CiADSP418]
"CANopen device profile for battery modules", CiA Draft Standard Proposal 418, Version 1.0

[CiADSP419]
"CANopen device profile for battery charger", CiA Draft Standard Proposal 419, Version 1.0

[CiADSP420]
"CANopen profiles for extruder downstream devices", CiA Draft Standard Proposal 420, Version 1.0

[CiATR308]
"CANopen performance testing", CiA Technical Recommendation 308, Version 1.0

[Comer00]
Comer. Douglas. *Internetworking with TCP/IP Vol.1: Principles, Protocols, and Architecture*, 4th Edition. 2000, Prentice Hall.

[Etschberger01]
Etschberger, Konrad. *Controller Area Network*. 2001, IXXAT Automation.

[Farsi99]
Farsi, Mohammad, Manuel Bernado and Martin Barbosa. *CANopen Implementation*. 1999, Research Studies Pr.

[Ganssle00]
Ganssle, Jack. *The Art of Designing Embedded Systems*. 2000, Newnes

[Ganssle03]
Ganssle, Jack and Michael Barr. *Embedded Systems Dictionary*. 2003, CMP Books.

[ISO7498]
ISO 7498-1:1994 Information technology - Open Systems Interconnection - Basic Reference Model: The Basic Model"

[Lawrenz97]
Lawrenz, Wolfhard. *CAN System Engineering: From Theory to Practical Applications*. 1997, Springer Verlag.

[Nolte]
Nolte, Thomas, Hans Hansson and Christer Norstrom. "Probabilistic Worst-Case Response-Time Analysis for the Controller Area Network"

[Pfeiffer01_1]
Pfeiffer, Olaf and John Rodrigues. "Internetworking Treads on MCU Turf" *EETimes*, May 21, 2001.

[Pfeiffer01_2]
Pfeiffer, Olaf. "Targeting Europe: Implementing CANopen" *Circuit Cellar Ink* #134, 2001

[Pfeiffer02_1]
Pfeiffer, Olaf. "Selecting the Best CAN Controller" *Circuit Cellar Ink #143*, 2002

[Pfeiffer02_2]
Pfeiffer, Olaf. "Making Medical Devices Smarter with CAN and CANopen Protocols" *Medical Electronics Manufacturing*, Fall, 2002

[Pfeiffer03]
Pfeiffer, Olaf and Paul Lukowicz. "Remote Access to Embedded Devices" Internet draft published 2003 with the RFC editor (www.rfc-editor.org).

[Rostan02]
Rostan, Martin and Josef Langermann. "High Precision Drive Synchronisation with CANopen" Proceedings of the 8th International CAN Conference, CAN in Automation.

[Smith01]
Smith, David. *Functional Safety.* 2001, Butterworth Heinemann.

[Stenerson02]
Stenerson, Jon. *Industrial Automation and Process Control.* 2002, Prentice Hall.

[Zuberi]
Zuberi, Khawar M. and Kang G. Shin. "Non-Preemptive Scheduling of Messages on Controller Area Network for Real-Time Control Applications"

K CANopen Glossary

A	
application layer	The application layer is the communication entity of the OSI (Open System Interface) reference model. It provides communication services to the application program.
application objects	Application objects are signals and parameters of the application program visible at the application layer API (application programming interface).
application profile	Application profiles define all communication objects and application objects in all devices that the network consists of.
asynchronous PDO	An asynchronous PDO is transmitted whenever a defined internal event occurs. This event may also be the elapsing of the PDO's event timer. If an asynchronous PDO is received the protocol software immediately updates the mapped objects in the Object Dictionary.

B	
boot-up message	CANopen communication service transmitted whenever a node enters the pre-operational state after initialization.
bus	Topology of a communication network, where all nodes are reached by passive links, which allows transmission in both directions.
bus analyzer	Tool, which monitors the bus and displays the transmitted bits. There are bus analyzers available on the physical layer, the data link layer, and different application layers (e.g. CANopen or DeviceNet.
bus arbitration	If at the very same moment several nodes try to access the bus, an arbitration process is necessary. At the end of this process, only one node has bus access. The bus arbitration process used in CAN protocol is CMSA/CD (Carrier Sense Multiple Access/Collision Detection) with AMP (Arbitration on Message Priority). This allows bus arbitration without destruction of messages.
bus length	The network cable length between the both termination resistors. The bus length of CANopen networks is limited by the used transmission rate. At 1 Mbps the maximum length is 25 m. When using lower transmission rates, longer bus lines may be used: at 50 kbps a length of 1 km is possible.
bus off state	The CAN controllers switch to bus off state when the TEC (transmit error counter) has reached 255. During bus off state, the CAN controller transmits recessive bits. When a CANopen device recovers from bus off state, it has to transmit the boot-up message and it is recommended to send an Emergency message with the appropriate error code.

C	
CAN	Controller Area Network (CAN) is a serial bus system originally developed by the Robert Bosch GmbH. It is internationally standardized by ISO 11898-1. CAN has been implemented by many semiconductor manufacturers.
CANopen	Family of profiles for embedded networking in industrial machinery, medical equipment, building automation (e.g. lift control systems, electronically controlled doors, integrated room control systems), railways, maritime electronics, truck-based superstructures, off-highway and off-road vehicles, etc.

CANopen application layer	The CANopen application layer and communication profile is standardized by EN 50325-4. It defines communication services and objects. In addition, it specifies the Object Dictionary and the network management (NMT).
CANopen Manager	The CANopen manager is responsible for the management of the network. The CANopen manager device shall include the NMT (network management) master, the SDO (service data object) manager, and the Configuration manager.
CANopen Safety	Communication protocol allowing transmission of safety-relevant data. The protocol requires just one physical CAN network. Redundancy is achieved by sending each message twice with bit-wise inverted content using two identifiers differing at least in two bits.
CAN protocol controller	The CAN protocol controller is part of a CAN module performing data en-/de-capsulation, bit-timing, CRC, bit-stuffing, error handling, failure confinement, etc.
CAN transceiver	The CAN transceiver is connected to the CAN controller and to the bus lines. It provides the line transmitter and the receiver. There are high-speed, fault-tolerant, and single-wire transceivers available as well as transceivers for power-line or fiber optic transmissions.
Certification	Official compliance test of components or devices to a specific standard. CiA officially certifies CANopen devices.
CiA DR 303	Draft recommendation for CANopen cabling and connector pin assignments, coding of prefixes and SI unit as well as LED usage.
CiA DS 102	Draft standard for high-speed transmission according to ISO 11898-2 using 9-pin D-sub connectors.
CiA DS 301	The CANopen application layer and communication profile specification covers the functionality of CANopen NMT (network management) slave devices.
CiA DSP 302	The draft standard proposal for programmable CANopen devices includes CANopen manager functions, dynamic SDO connections, standardized boot-up procedure for NMT slaves as well as program download.
CiA DSP 304	The CANopen safety protocol specification is approved by German authorities and is compliant to SIL class 3 applications.
CiA DSP 305	The Layer Setting Services (LSS) specify how to set node-ID and transmission rate via the CANopen network.
CiA DSP 306	This draft standard proposal defines format and content of Electronic Data Sheets (EDS) to be used in configuration tools.

CiA DSP 308	The CANopen framework for maritime applications defines redundancy of networks including swapping mechanism for SDOs and PDOs.
CiA DSP 309	Set of gateway specifications for CANopen to Ethernet-based networks (e.g. Modbus TCP(IP).
CiA DS 401	The CANopen device profile for generic I/O modules covers the definition of digital and analog input and output devices.
CiA DSP 402	The CANopen device profile for drives and motion controllers defines the interface to frequency inverters, servo controllers as well as stepper motors.
CiA DS 404	The CANopen device profile for measuring devices and closed-loop controllers supports also multi-channel devices.
CiA DSP 405	The CANopen device and interface profile for IEC 61131-3 compatible controllers is based on the CiA DSP 302 specification using network variables to be mapped into PDOs, and function blocks for SDO services, etc.
CiA DS 406	The CANopen device profile for encoders defines the communication of rotating as well as linear sensors.
CiA DSP 407	The CANopen application profile for passenger information systems developed in cooperation with the German VDV specifies interfaces for a range of devices including displays, ticket printers, passenger counting units, main onboard computer, etc.
CiA DSP 408	The CANopen device profile for hydraulic controllers and proportional valves is compliant to the bus-independent VDMA device profile.
CiA DSP 410	The CANopen device profile for inclinometer supports 16-bit as well as 32-bit sensors.
CiA DSP 412	The CANopen device profiles for medical equipment specify the interfaces for x-ray collimators, x-ray generators, stands and tables.
CiA DSP 413	The CANopen interface profiles for in-vehicle truck gateways specify gateways to ISO 11992, J1939, and other in-vehicle networks. The CANopen network is mainly used for truck- or trailer-based superstructures, e.g. as in garbage trucks, truck-mounted cranes, and concrete mixers.
CiA DSP 414	The CANopen device profile for weaving machines specifies the interface for feeder sub-systems.

CiA DSP 415	The CANopen application profile for asphalt pavers specifies interfaces to different devices used in road construction machinery.
CiA DSP 416	The CANopen application profile for building doors specifies interfaces for locks, sensors, and other devices used in electronically controlled building doors.
CiA DSP 417	The CANopen application profile for lift control specifies the interfaces for car controller, door controller, call controller and other controllers as well as for car units, door units, input panels, and display units, etc.
CiA DSP 418	The CANopen device profile for battery modules specifies the interface to communicate with battery chargers.
CiA DSP 419	The CANopen device profile for battery charger specifies the interface to communicate with the battery module.
CiA DSP 420	The CANopen device profile family for extruder downstream devices defines interfaces for puller, corrugator and saw devices.
CiA DSP 421	The CANopen device profile for railways specifies interfaces to sub-systems such as diesel engines, brake controllers, door controllers, etc.
CiA DSP 422	The CANopen application profile for municipal vehicles defines the communication of sub-systems used in garbage trucks.
CiA TR 308	This technical report specifies some timings for CANopen performance testing tools.
Client SDO	The Client SDO initiates the SDO communication by means of reading or writing to the Object Dictionary of the server device.
Client/server communication	In a client/server communication the client initiates the communication with the server. It is always a point-to-point communication.
COB ID	The COB ID is the object specifying the CAN message identifier and additional parameters such as valid/invalid and remote frame support.
communication object (COB)	A communication object is one or more CAN messages with a specific functionality, e.g. PDO, SDO, Emergency, Time, or Error Control.
communication profile	A communication profile defines the content of communication objects such as Emergency, Time, Sync, Heartbeat, NMT, etc. in CANopen.
Configuration Manager	The Configuration Manager (CMT) provides mechanisms for configuration of CANopen devices during boot-up.

confirmed communication	Confirmed communication services requires a bi-directional communication, meaning that the receiving node sends a confirmation that the message has been received successfully.
conformance test plan	Definitions of test cases that have to be passed successfully in order to achieve conformance to a communication standard. The conformance test plan for CAN is standardized by ISO 16845.
conformance test tool	A conformance test tool is the implementation of a conformance test plan.
consumer	In CAN networks a receiver of messages is called a consumer meaning the acceptance filter is opened.

D	
data type	Object attribute in CANopen defining the format, e.g. UNSIGNED8, INTEGER16, BOOLEAN, etc.
data link layer	Second layer in the OSI reference model providing basic communication services. The CAN data link layer defines data, remote, error, and overload frames.
default value	Object attribute in CANopen defining the pre-setting of not user-configured objects after power-on or application reset.
device profile	A device profile defines the device-specific communication services including the configuration services in all details.
Draft Recommendation (DR)	This kind of recommendation is not fixed, but it is published. CiA's draft recommendations are not changed within one year.
Draft Standard (DS)	This kind of standard is not fixed, but it is published. CiA's draft standards are not changed within one year.
Draft Standard Proposal (DSP)	This kind of standard is a proposal, but it is published. CiA's draft standard proposals may be changed anytime without notification.
D-sub connector	Standardized connectors. Most common in use is the 9-pin D-sub connector (DIN 41652); its pin-assignment for CAN networks is specified in CiA DS 102.

E	
EDS checker	Software tool that checks the conformity of electronic data sheets. The CANopen EDS checker is available on CiA's website to be downloaded..

EDS generator	Software tool that generates CANopen electronic data sheets.
Electronic Data Sheet (EDS)	Electronic data sheets describe the functionality of a device in a standardized manner.
Emergency message	Pre-defined communication service in CANopen mapped into a single 8-byte data frame containing a 2-byte standardized error code, the 1-byte error register, and 5-byte manufacturer-specific information. It is used to communicate device and application failures.
EN 50325-4	CENELEC standard defining the CANopen application layer (version 4.0).
Entry category	Object attribute in CANopen defining if this object is mandatory or optional.
Error code	CANopen specifies standardized error codes transmitted in emergency messages.
Error control message	The CANopen error control messages are mapped to a single 1-byte CAN data frame assigned with a fixed identifier that is derived from the device's Node ID. It is transmitted as boot-up message before entering pre-operational state after inititialization, and it is transmitted if remotely requested by the NMT Master (node guarding) or periodically by the device (heartbeat).
event driven	Event driven messages are transmitted when a defined event occurs in the node. This may be a change of input states, elapsing of a local timer, or any other local event.
event timer	The event timer is assigned in CANopen to one PDO. It defines the frequency of transmission.
expedited SDO	This is a confirmed communication service of CANopen (peer-to-peer). It is made up by one SDO initiate message of the client node and the corresponding confirmation message of the server node. Expedited SDOs are used if not more than 4 byte of data has to be transmitted.

F	
flying master	In safety-critical applications, it may be required that a missing NMT Master is substituted automatically by another stand-by NMT Master. This concept of redundancy is called flying master.

form error	A corruption of one of the pre-defined recessive bits (CRC delimiter, ACK delimiter and EOF) is regarded as a form error condition that will cause the transmission of an error frame in the very next bit-time.
function code	First four bits of the CAN identifier in the CANopen pre-defined identifier set indicating the function of the communication object (e.g. TPDO_1 or Error Control message).

G	
galvanic isolation	Galvanic isolation in CAN networks is performed by optocouplers or transformers placed between CAN controller and CAN transceiver chip.
gateway	Device with at least two network interfaces transforming all seven OSI (open system interconnection) protocol layers, e.g. CANopen-to-Ethernet gateway.

H	
heartbeat	CANopen uses heartbeat message to indicate that a node is still alive. This message is transmitted periodically.
heartbeat consumer time	The heartbeat consumer time defines the time when a node is regarded as no longer alive due to a missing heartbeat message.
heartbeat producer time	The heartbeat producer time defines the transmission frequency of a heartbeat message.

I	
identifier	In general, the term identifier refers to a CAN message identifier. The CAN message identifier identifies the content of a data frame. The identifier of a remote frame corresponds to the identifier of the requested data frame. The identifier includes implicitly the priority for the bus arbitration.
Index	16-bit address to access the CANopen dictionary; for array and records the address is extended by an 8-bit Subindex.
line topology	Networks, where all nodes are connected directly to one bus line. CAN networks use theoretically just line topologies without any stub cable. However in practice you find tree and star topologies as well.

inhibit timer	Object in CANopen for PDOs and Emergency messages that forbids for the specified time (inhibit time) a transmission of this communication object.
Initialization state	NMT slave state in CANopen that is reached automatically after power on and communication or application reset.
interface profile	CANopen profile that describes just the interface and not the application behavior of device, e.g. gateway and bridge devices.
ISO 11898-1	International standard defining the CAN data link layer including LLC, MAC and PLS sub-layers.
ISO 11898-2	International standard defining the CAN high-speed MAU.

L	
Life guarding	Method in CANopen to detect that the NMT Master does not guard the NMT slave anymore. This not recommended for new systems designs.

M	
master	Communication or application entity that is allowed to control a specific function. In networks this is for example the initialization of a communication service.
Multiplexed PDO (MPDO)	The MPDO is made of 8 byte including one control byte, three multiplexer bytes (containing the 24-bit Index and Subindex), and four bytes of object data.

N	
network length	Bus length. The network cable length between the both termination resistors. The bus length of CANopen networks is limited by the used transmission rate. At 1 Mbps the maximum length is 25 m. When using lower transmission rates, longer bus lines may be used: at 50 kbps a length of 1 km is possible.
network management	Entity responsible for the network boot-up procedure and the optional configuration of nodes. It also may include node-supervising functions such as node guarding.
network variables	Network variables are used in programmable CANopen devices to be mapped into PDOs after programming the device.

NMT	Network management in CANopen.
NMT Master	The NMT Master device performs the network management by means of transmitting the NMT message. With this message, it controls the state machines of all connected NMT Slave devices.
NMT Slave	The NMT Slaves receive the NMT message, which contains commands for the NMT state machine implemented in CANopen devices.
NMT state machine	The NMT state machines support different states and the highest prior CAN message transmitted controls the transition to the states by the NMT Master.
node guarding	Mechanism used in CANopen and CAL to detect bus off or disconnected devices. The NMT Master sends a remote frame to the NMT slave that is answered by the corresponding error control message.
Node ID	Unique identifier for a device required by different CAN-based higher-layer protocols in order to assign CAN identifiers to this device, e.g. in CANopen and DeviceNet. In the pre-defined connection set of CANopen some of the CAN message identifier are derived from the assigned Node ID.

O	
Object Dictionary	Heart of each CANopen device containing all communication and application objects.
operational state	In the NMT operational state all CANopen communication services are available.

P	
PDO mapping	In PDOs, there may be mapped up to 64 objects. The PDO mapping is described in the PDO mapping parameters.
pin assignment	Definition of the use of connector pins.

pre-defined connection set	The pre-defined connection set is a default assignment of CAN message identifiers to CANopen communication objects. Some CANopen communication objects are distributed in broadcast (NMT message, Sync message, Time message) and others are transmitted between NMT Master device and dedicated NMT slave devices (PDO, SDO, Emergency, and Error Control). This default assignment guarantees that the CAN message identifiers are uniquely assigned in the network, if the node-ID has been assigned uniquely.
pre-operational state	In the NMT pre-operational state no CANopen PDO communication is allowed.
Process Data Object (PDO)	Communication object defined by the PDO communication parameter and PDO mapping parameter objects. It is an unconfirmed communication service without protocol overhead.
producer	In CAN networks a transmitter of messages is called a producer.
protocol	Formal set of conventions and rules for the exchange of information between nodes, including the specification of frame administration, frame transfer and physical layer.

R	
receiver	A CAN node is called receiver or consumer, if it is not transmitter and the bus is not idle.
redundant networks	In some safety-critical applications (e.g. maritime systems), redundant networks may be required that provide swapping capability in case of detected communication failures.
remote frame	With a remote frame another node is requested to transmit the corresponding data frame identified by the very same identifier. The remote frame's DLC has the value of the corresponding data frame DLC. The data field of the remote frame has a length of 0 byte.
remote transmission request (RTR)	Bit in the arbitration field indicating if the frame is a remote frame (recessive value) or a data frame (dominant value).
repeater	Passive component that refreshes CAN bus signals. It is used to increase the maximum number of nodes, or to achieve longer networks (>1 km), or to implement tree or meshed topologies.
reset application	This NMT command resets all objects in CANopen devices to the default values or the permanently stored configured values.

reset communication	This NMT command resets only the communication objects in CANopen devices to the default values or the permanently stored configured values.
RPDO	The Receive Process Data Object (RPDO) is a communication object that is received by a CANopen device.

S	
SDO block transfer	SDO block transfer is an CANopen communication services for increasing downloading In SDO block transfer, the confirmation is send after the reception of a number of SDO segments.
SDO Manager	The SDO Manager handles the dynamic establishment of SDO connections. It resides on the very same node as the NMT Master.
segmented SDO	If objects longer than 4 byte are transmitted by means of SDO services, a segmented transfer is used. The number of segments is theoretically not limited.
Server SDO	The Server SDO receives the SDO messages from the corresponding SDO Client and responses each SDO message or a block of SDO messages (SDO block transfer).
Service Data Object (SDO)	SDOs provide the access to entries in the CANopen Object Dictionary. An SDO is made up of at least two CAN messages with different identifiers. SDOs are always confirmed point-to-point communication services.
SI unit	International system of units for physical values as specified in ISO 1000:1983.
stopped state	NMT state in which only NMT messages are performed and under some conditions error control messages are transmitted.
sub-index	8-bit sub-address to access the sub-objects of arrays and records. Note: In this book Subindex is used instead of sub-index and Subentry instead of sub-object.
suspend transmission	CAN controllers in error passive mode have to wait additional 8 bit-times before the next data or remote frame may be transmitted.
SYNC message	Dedicated CANopen message forcing the receiving nodes to sample the inputs mapped into synchronous TPDOs. Receiving this message causes the node to set the outputs to values received in the previous synchronous RPDO.

T	
termination resistor	In CAN high-speed networks with bus topology, both ends are terminated with resistors in order to suppress reflections.
TIME message	Standardized message in CANopen containing the time as a 6-byte value given as ms after midnight and days after 1st January 1984.
TPDO	The Transmit Process Data Object (TPDO) is a communication object that is transmitted by a CANopen device.
transmission type	CANopen object defining the scheduling of a PDO.

V	
value definition	Detailed description of the value range in CANopen profiles.
value range	Object attribute in CANopen defining the allowed values that this object supports.

Index

Symbols

Numerics

A

B

C

D

E

N

O

P

R

S

T

U

V

W

X-Z

About Embedded Systems Academy

When founded in 2000 in California, the Embedded Systems Academy Inc. focused on training for new microcontroller architectures and communication aspects of Embedded Systems. Gradually, together with its soon thereafter founded partner Embedded Systems Academy GmbH of Germany, the focus shifted towards CAN, CANopen and J1939 communication protocols. Today, their products include hardware, software and libraries to implement, debug and test devices using these protocols. For further information please visit www.esacademy.com.